Journal of Homotopy and Related Structu

Volume 6(1&2), 2011

Aims and Scope

Homotopy is a basic discipline of mathematics having fundamental and various applications to important fields of mathematics.

The Journal has a wide scope which ranges from homotopical algebra and algebraic topology to algebraic number theory and functional analysis. Diverse algebraic, geometric, topological and categorical structures are closely related to mathematics such as general algebra, algebraic topology, algebraic geometry, category theory, differential geometry, computer science, K-theory, functional analysis, Galois theory and in physical sciences as well.

Journal of Homotopy and Related Structures intends to develop its vision on the determining role of homotopy in mathematics. The aim of the Journal is to show the importance, merit and diversity of homotopy in mathematical sciences.

Journal of Homotopy and Related Structures is primarily concerned with publishing carefully refereed significant and original research papers. However, a limited number of carefully selected survey and expository papers are also included, and special issues devoted to proceedings of meetings in the field and to Festschrifts will also be published.

Published jointly by College Publications and Tbilisi Center for Mathematical Sciences

ISBN 978-1-84890-118-6

College Publications
Scientific Director: Dov Gabbay
Managing Director: Jane Spurr
Department of Computer Science
King's College London, Strand, London WC2R 2LS, UK

http://www.collegepublications.co.uk

Original cover design by Laraine Welch
Printed by Lightning Source, Milton Keynes, UK

Journal of Homotopy and Related Structures

Volume **6**(1), 2011

Table of Contents

Journal of Homotopy and Related Structures

Volume **6**(2), 2011

Table of Contents

Journal of Homotopy and Related Structures, vol. 6(1), 2011, pp.1–38

CLASSIFYING SPACES AND FIBRATIONS OF SIMPLICIAL SHEAVES

MATTHIAS WENDT

(*communicated by Paul Goerss*)

Abstract

In this paper, we discuss the construction of classifying spaces of fibre sequences in model categories of simplicial sheaves. One construction proceeds via Brown representability and provides a classification in the pointed model category. The second construction is given by the classifying space of the monoid of homotopy self-equivalences of a simplicial sheaf and provides the unpointed classification.

Contents

1. Introduction

In this paper, we discuss the classification of fibrations in categories of simplicial sheaves. As usual, the results are modelled on the corresponding results for simplicial sets or topological spaces which we first discuss.

For simplicial sets, there are two approaches to the construction of classifying spaces. The first approach uses Brown representability to classify rooted fibrations, yielding a classification in the pointed category. This line of construction

The results presented here are taken from my PhD thesis [**Wen07**] which was supervised by Annette Huber-Klawitter. I would like to use the opportunity to thank her for her encouragement and interest in my work. I would also like to thank the following people: Fernando Muro and Jiří Rosický for explanations on Brown representability, the referee for pointing out some mistakes concerning pointed vs. unpointed classification, and Paul Goerss, Jim Stasheff and an editor of JHRS for helpful comments.
Received May 12, 2010, revised November 9, 2010; published on January 18, 2011.
2000 Mathematics Subject Classification: 18F20, 55R15.
Key words and phrases: classifying spaces, fibre sequence, simplicial sheaves

has been pursued in the work of Allaud [**All66**], Dold [**Dol66**], and Schön [**Sch82**]. The second approach applies the bar construction to the monoid of homotopy self-equivalences of the fibre. This is a generalization of the classifying space of a topological group, cf. [**Mil56**], further developed by Dold and Lashof for associative H-spaces, cf. [**DL59**], and applied to the classification of fibrations in [**Sta63**] and [**May75**]. This approach yields a classification in the unpointed category. The two approaches do not yield equivalent classifying spaces: the rooted fibrations carry an action of the group of homotopy self-equivalences, and dividing out this action yields the unpointed classifying space of the second approach. A survey on construction of classifying spaces and classification of fibrations can be found in [**Sta70**] or [**May75**].

Now we want to explain why this theory works in the general setting of simplicial sheaves. On the one hand, *fibrations of simplicial sheaves can be glued.* This is of course not true on the nose, but as for simplicial sets there is a way around this problem. The essence of the solution is that some kind of "homotopy distributivity" holds – in some situations it is possible to interchange homotopy limits and homotopy colimits. The notion of homotopy distributivity is due to Rezk [**Rez98**] and can be used to generalize various classical results on homotopy pullbacks and homotopy colimits, such as Puppe's theorem or Mather's cube theorem. This theory is developed in Section 2. Once such a glueing for fibrations of simplicial sheaves is developed, it is a simple matter to prove that the conditions for a version of Brown representability are satisfied, yielding classifying spaces for analogs of rooted fibrations of simplicial sheaves. On the other hand, *fibrations of simplicial sheaves correspond to principal bundles under homotopy self-equivalences.* Suitably formulated, we can associate to a simplicial sheaf X a simplicial sheaf of monoids consisting of homotopy self-equivalences of X. To this monoid we can apply the bar construction. One can prove that the resulting space classifies fibre sequences of simplicial sheaves.

In our approach to the construction of classifying spaces, we introduce a notion of local triviality of fibrations in the Grothendieck topology. This condition is one possible generalization of the usual condition that all fibres of the fibration should have the homotopy type of the given fibre F. In the first approach via Brown representability, this condition ensures that the fibre functor is indeed set-valued. In the second approach using the bar construction, it comes in naturally because we can not talk about fibre sequence if the base is not pointed.

The main result of the paper is the following theorem:

Theorem 1. *Let T be a site.*

(i) Assume that the category $\Delta^{op}Shv(T)$ of simplicial sheaves on T is compactly generated, cf. Definition 4.8. Let F be a pointed simplicial sheaf on T. There exists a pointed simplicial sheaf B^fF which classifies locally trivial fibrations with fibre F up to (rooted) equivalence, i.e. for each pointed simplicial sheaf X there is a bijection between the set of equivalence classes of fibre sequences over X with fibre F and the set of pointed homotopy classes of maps $X \to BF$.

(ii) Let F be a simplicial sheaf on T. There exists a simplicial sheaf denoted by $B(*, \mathrm{hAut}_\bullet(F), *)$ which classifies locally trivial morphisms with fibre F up to

equivalence, i.e. for each simplicial sheaf there is a bijection between the set of equivalence classes of locally trivial morphisms over X with fibre F and the set of unpointed homotopy classes of maps $X \to B(*, \mathrm{hAut}_\bullet(F), *)$.

The two classification results can be found in Theorem 4.14 and Theorem 5.10. The main input in both of them is homotopy distributivity which originally is a result of Rezk [**Rez98**]. We give a short proof for topoi with enough points in Proposition 2.15.

One word on the relation between our approach and the classification results in [**DK84**]: given fixed simplicial sheaves B and F, analogs of the classification results of [**DK84**] can be used to construct a simplicial set whose components are in one-to-one correspondence with fibre sequences over B with fibre F. However, these results do not imply that the various simplicial sets are the sections of one simplicial sheaf. It is exactly this *internal* classification that we are after. For this, some sort of homotopy distributivity is needed, as we discuss in Section 2.

Finally, a short sketch of the envisioned applications is in order. The main motivation for the research reported in this paper comes from $\mathbb{A}^1$-homotopy theory, which is a homotopy theory for algebraic varieties defined by Morel and Voevodsky [**MV99**]. On the one hand, the homotopy distributivity results from Section 2 have been used in [**Wen10**] to give descriptions of $\mathbb{A}^1$-fundamental groups of smooth toric varieties. On the other hand, the theory of classifying spaces developed here allows several results on unstable localization of fibre sequences for simplicial sets to be carried over to simplicial sheaves. This is discussed in [**Wen07**, Chapter 4] and will be further elaborated in a forthcoming paper. The most interesting application, however, is in $\mathbb{A}^1$-homotopy theory. The results presented here allow the construction of classifying spaces, and the localization theory of [**Wen07**] allows us to obtain checkable conditions under which fibrations which are locally trivial in the Nisnevich topology are indeed $\mathbb{A}^1$-local. This will be discussed in [**Wen09**].

Structure of the Paper: In Section 2, we develop the necessary preliminaries for homotopy distributivity which will be needed. Section 3 we discuss locally trivial fibrations in categories of simplicial sheaves. Then the two classification results are proved in Section 4 and Section 5.

2. Homotopy Limits and Colimits of Simplicial Sheaves

2.1. Model Structures for Simplicial Sheaves

The global pattern in the theory of model structures on categories of simplicial sheaves is always the same: a category of simplicial sheaves behaves in many aspects like the category of simplicial sets. This is also evident in the proofs, which reduce statements about simplicial sheaves to known statements about simplicial sets.

The basic definitions of sites and categories of sheaves on them can be found in [**MM92**]. We will freely use these as well as the notions of homotopical algebra. For the definition of model categories, see [**GJ99**] with a particular focus on simplicial sets, as well as [**Hov98**] and [**Hir03**].

We denote by $\Delta^{op}\mathcal{C}$ the category of simplicial objects in the category $\mathcal{C}$. In particular, the category of simplicial sheaves on a site T is denoted by $\Delta^{op}\mathcal{S}hv(T)$.

The following comprises the main facts about model structures on simplicial sheaves.

Theorem 2.1. *Let $\mathcal{E}$ be a topos. Then the category $\Delta^{op}\mathcal{E}$ of simplicial objects in $\mathcal{E}$ has a model structure, where the*

(i) cofibrations are monomorphisms,

(ii) weak equivalences are detected on a fixed Boolean localization,

(iii) fibrations are determined by the right lifting property.

The above definition of weak equivalences does not depend on the Boolean localization.

The injective model structure of Jardine on the category of (pre-)sheaves of simplicial sets on T is a proper simplicial and cellular model structure.

Existence is proved in [**Jar96**, Theorems 18 and 27]. Properness and simpliciality are proven in [**Jar96**, Theorem 24]. The fact that the model categories are cofibrantly generated is implicit in Jardine's proofs, though not explicitly stated. The combinatoriality follows since categories of sheaves on a Grothendieck site are locally presentable. Cellularity is proven in [**Hor06**, Theorem 1.4].

2.2. Recollection on Homotopy Limits and Colimits

Homotopy colimits and limits are homotopy-invariant versions of the ordinary colimits and limits for categories. Abstractly, one can define the ordinary colimit of a diagram $\mathcal{X} : \mathcal{I} \to \mathcal{C}$ as left adjoint of the diagonal functor $\Delta_{\mathcal{I}} : \mathcal{C} \to \hom(\mathcal{I}, \mathcal{C})$, where $\hom(\mathcal{I}, \mathcal{C})$ is the category of $\mathcal{I}$-diagrams in $\mathcal{C}$. Similarly, the ordinary limit is the right adjoint of the diagonal, cf. [**Mac98**, Section X.1]. Homotopy colimits and limits are then defined as suitable derived functors of the ordinary colimit and limit functors.

A general reference for homotopy limits and colimits is [**Hir03**], in the context of simplicial sheaves see also [**MV99**]. We shortly recall the definition of homotopy limits and colimits.

Definition 2.2. *Let $\mathcal{C}$ be a cofibrantly generated simplicial model category, and $\mathcal{I}$ be any small category.*

Colimits *The category $\mathrm{Hom}(\mathcal{I}, \mathcal{C})$ of $\mathcal{I}$-indexed diagrams in $\mathcal{C}$ has the structure of a simplicial model category by taking the weak equivalences and fibrations to be the pointwise ones. Then the diagonal $\Delta : \mathcal{C} \to \mathrm{Hom}(\mathcal{I}, \mathcal{C})$ preserves fibrations and weak equivalences, and therefore is a right Quillen functor. Its left adjoint $\mathrm{colim} : \mathrm{Hom}(\mathcal{I}, \mathcal{C}) \to \mathcal{C}$ is thus a left Quillen functor, and we can define its derived functor*

$$\operatorname*{hocolim}_{\mathcal{I}} = L \operatorname*{colim}_{\mathcal{I}} : \mathcal{X} \mapsto \operatorname*{colim}_{\mathcal{I}} Q\mathcal{X},$$

where Q is a cofibrant replacement in the model category $\mathrm{Hom}(\mathcal{I}, \mathcal{C})$.

Limits *Dually, the category $\mathrm{Hom}(\mathcal{I}, \mathcal{C})$ also has a simplicial model structure where the weak equivalences and cofibrations are the pointwise ones. Then the diagonal $\Delta : \mathcal{C} \to \mathrm{Hom}(\mathcal{I}, \mathcal{C})$ preserves cofibrations and weak equivalences, and* *therefore is a left Quillen functor. Its right adjoint* $\lim : \mathrm{Hom}(\mathcal{I}, \mathcal{C}) \to \mathcal{C}$ *is thus a right Quillen functor, and we can define its derived functor*

$$\operatorname{holim}_{\mathcal{I}} = R \lim_{\mathcal{I}} : \mathcal{X} \mapsto \lim_{\mathcal{I}} R\mathcal{X},$$

where R *is a fibrant replacement in the model category* $\mathrm{Hom}(\mathcal{I}, \mathcal{C})$.

We usually denote the homotopy colimit of an $\mathcal{I}$*-diagram* $\mathcal{X}$ *by* $\operatorname{hocolim}_{\mathcal{I}} \mathcal{X}$*, the special case of a homotopy pushout is denoted by* $A \cup^h_B C$*. Similarly, homotopy limits are usually denoted by* $\operatorname{holim}_{\mathcal{I}} \mathcal{X}$*, and the homotopy pullbacks by* $A \times^h_B C$.

There are also more concrete constructions of homotopy limits and colimits. Since we are not going to need these descriptions, we just refer to [**Hir03**, Chapter 18].

The fact that homotopy colimits resp. limits can be defined as left resp. right derived functors of colimits resp. limits implies that they are homotopy invariant [**Hir03**, Theorem 18.5.3].

Proposition 2.3. *Let* $\mathcal{C}$ *be a simplicial model category, and let* $\mathcal{I}$ *be a small category. If* $f : \mathcal{X} \to \mathcal{Y}$ *is a morphism of* $\mathcal{I}$*-diagrams of cofibrant objects in* $\mathcal{C}$ *which is an objectwise equivalence, then*

$$\operatorname{hocolim}_{\mathcal{I}} f : \operatorname{hocolim}_{\mathcal{I}} \mathcal{X} \to \operatorname{hocolim}_{\mathcal{I}} \mathcal{Y}$$

is a weak equivalence of cofibrant objects.

Dually, if $f : \mathcal{X} \to \mathcal{Y}$ *is a morphism of* $\mathcal{I}$*-diagrams in* $\mathcal{C}$ *which is an objectwise equivalence of fibrant objects, then*

$$\operatorname{holim}_{\mathcal{I}} f : \operatorname{holim}_{\mathcal{I}} \mathcal{X} \to \operatorname{holim}_{\mathcal{I}} \mathcal{Y}$$

is a weak equivalence of fibrant objects.

Moreover, homotopy colimits and limits interact nicely with the corresponding left resp. right Quillen functors.

Proposition 2.4. *Let* $F : \mathcal{C} \to \mathcal{D}$ *be a left Quillen functor. Then the following diagram commutes up to isomorphism:*

$$\begin{array}{ccc} \mathrm{Ho}\,\mathrm{Hom}(\mathcal{I}, \mathcal{C}) & \xrightarrow{\text{hocolim}} & \mathrm{Ho}\,\mathcal{C} \\ {\scriptstyle LF}\downarrow & & \downarrow{\scriptstyle LF} \\ \mathrm{Ho}\,\mathrm{Hom}(\mathcal{I}, \mathcal{D}) & \xrightarrow[\text{hocolim}]{} & \mathrm{Ho}\,\mathcal{D}, \end{array}$$

One example of this situation is the relation between homotopy colimits and hom-functors as stated in [**MV99**, Lemma 2.1.19].

Finally, we state a standard fact on homotopy pullbacks, cf. also [**GJ99**, Lemma II.8.22]:

Lemma 2.5. *Let* $\mathcal{C}$ *be a proper model category, and let the following commutative diagram be given:*

$$\begin{array}{ccccc} X_1 & \longrightarrow & X_2 & \longrightarrow & X_3 \\ \downarrow & & \downarrow & & \downarrow \\ Y_1 & \longrightarrow & Y_2 & \longrightarrow & Y_3. \end{array}$$

If the inner squares are homotopy pullback squares, then so is the outer. If the outer square and the right inner square are homotopy pullback squares, then so is the left inner square.

2.3. Functorialities

We first recall the basic result that geometric morphisms of Grothendieck topoi induce Quillen functors. This is basically a reformulation of [**MV99**, Proposition 2.1.47].

Proposition 2.6. *Let $f : \mathcal{F} \to \mathcal{E}$ be a geometric morphism of Grothendieck topoi. We also denote by $f^* : \Delta^{op}\mathcal{E} \to \Delta^{op}\mathcal{F}$ and $f_* : \Delta^{op}\mathcal{F} \to \Delta^{op}\mathcal{E}$ the induced functors on the categories of simplicial sheaves. Then (f^*, f_*) is a Quillen pair, i.e. f^* preserves cofibrations and trivial cofibrations and f_* preserves fibrations and trivial fibrations.*

Finally, we recall that weak equivalences are reflected along surjective geometric morphisms.

Proposition 2.7. *Let $f : \mathcal{E}' \to \mathcal{E}$ be a surjective geometric morphism, and let $g : A \to B$ be a morphism in $\mathcal{E}$. Then g is a weak equivalence if $f^*g : f^*A \to f^*B$ is a weak equivalence in $\mathcal{E}'$.*

Proof. If f is surjective, then any Boolean localization of $\mathcal{E}'$ is a Boolean localization of $\mathcal{E}$, because a Boolean localization of $\mathcal{E}$ is simply a surjective geometric morphism $\mathcal{B} \to \mathcal{E}$, where $\mathcal{B}$ is the topos of sheaves on a complete Boolean algebra. In [**Jar96**], it was proved that the weak equivalences which are defined via Boolean localizations are independent of the Boolean localization.

A morphism $f : A \to B$ is thus a weak equivalence in $\mathcal{E}$ if it is a morphism after pullback along $f^* : \mathcal{E} \xrightarrow{g^*} \mathcal{E}' \to \mathcal{B}$, where the latter morphism is a chosen Boolean localization of $\mathcal{E}'$. But by definition, this is equivalent to the fact that g^*f is a weak equivalence in $\mathcal{E}'$. This proves the claim. □

2.4. Homotopy Colimits

In this subsection, we recall the behaviour of homotopy colimits under the inverse image part of a geometric morphism. The inverse image preserves homotopy colimits, and reflects them if the geometric morphism is surjective.

Proposition 2.8. *Let $\mathcal{E}$ be a topos, and let $f : \mathcal{E}' \to \mathcal{E}$ be a geometric morphism. Then $f^* : \Delta^{op}\mathcal{E} \to \Delta^{op}\mathcal{E}'$ preserves homotopy colimits.*

Proof. f^* is a left Quillen functor, cf. Proposition 2.6. The result follows from Proposition 2.4. □

Proposition 2.9. *Let $\mathcal{E}$ be a topos, let $\mathcal{I}$ be a small category, and let $f : \mathcal{E}' \to \mathcal{E}$ be a geometric morphism. If f is surjective, then $f^* : \Delta^{op}\mathcal{E} \to \Delta^{op}\mathcal{E}'$ reflects homotopy colimits. In other words, $\mathcal{X} : \mathcal{I} \to \Delta^{op}\mathcal{E}$ is a homotopy colimit diagram if and only if $f^*\mathcal{X} : \mathcal{I} \to \Delta^{op}\mathcal{E}'$ is a homotopy colimit diagram.*

Proof. Recall that $\mathcal{X}$ is a homotopy colimit diagram if the natural map

$$\Psi : \operatorname*{hocolim}_{\mathcal{I}} \mathcal{X} \to \operatorname*{colim}_{\mathcal{I}} \mathcal{X}$$

is a weak equivalence.

We have a diagram

$$\begin{array}{ccc}
f^* \operatorname{hocolim} \mathcal{X} & \longrightarrow & f^* \operatorname{colim} \mathcal{X} \\
\vdots\uparrow & & \uparrow \\
\operatorname{hocolim} f^*\mathcal{X} & \longrightarrow & \operatorname{colim} f^*\mathcal{X}
\end{array}$$

The left arrow exists because to compute $\operatorname{hocolim} \mathcal{X}$, we use a cofibrant replacement which is preserved by the left Quillen functor f^*. Therefore there is a cone from the cofibrant diagram $\mathcal{X}$ to $f^* \operatorname{hocolim} \mathcal{X}$ which has to factor through the colimit, which is also the homotopy colimit since the diagram is cofibrant. The vertical morphisms are weak equivalences by Proposition 2.4, hence $f^*\Psi$ can be identified up to weak equivalence with the map

$$\operatorname*{hocolim}_{\mathcal{I}} f^*\mathcal{X} \to \operatorname*{colim}_{\mathcal{I}} f^*\mathcal{X},$$

which is a weak equivalence if $f^*\mathcal{X}$ is a homotopy colimit diagram.

If f is surjective, it reflects weak equivalences, cf. Proposition 2.7. This proves the claim. □

This implies that homotopy colimits in a model category of simplicial sheaves can be checked on points, provided there are enough points, cf. [**Wen07**, Proposition 3.1.10].

Corollary 2.10. *Let $\mathcal{E}$ be a topos with enough points, let $\mathcal{I}$ be a small category, and let $\mathcal{X} : \mathcal{I} \to \Delta^{op}\mathcal{E}$ be a diagram. Then $\mathcal{X}$ is a homotopy colimit diagram if and only if for each point p of $\mathcal{E}$ in a conservative set of points, the corresponding diagram $p^*(\mathcal{X}) : \mathcal{I} \to \Delta^{op}\mathcal{S}et$ is a homotopy colimit diagram.*

Proof. This follows from Proposition 2.9: if $\mathcal{E}$ has enough points, we can choose a conservative set C of points, and then the geometric morphism

$$\prod_{p \in C} \mathcal{S}et \to \mathcal{E}$$

is surjective. □

2.5. Homotopy Pullbacks

Finally, we recall the behaviour of homotopy pullbacks under inverse images of geometric morphisms. As for homotopy colimits, they are preserved by inverse images and reflected, provided the geometric morphism is surjective. The argument does however not work for arbitrary homotopy limits, since the inverse image fails to be a right Quillen functor.

Proposition 2.11. *Let $\mathcal{E}$ be a topos, let $f : \mathcal{E}' \to \mathcal{E}$ be a geometric morphism, and let the following commutative diagram $\mathcal{X}$ in $\Delta^{op}\mathcal{E}$ be given:*

$$\begin{array}{ccc} A & \longrightarrow & B \\ \downarrow & & \downarrow \\ C & \longrightarrow & D. \end{array}$$

If $\mathcal{X}$ is a homotopy pullback diagram in $\Delta^{op}\mathcal{E}$, then $f^*\mathcal{X}$ is a homotopy pullback diagram in $\Delta^{op}\mathcal{E}'$. If moreover f is surjective, and $f^*\mathcal{X}$ is a homotopy pullback diagram in $\Delta^{op}\mathcal{E}'$, then $\mathcal{X}$ is a homotopy pullback diagram in $\Delta^{op}\mathcal{E}$.

Proof. The first assertion, i.e. that homotopy pullback squares are preserved by the inverse image part of a geometric morphism is proved in [**Rez98**, Theorem 1.5].

Recall that $\mathcal{X}$ is a homotopy pullback diagram if there exists a factorization of $f : B \to D$ into a trivial cofibration $i : B \to \tilde{B}$ and a fibration $g : \tilde{B} \to D$, such that the induced morphism $A \to C \times_D \tilde{B}$ is a weak equivalence. Since f is surjective, it suffices to show that the induced morphism $f^*(A) \to f^*(C \times_D \tilde{B}) \cong f^*(C) \times_{f^*(D)} f^*(\tilde{B})$ is a weak equivalence. Note that geometric morphisms preserve finite limits by definition, which explains the last isomorphism.

Consider the diagram

$$\begin{array}{ccc} f^*(A) & \longrightarrow & f^*(B) \\ \downarrow & & \downarrow \\ f^*(C) \times_{f^*(D)} f^*(\widetilde{B}) & \longrightarrow & f^*(\widetilde{B}) \\ \downarrow & & \downarrow \\ f^*(C) & \longrightarrow & f^*(D). \end{array}$$

Since homotopy pullbacks are preserved by geometric morphisms, the lower square is a homotopy pullback. By assumption, the outer square is also homotopy pullback square, therefore the upper square is a homotopy pullback, cf. Lemma 2.5. Since f preserves weak equivalences, $f^*(B) \to f^*(\widetilde{B})$ is a weak equivalence. Therefore, the morphism $f^*(A) \to f^*(C) \times_{f^*(D)} f^*(\widetilde{B})$ is also a weak equivalence. This proves the result.

□

As for homotopy colimits, we find that homotopy pullbacks in a category of simplicial sheaves can be checked on points, provided there are enough points, cf. [**Wen07**, Proposition 3.1.11].

Corollary 2.12. *Let $\mathcal{E}$ be a topos with enough points, and let the following commutative diagram $\mathcal{X}$ of simplicial sheaves in $\Delta^{op}\mathcal{E}$ be given:*

$$\begin{array}{ccc} A & \longrightarrow & B \\ \downarrow & & \downarrow{\scriptstyle f} \\ C & \longrightarrow & D. \end{array}$$

This is a homotopy pullback diagram iff for each point p of T in a conservative set of points, the diagram $p^*(\mathcal{X})$ of simplicial sets is a homotopy pullback diagram.

2.6. Homotopy Distributivity

The results on homotopy limits and colimits from the previous section can be used to give a simple proof of the following result of Rezk on homotopy distributivity in categories of simplicial sheaves, cf. [**Rez98**, Theorem 1.4]. These results generalize various results on commuting homotopy pullbacks and homotopy colimits known to hold for simplicial sets, such as Mather's cube theorem and Puppe's theorem, cf. Corollary 2.16 and Proposition 2.17. Moreover, homotopy distributivity allows the construction of classifying spaces for fibre sequences, cf. [**Wen07**].

We begin by explaining the precise definition of homotopy distributivity, which is a homotopical generalization of the usual infinite distributivity law which holds for topoi. It is a statement about commutation of arbitrary small homotopy colimits with finite homotopy limits. Since any finite homotopy limit can be constructed via homotopy pullbacks, it suffices to check that homotopy pullbacks distribute over arbitrary homotopy colimits. Most of the work on homotopy distributivity is due to Rezk [**Rez98**].

The situation is the following. Let $\mathcal{C}$ be a simplicial model category, let $\mathcal{I}$ be a small category, and let $f : \mathcal{X} \to \mathcal{Y}$ be a morphism of $\mathcal{I}$-diagrams in $\mathcal{C}$. The diagrams we are most interested in are the following:

For any $i \in \mathcal{I}$, we have a commutative square

$$\begin{array}{ccc} \mathcal{X}(i) & \longrightarrow & \operatorname{colim}_{\mathcal{I}} \mathcal{X} \\ {\scriptstyle f(i)}\downarrow & & \downarrow \\ \mathcal{Y}(i) & \longrightarrow & \operatorname{colim}_{\mathcal{I}} \mathcal{Y}. \end{array} \qquad (1)$$

Moreover, for any $\alpha : i \to j$ in $\mathcal{I}$ we have a commutative square

$$\begin{array}{ccc} \mathcal{X}(i) & \xrightarrow{\mathcal{X}(\alpha)} & \mathcal{X}(j) \\ {\scriptstyle f(i)}\downarrow & & \downarrow{\scriptstyle f(j)} \\ \mathcal{Y}(i) & \xrightarrow[\mathcal{Y}(\alpha)]{} & \mathcal{Y}(j). \end{array} \qquad (2)$$

Now we are ready to state the definition of homotopy distributivity, following [**Rez98**].

Definition 2.13 (Homotopy Distributivity). *In the above situation, we say that $\mathcal{C}$ satisfies* homotopy distributivity *if for any morphism $f : \mathcal{X} \to \mathcal{Y}$ of $\mathcal{I}$-diagrams in $\mathcal{C}$ for which $\mathcal{Y}$ is a homotopy colimit diagram, i.e.* $\operatorname{hocolim}_{\mathcal{I}} \mathcal{Y} \to \operatorname{colim}_{\mathcal{I}} \mathcal{Y}$ *is a weak equivalence, the following two properties hold:*

(HD i) If each square of the form (1) is a homotopy pullback, then $\mathcal{X}$ is a homotopy colimit diagram.

(HD ii) If $\mathcal{X}$ is a homotopy colimit diagram, and each diagram of the form (2) is a homotopy pullback, then each diagram of the form (1) is also a homotopy pullback.

Example 2.14. *The category $\Delta^{op}\mathcal{S}et$ of simplicial sets satisfies homotopy distributivity. This follows e.g. from the work of Puppe [**Pup74**] and Mather [**Mat76**].* □

More generally, homotopy distributivity holds for all model categories of simplicial sheaves on a Grothendieck site and can be proven quite easily if the site has enough points. We give a short and simple proof of homotopy distributivity, based on the reflection of homotopy colimits and pullbacks proved earlier. The general statement and proof using Boolean localizations can be found in [**Rez98**].

Proposition 2.15. *Let $\mathcal{E}$ be a Grothendieck topos with enough points. Then homotopy distributivity holds for the injective model structure on $\Delta^{op}\mathcal{E}$.*

Proof. Since there are enough points, there exists a surjective geometric morphism

$$f : \prod_{p \in C} \mathcal{S}et \to \mathcal{E},$$

where C is a conservative set of points. By Propositions 2.9 and 2.11 the properties of homotopy colimit resp. homotopy pullback diagrams can be checked locally. The assertion then follows from homotopy distributivity for simplicial sets. □

We next discuss two important consequences of homotopy distributivity for model categories of simplicial sheaves. One is a generalization of Mather's cube theorem [**Mat76**]. The other generalizes a theorem of Puppe [**Pup74**] on commuting homotopy fibres and homotopy pushouts to simplicial sheaves.

Corollary 2.16 (Mather's Cube Theorem). *Let $\mathcal{E}$ be any Grothendieck topos. Consider the following diagram of simplicial objects in $\mathcal{E}$:*

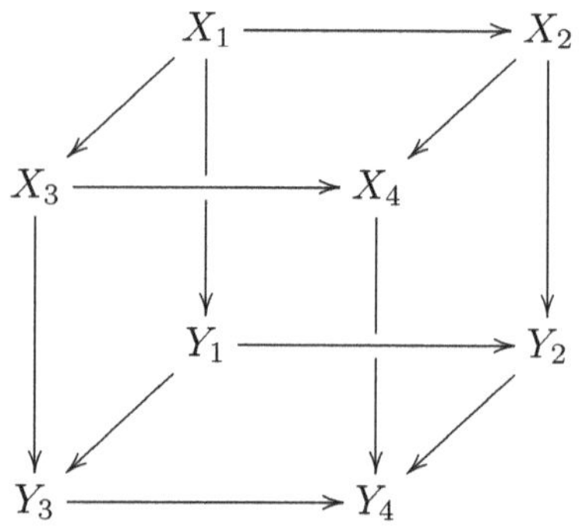

Assume that the bottom face, i.e. the one consisting of the spaces Y_i, is a homotopy pushout, and that all the vertical faces are homotopy pullbacks. Then the top face is a homotopy pushout.

Moreover, taking the homotopy fibre commutes with homotopy pushouts: for a commutative diagram

$$\begin{array}{ccccc} E_2 & \longleftarrow & E_0 & \longrightarrow & E_1 \\ {\scriptstyle p_2}\downarrow & & \downarrow{\scriptstyle p_0} & & \downarrow{\scriptstyle p_1} \\ B_2 & \longleftarrow & B_0 & \longrightarrow & B_1 \end{array}$$

in which the squares are homotopy pullbacks, we have weak equivalences

$$\operatorname{hofib} p_i \xrightarrow{\cong} \operatorname{hofib}(p : E_1 \cup^h_{E_0} E_2 \to B_1 \cup^h_{B_0} B_2).$$

Proof. This is a consequence of homotopy distributivity, cf. Proposition 2.15, applied to homotopy pushout diagrams. The assumption in the definition of homotopy distributivity is that the bottom face is a homotopy colimit diagram, i.e. a homotopy pushout.

For the first assertion, we note that since all the vertical faces are homotopy pullbacks, the diagonal square in the cube consisting of X_1, Y_1, X_4 and Y_4 is also a homotopy pullback, by the homotopy pullback lemma 2.5. By (HD i) we conclude that the top square is a homotopy colimit diagram, i.e. X_4 is weakly equivalent to the homotopy pushout $X_2 \cup^h_{X_1} X_3$. The restriction in the definition of homotopy distributivity that X_4 be the point-set pushout of X_2 and X_3 along X_1 is not essential. Without loss of generality we can assume that the morphisms $X_1 \to X_3$ resp. $Y_1 \to Y_3$ are cofibrations, and that X_4 resp. Y_4 are point-set pushouts. If this is not the case, just replace the morphism by cofibrations, and obtain a cube which is weakly equivalent to the cube we started with.

For the second statement note that since the squares are homotopy pullbacks, we have $\operatorname{hofib} p_0 \cong \operatorname{hofib} p_1 \cong \operatorname{hofib} p_2$. Factoring $E_0 \to E_1$ resp. $B_0 \to B_1$ as a cofibration followed by a trivial fibration, we can assume that these morphisms are cofibrations. Denote $E = E_1 \cup^h_{E_0} E_2$ and $B = B_1 \cup^h_{B_0} B_2$. Then we are in the situation to apply (HD ii). This implies that all the squares

$$\begin{array}{ccc} E_i & \longrightarrow & E \\ \downarrow & & \downarrow \\ B_i & \longrightarrow & B \end{array}$$

are homotopy pullback squares. In particular, we get the desired weak equivalences $\operatorname{hofib} p_i \cong \operatorname{hofib} p$. □

The following is a version of Puppe's Theorem [**Pup74**] for simplicial sheaves:

Proposition 2.17 (Puppe's Theorem). *Let $\mathcal{E}$ be a Grothendieck topos, and let $\mathcal{X} : \mathcal{I} \to \Delta^{op}\mathcal{E}$ be a diagram of simplicial objects over a fixed base simplicial object Y, i.e. the following diagram commutes for every $\alpha : i \to j$ in $\mathcal{I}$:*

$$\begin{array}{ccc} \mathcal{X}(i) & \xrightarrow{\mathcal{X}(\alpha)} & \mathcal{X}(j) \\ & \searrow \quad \swarrow & \\ & Y & \end{array}$$

There is an associated diagram of homotopy fibres

$$\mathcal{F} : \mathcal{I} \to \Delta^{op}\mathcal{E} : i \mapsto \mathrm{hofib}(\mathcal{X}(i) \to Y)$$

Denoting $X = \mathrm{hocolim}_{\mathcal{I}} \mathcal{X}$ *and* $F = \mathrm{hocolim}_{\mathcal{I}} \mathcal{F}$, *we have a weak equivalence* $\mathrm{hofib}(X \to Y) \simeq F$.

Proof. We construct a new morphism of diagrams $\mathcal{G} \to \mathcal{X}$, where the diagram $\mathcal{G}$ is defined by

$$\mathcal{G} : \mathcal{I} \to \Delta^{op}\mathcal{E} : i \mapsto \mathcal{X}(i) \times^h_X \mathrm{hofib}(X \to Y).$$

Without loss of generality we can assume $\mathrm{hofib}(X \to Y) \to X$ is a fibration. Then the homotopy pullbacks above are ordinary pullbacks, and $\mathrm{colim}\, \mathcal{G} \cong \mathrm{hofib}(X \to Y)$. We apply the homotopy pullback lemma to the following diagram:

$$\begin{array}{ccc} \mathcal{X}(i) \times^h_X \mathrm{hofib}(X \to Y) & \longrightarrow & \mathcal{X}(i) \\ \downarrow & & \downarrow \\ \mathrm{hofib}(X \to Y) & \longrightarrow & X \\ \downarrow & & \downarrow \\ * & \longrightarrow & Y \end{array}$$

This implies the following weak equivalence

$$\mathcal{X}(i) \times^h_X \mathrm{hofib}(X \to Y) \simeq \mathrm{hofib}(\mathcal{X}(i) \to Y) = \mathcal{F}(i).$$

Invariance of homotopy colimits under weak equivalence, cf. Proposition 2.3, implies a weak equivalence $\mathrm{hocolim}\, \mathcal{G} \cong \mathrm{hocolim}\, \mathcal{F}$. Homotopy distributivity applied to the projection morphism $\mathcal{G} \to \mathcal{X}$ implies that $\mathcal{G}$ is a homotopy colimit diagram. Putting everything together we obtain weak equivalences $\mathrm{hofib}(X \to Y) \cong \mathrm{colim}\, \mathcal{G} \simeq \mathrm{hocolim}\, \mathcal{G} \simeq \mathrm{hocolim}\, \mathcal{F}$, whence the desired statement follows. □

2.7. Ganea's Theorem

It is now possible to obtain some fibre sequences for simplicial sheaves, which are known to hold for simplicial sets by homotopy distributivity. In the case of simplicial sets, these fibre sequences are all more or less consequences of Ganea's work [**Gan65**]. Their simplicial sheaf analogues have been used in [**Wen10**] to provide partial descriptions of the $\mathbb{A}^1$-fundamental group of smooth toric varieties.

We start out with a theorem describing the homotopy fibre of the fold map. The proof is essentially the one given in [**DF96**, Appendix HL], which simply applies homotopy distributivity to one of the simplest situations possible:

Proposition 2.18 (Ganea's Theorem). *Let $\mathcal{E}$ be a Grothendieck topos, and let X be a simplicial object in $\mathcal{E}$. The sequence $\Sigma\Omega X \to X \vee X \to X$ is a fibre sequence in $\Delta^{op}\mathcal{E}$.*

Proof. This is an instance of Proposition 2.17 applied to the diagram:

$$\begin{array}{ccccc} X & \longleftarrow & * & \longrightarrow & X \\ {\scriptstyle =}\downarrow & & \downarrow & & \downarrow{\scriptstyle =} \\ X & \underset{=}{\longleftarrow} & X & \underset{=}{\longrightarrow} & X \end{array}$$

We are taking the homotopy colimit of the diagram over the fixed base space X, and the homotopy colimit of the upper line yields $X \vee X$. The map to X is the fold map $\vee : X \vee X \to X$. Then Proposition 2.17 shows that the fibre is given by the homotopy colimit of the diagram of fibres:

$$* \longleftarrow \Omega X \longrightarrow *.$$

This is by definition $\Sigma\Omega X$. □

Example 2.19. *A particular topological instance of the above is the fibre sequence*

$$S^2 \to \mathbb{C}P^\infty \vee \mathbb{C}P^\infty \to \mathbb{C}P^\infty.$$

A similar fibre sequence exists in $\Delta^{op}\mathcal{S}hv(\mathrm{Sm}_S)$ with any of the usual topologies. This implies that there is a fibre sequence

$$\Sigma^1_s \mathbb{G}_m \to B\mathbb{G}_m \vee B\mathbb{G}_m \to B\mathbb{G}_m.$$

$\mathbb{A}^1$-locally, this yields a fibre sequence

$$\mathbb{P}^1 \to \mathbb{P}^\infty \vee \mathbb{P}^\infty \to \mathbb{P}^\infty.$$

□

There are also other fibre sequences one can obtain: By considering similar diagrams as in [**DF96**, Appendix HL] we get the following fibration sequences in any model category of simplicial sheaves. In the next proposition, $X * Y$ denotes the join of X and Y which is defined as the homotopy pushout of the diagram $X \leftarrow X \times Y \to Y$.

Proposition 2.20. *Let $\mathcal{E}$ be a Grothendieck topos, and X be a simplicial object in $\mathcal{E}$. The sequence $\Omega X_0 * \Omega X_1 \to X_0 \vee X_1 \to X_0 \times X_1$ is a fibre sequence in $\Delta^{op}\mathcal{E}$.*

Proof. Apply Puppe's theorem 2.17 to the following diagram, the horizontal lines are the pushout diagrams and the vertical lines are fibre sequences:

$$\begin{array}{ccccc} \Omega X_0 \times * & \longleftarrow & \Omega X_0 \times \Omega X_1 & \longrightarrow & * \times \Omega X_1 \\ \downarrow & & \downarrow & & \downarrow \\ * \times X_1 & \longleftarrow & * & \longrightarrow & X_0 \times * \\ \downarrow & & \downarrow & & \downarrow \\ X_0 \times X_1 & \longleftarrow & X_0 \times X_1 & \longrightarrow & X_0 \times X_1. \end{array}$$

□

Example 2.21. *An instantiation of the above fibre sequence similar to the one given in Example 2.19 is the following fibre sequence in* $\Delta^{op}\mathcal{S}hv(\mathrm{Sm}_S)$*:*

$$\mathbb{G}_m * \mathbb{G}_m \to B\mathbb{G}_m \vee B\mathbb{G}_m \to B\mathbb{G}_m \times B\mathbb{G}_m.$$

$\mathbb{A}^1$*-locally, this yields a fibre sequence*

$$\mathbb{A}^2 \setminus \{0\} \to \mathbb{P}^\infty \vee \mathbb{P}^\infty \to \mathbb{P}^\infty \times \mathbb{P}^\infty.$$

□

As a final example, we restate yet another theorem of Ganea [**Gan65**]. It should by now be obvious, which diagram to apply Puppe's theorem to.

Proposition 2.22. *Let* $\mathcal{E}$ *be a Grothendieck topos, and let* $F \to E \to B$ *be any fibre sequence of simplicial objects. Then there is another fibre sequence*

$$F * \Omega B \longrightarrow E \cup CF = E/F \longrightarrow B.$$

2.8. Canonical Homotopy Colimit Decomposition

Let $p : E \to B$ be a fibration of fibrant simplicial sets. Then the canonical homotopy colimit decomposition of B allows to write B as homotopy colimit of standard simplices $\Delta^n \to B$. Then we can pull back the fibration p to these simplices and obtain the homotopy fibres. By homotopy distributivity, E can be written as the homotopy colimit over the simplex category $\Delta \downarrow B$ of the homotopy fibres. The same statement works for simplicial sheaves: The right notion to formulate it is the canonical homotopy colimit decomposition for objects in a combinatorial model category, which was described in detail in [**Dug01**].

Let $\mathcal{M}$ be a combinatorial model category, $\mathcal{C}$ be a small category. For any functor $I : \mathcal{C} \to \mathcal{M}$ and a fixed cosimplicial resolution $\Gamma_I : \mathcal{C} \to \Delta\mathcal{M}$, we obtain a functor $\mathcal{C} \times \Delta \to \mathcal{M} : (U, [n]) \mapsto \Gamma(n)(U)$. For any object X, we can consider the over-category (resp. comma category in Mac Lane's terminology [**Mac98**, Section II.6]) $(\mathcal{C} \times \Delta \downarrow X)$ and the canonical diagram $(\mathcal{C} \times \Delta \downarrow X) \to \mathcal{M} : \Gamma(n)(U) \mapsto U \times \Delta^n$.

Lemma 2.23. *Let* T *be a site, and let* $p : E \to B$ *be a fibration of fibrant simplicial sheaves. Then* p *is weakly equivalent to the morphism of simplicial sheaves*

$$\operatorname{hocolim} \mathcal{F} \to \operatorname{hocolim}(T \times \Delta \downarrow B),$$

where $(T \times \Delta \downarrow B)$ *is the canonical diagram associated to some fixed cosimplicial resolution, and the diagram* $\mathcal{F}$ *is the diagram of homotopy fibres: the index category is still* $(T \times \Delta \downarrow B)$*, but an object* $U \times \Delta^n \to B$ *is mapped to the pullback* $(U \times \Delta^n) \times_B E$*, which is the fibre of* p *over* U*.*

This is not as useful as the same construction for simplicial sets, since the homotopy types of the various $U \in T$ are different, which is the same as saying that a simplicial sheaf is not locally contractible. Therefore, not all of the simplicial sheaves $(U \times \Delta^n) \times_B E$ are weakly equivalent.

3. Preliminaries on Fibre Sequences

We first repeat the definition of fibre sequences in model categories, taken from [**Hov98**]. For details of the proof see [**Hov98**, Theorem 6.2.1].

Definition 3.1. *Given a fibration $p : E \to B$ of fibrant objects with fibre $i : F \to E$. There is an action of ΩB on F, given as follows. Let $h : A \times I \to B$ represent $[h] \in [A, \Omega B]$ and let $u : A \to F$ represent $[u] \in [A, F]$. We define $\alpha : A \times I \to E$ as the lift in the following diagram:*

$$\begin{array}{ccc} A & \xrightarrow{i \circ u} & E \\ {\scriptstyle i_0}\downarrow & & \downarrow{\scriptstyle p} \\ A \times I & \xrightarrow[h]{} & B \end{array}$$

Then define $[u].[h] = [w]$ with $w : A \to F$ to be the unique map satisfying $i \circ w = \alpha \circ i_1$.

This defines a natural right action of $[A, \Omega B]$ on $[A, F]$ for any A, which suffices to provide an action of ΩB on F.

Note that the action of ΩB on F is an action in the homotopy category: ΩB acts on F only up to homotopy since the action is defined by using the homotopy lifting property, cf. [**Sta74**].

This motivates the definition of fibre sequences [**Hov98**, Definition 6.2.6], given as follows:

Definition 3.2. *Let $\mathcal{C}$ be a pointed model category. A* fibre sequence *is a diagram $X \to Y \to Z$ together with a right action of ΩZ on X that is isomorphic in* Ho $\mathcal{C}$ *to a diagram $F \xrightarrow{i} E \xrightarrow{p} B$ where p is a fibration of fibrant objects with fibre i and F has the right ΩB-action of Definition 3.1.*

The following proposition shows that fibrations induce fibre sequences in the sense of of Definition 3.2.

Proposition 3.3. *Let $\mathcal{C}$ be a proper pointed model category. Let $p : E \to B$ be a fibration, and denote by F a cofibrant replacement of $p^{-1}(*)$. Then $F \to E \xrightarrow{p} B$ is a fibre sequence.*

3.1. Locally trivial morphisms

Already in the case of simplicial sets, one has to restrict the classification problem for fibrations to obtain a classifying space. One possible such restriction is to consider only base spaces B which are connected. Another approach is to consider only fibrations $p : E \to B$ for which the fibres $p^{-1}(b)$ have the weak homotopy type of F for all $b \in B$.

Also in the simplicial sheaf case, we need such a restriction. The obvious way to define connectedness for simplicial sheaves is the one used e.g. in [**MV99**, Corollary 2.3.22].

Definition 3.4. *Let X be a pointed simplicial sheaf on a Grothendieck site T. We say that X is* connected *if $L^2\pi_0 X = *$, where L^2 denotes sheafification. In other words, for any point x of the topos $\mathcal{S}hv(T)$, we require that the simplicial set $x^*(X)$ is connected.*

The main difference to the topological notion of connectedness is that a topological space is always the disjoint union of its connected components. This is no longer true for simplicial sheaves. The representable sheaves of a site can be viewed as constant simplicial sheaves; usually they are neither connected in the above sense nor decomposable into a direct sum of connected sheaves.

The topological way out of the connectivity problem therefore becomes a little awkward. We will consider a different type of condition which makes sure that the fibre sequences over a general simplicial sheaf form a set (at least after passing to equivalence classes). This is done by introducing local triviality with respect to a Grothendieck topology – the least common denominator of the algebraic topology and algebraic geometry usage of terms like fibration.

Definition 3.5. *Let T be a Grothendieck site. We say that a morphism $p : E \to B$ of simplicial sheaves is* locally trivial with fibre *F, if for each object U in T and each morphism $U \times \Delta^n \to B$, there exists a covering $\bigsqcup U_i \to U$ such that there are weak equivalences*

$$E \times_B (U_i \times \Delta^n) \simeq F \times (U_i \times \Delta^n).$$

Example 3.6. *As an example, consider the category of smooth manifolds with the Grothendieck topology generated by the open coverings. A fibre sequence $F \to E \to B$ is locally trivial if for each pullback $E \times_B M \to M$ of this sequence to a smooth manifold M, there exists a covering $\bigsqcup U_i \to M$ of M by open submanifolds such that $U_i \times_B E \simeq F$. But a fibration $E \to M$ over a smooth (connected) manifold M is always locally trivial in this sense: for each contractible open submanifold U of M, then $U \times_M E$ is weakly equivalent to the point set fibre over any point of U. Therefore, fibrations of connected topological spaces are indeed locally trivial in the above sense.*

*Note also that the local triviality condition forces all points to have fibres weakly equivalent to X. This shows that the above local triviality condition reduces to the usual assumptions used e.g. in [**All66**].* □

We remark that the results discussed in Section 2 are an analogue of the theory of quasi-fibrations, cf. [**DT58, DL59**]. In fact, we have the following:

Proposition 3.7. *Let $p : E \to B$ be a locally trivial morphism of pointed simplicial sheaves with fibre $F = p^{-1}(*)$. Then $F \to E \to B$ is a fibre sequence in the sense of Definition 3.2.*

4. First Variant: Brown Representability

In this section, we will construct classifying spaces of fibre sequences via the Brown representability theorem. For topological spaces, this approach was used by

Allaud [**All66**], Dold [**Dol66**], and Schön [**Sch82**]. A textbook treatment of this approach can be found in [**Rud98**].

4.1. Fibre Sequences Functor

We now define the functor mapping a simplicial sheaf to the *set* of fibre sequences with fixed fibre over this simplicial sheaf. We will work in the injective model category of *pointed* simplicial sheaves on some site T. This is due to the fact that fibre sequences as in Definition 3.2 are only defined in pointed model categories. Moreover, the Brown representability theorem also requires pointed model categories. There are examples in [**Hel81**] showing that Brown representability might fail already for unpointed topological spaces.

Definition 4.1. *Recall from Definition 3.2 that a fibre sequence over X with fibre F is a diagram $F \to E \xrightarrow{p} X$ with an ΩX-action on F which is isomorphic in the homotopy category to the fibre sequence associated to a fibration $p : \tilde{E} \to \tilde{X}$ of fibrant replacements $\tilde{X}$ of X and $\tilde{E}$ of E. Up to isomorphism in the homotopy category, we will usually assume that our fibre sequence $F \to E \xrightarrow{p} X$ is represented by some actual fibration over some fibrant replacement $\tilde{X}$ of X.*

A morphism of fibre sequences *is a diagram in $\Delta^{op}Shv(T)$*

$$\begin{array}{ccccc} F_1 & \longrightarrow & E_1 & \xrightarrow{p_1} & B_1 \\ {\scriptstyle f}\downarrow & & {\scriptstyle g}\downarrow & & \downarrow{\scriptstyle h} \\ F_2 & \longrightarrow & E_2 & \xrightarrow[p_2]{} & B_2, \end{array}$$

such that the left square commutes up to homotopy, and the right square is commutative, and f is Ωh-equivariant, i.e. the following diagram is homotopy commutative:

$$\begin{array}{ccc} \Omega B_1 \times F_1 & \longrightarrow & F_1 \\ {\scriptstyle \Omega h \times f}\downarrow & & \downarrow{\scriptstyle f} \\ \Omega B_2 \times F_2 & \longrightarrow & F_2. \end{array}$$

This in particular allows to define what an equivalence of fibre sequences over X is: Two fibre sequences over X with fibre F are equivalent if there is an isomorphism of fibre sequences

$$\begin{array}{ccccc} F & \longrightarrow & E_1 & \longrightarrow & X \\ {\scriptstyle \mathrm{id}}\downarrow & & \downarrow & & \downarrow{\scriptstyle \mathrm{id}} \\ F & \longrightarrow & E_2 & \longrightarrow & X, \end{array}$$

in the homotopy category $\mathrm{Ho}\,\Delta^{op}Shv(T)$. *We denote this by $E_1 \sim E_2$.*

Remark 4.2. *(i) The following can be assumed without loss of generality: we can assume that the base B is fibrant, that the morphism p is a fibration, and that F is the point-set fibre of p over $* \hookrightarrow B$. This basically follows from Proposition 3.3.*

(ii) *Note that in the definition of a morphism of fibre sequences we can always arrange for the right square to be commutative on the nose. We just lift the morphism $h \circ p_1$ along the fibration p_2. This makes the right square commutative, and leaves the left square commutative up to homotopy.*

(iii) *In the case of topological spaces, the above definition was used by Allaud, cf. [**All66**]. It coincides with the notion of fibre homotopy equivalence by a theorem of Dold, cf. [**Dol63**, Theorem 6.3].*

Lemma 4.3. *Equivalence of fibre sequences is an equivalence relation.*

Proof. This is clear since equivalence was defined by isomorphism in the homotopy category, which implies reflexivity, symmetry and transitivity. □

Definition 4.4 (Pullback of Fibre Sequences)**.** *Let $f : B_1 \to B_2$ be a pointed map, and let $F \to E_2 \to B_2$ be a fibre sequence. We define a fibre sequence with fibre F over B_1 as follows:*

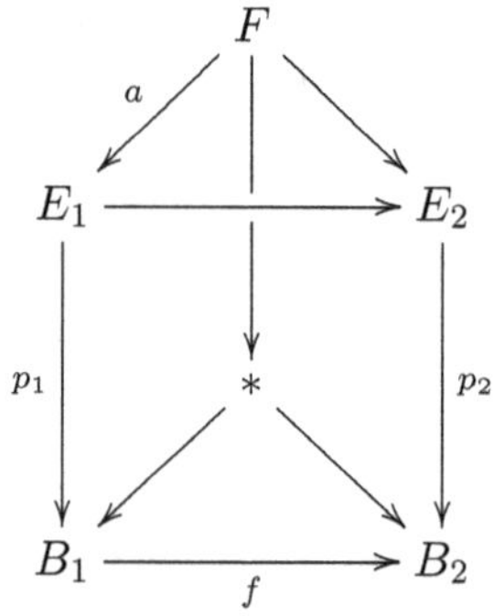

We assume that p_2 is a fibration, and define E_1 as the pullback $E_2 \times_{B_2} B_1$ of p_2 along f. Note that E_1 is therefore also the homotopy pullback of p_2 along f, and p_1 is a fibration. By the universal property of pullbacks, we have a morphism $a : F \to E_1$. Moreover, by the pullback lemma we have $p_1^{-1}(*) = F$, and since p_1 is a fibration, this is also the homotopy fibre.

Let T be a Grothendieck site. For given pointed simplicial sheaves X and F on T, let $\mathcal{H}^{\mathrm{pt}}(X, F)$ denote the collection of equivalence classes of locally trivial fibre sequences over X with fibre F modulo the equivalence $\sim$. We want to show that this is a set.

Proposition 4.5. *For any $X, F \in \Delta^{op}Shv(T)_*$, the collection $\mathcal{H}^{\mathrm{pt}}(X, F)$ is a set. Hence, with the pullbacks as in Definition 4.4, we have a functor*

$$\mathcal{H}^{\mathrm{pt}}(-, F) : \Delta^{op}Shv(T)_* \to \mathcal{S}et_*.$$

The natural base point of $\mathcal{H}^{\mathrm{pt}}(X, F)$ is given by the trivial fibre sequence $F \to X \times F \to X$, where the first map is inclusion via the base point $* \to X$, and the second is the product projection.

Proof. We first show that for every simplicial sheaf X there is only a set of equivalence classes of fibre sequences $F \to E \to X$. We follow the lines of [**Rud98**, Theorem IV.1.55]. Note that this includes forward references to Proposition 4.11, Proposition 4.13 and Proposition 4.12.

We start with the case of fibre sequences $F \to E \to U$ for $U \in T$ viewed as constant simplicial sheaf. We assume $E \to U$ is actually a fibration. For the above fibre sequence, there exists a covering U_i of U such that $F \to E \times_U U_i \to_U U_i$ is a trivial fibre sequence with given trivializations $E \times_U U_i \simeq F \times U_i$ and transition morphisms $F \times (U_i \times_U U_j) \to F \times (U_i \times U_j)$ which are weak equivalences. Now Proposition 2.15 implies that the original fibre sequence $F \to E \to U$ can be reconstructed up to equivalence as the homotopy colimit

$$F \to \operatorname{hocolim} E_i \to \operatorname{hocolim} U_i.$$

The Grothendieck site is (essentially) small, so there is only a set of coverings, and for a given covering, there is only a set of possible transition morphisms. The set of all locally trivial fibre sequences up to equivalence is therefore contained in the product of the sets of all possible transition morphisms (indexed by the possible coverings of U). It is therefore a set.

Next, we extend this result to simplicial sheaves of the form $U \times \Delta^n$ for $U \in T$. The argument in Proposition 4.11 is independent of the $\mathcal{H}^{\mathrm{pt}}(-, F)$ being sets. It therefore shows that for a weak equivalence $f : X \to Y$, if $\mathcal{H}^{\mathrm{pt}}(X, F)$ is a set, then so is $\mathcal{H}^{\mathrm{pt}}(Y, F)$. We find that for any simplicial sheaf of the form $U \times \Delta^n$ with $U \in T$, $\mathcal{H}^{\mathrm{pt}}(U \times \Delta^n, F)$ is a set.

Finally, we use the decomposition of fibrations over the canonical homotopy colimit presentation of the base simplicial sheaf, cf. Lemma 2.23. We consider F-fibre sequences over B, and decompose B as a homotopy colimit over the category of simplices $(T \times \Delta \downarrow B)$. The simplicial sheaves indexed by this diagram are of the form $U \times \Delta^n$ for $U \in T$, and we have already shown that fibre sequences over these form a set. Moreover, the site T is (essentially) small, therefore the diagram is set-indexed.

We now have to show that for any set-indexed homotopy colimit $\operatorname{hocolim}_\alpha X_\alpha$ of spaces X_α for which $\mathcal{H}^{\mathrm{pt}}(X_\alpha, F)$ is a set, the collection

$$\mathcal{H}^{\mathrm{pt}}(\operatorname*{hocolim}_{\alpha} X_\alpha, F)$$

is also a set. Since all homotopy colimits can be decomposed into homotopy pushouts and wedges, it suffices to show this assertion for these special homotopy colimits.

The proof of Proposition 4.13 shows that if $\mathcal{H}^{\mathrm{pt}}(X_\alpha, F)$ is a set for a set-indexed collection X_α, then $\mathcal{H}^{\mathrm{pt}}(\bigvee_\alpha X_\alpha, F)$ is also a set.

For the homotopy pushouts, we use the proof of Proposition 4.12. We get a surjective morphism of classes

$$d : \mathcal{H}^{\mathrm{pt}}(B_1 \cup_{B_0} B_2, F) \twoheadrightarrow \mathcal{H}^{\mathrm{pt}}(B_1, F) \times_{\mathcal{H}^{\mathrm{pt}}(B_0,F)} \mathcal{H}^{\mathrm{pt}}(B_2, F).$$

By assumption, $\mathcal{H}^{\mathrm{pt}}(B_1, F) \times_{\mathcal{H}^{\mathrm{pt}}(B_0,F)} \mathcal{H}^{\mathrm{pt}}(B_2, F)$ is a set. The morphism d decomposes a fibre sequence E over $B_1 \cup_{B_0} B_2$ into the pullbacks of the fibre sequence E to B_1 resp. B_2. These fibre sequences are remembered in the element in $\mathcal{H}^{\mathrm{pt}}(B_1, F) \times_{\mathcal{H}^{\mathrm{pt}}(B_0,F)} \mathcal{H}^{\mathrm{pt}}(B_2, F)$. What is forgotten, and what constitutes the kernel of d is the isomorphism between the pullbacks of E to B_1 resp. B_2. Since there is only a set of automorphisms for any given fibre sequence, the kernel of d is also a set. This implies that $\mathcal{H}^{\mathrm{pt}}(B_1 \cup_{B_0} B_2, F)$ is also a set.

Therefore, $\mathcal{H}^{\mathrm{pt}}(B, F)$ is a set for any simplicial sheaf B.

We still need to check that the pullback is really well-defined. This is a simple diagram check, using the cogluing lemma and therefore needing properness: assume we have two fibre sequences E_1 and E_2 over B, which are isomorphic in the homotopy category. We may assume that $p_i : E_i \to B$ are fibrations. If not, we choose factorizations. The independence of the choice of such fibrant replacements is proven in Proposition 4.6. We consider the pullback $E_i \times_B A$ of the fibre sequence E_i along the morphism $f : A \to B$. The isomorphism in the homotopy category lifts to a zig-zag of weak equivalences, so it suffices to show that a weak equivalence $g : E_1 \to E_2$ pulls back to a weak equivalence $f : E_1 \times_B A \to E_2 \times_B A$. This follows from the cogluing lemma [**GJ99**, Corollar II.8.13].

Finally, for any fibre sequence $F \to E \to B$, which is locally trivial in the T-topology the pullback $F \to E \times_B B' \to B'$ along any morphism B' is again locally trivial. This follows by a simple argument from the pullback lemma: the pullback of $E' = E \times_B B'$ along any morphism $U \to B'$ for $U \in T$ is also the pullback of E along $U \to B' \to B$. □

Proposition 4.6. *Let T be a Grothendieck site. All spaces and maps appearing below are in the category $\Delta^{\mathrm{op}}Shv(T)$ of simplicial sheaves on T.*

(i) For any commutative diagram

$$\begin{array}{ccc} E_1 & \xrightarrow{\simeq} & E_2 \\ & {\scriptstyle p_1}\searrow \quad \swarrow{\scriptstyle p_2} & \\ & B & \end{array}$$

with p_1 and p_2 fibrations, the induced weak equivalence on the fibres is equivariant for the ΩB-action in the homotopy category.

(ii) Let $p : E \to B$ be any morphism with homotopy fibre $F = \mathrm{hofib}\, p$. Then the class $[\tilde{p}] \in \mathcal{H}^{\mathrm{pt}}(B, F)$ of a fibrant replacement $\tilde{p} : \tilde{E} \to B$ of p is independent of the choice of fibrant replacement.

(iii) For any homotopy pullback

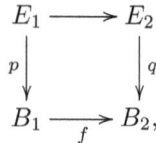

$$\begin{array}{ccc} E_1 & \longrightarrow & E_2 \\ {\scriptstyle p}\downarrow & & \downarrow{\scriptstyle q} \\ B_1 & \xrightarrow[f]{} & B_2, \end{array}$$

we have $f^*[q] = [p]$.

Proof. (i) This follows since the action as in Definition 3.1 is given by liftings in a diagram:

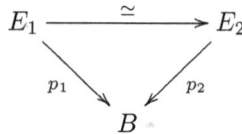

$$
\begin{array}{ccc}
A & \xrightarrow{\;iou\;} & E_1 \\
& \nearrow & \\
io \downarrow \quad \theta & & \downarrow \simeq \\
& & E_2 \\
& & \downarrow p_2 \\
A \times I & \xrightarrow{\;h\;} & B
\end{array}
$$

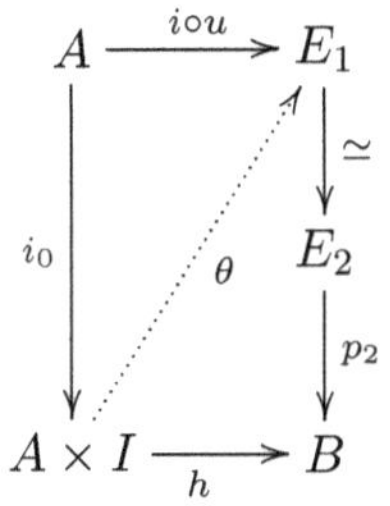

The action of ΩB on u is given by the lift θ which factors through the fibre. Lifting to E_1 and composing with the weak equivalence $E_1 \to E_2$ yields a lift for E_2. Therefore, the induced weak equivalence of the fibres is equivariant for the ΩB-action.

(ii) Note that the injective model structure on simplicial sheaves is a proper model category, see Theorem 2.1. By Proposition 3.3, $F \to E \to B$ is a fibre sequence for p a fibration. Now for an arbitrary map $p : E \to B$, we define $[p]$ as the fibre sequence associated to the fibration in a factorization

$$
\begin{array}{ccc}
E & \xrightarrow{\;\simeq\;}_{i} & \tilde{E} \\
& {\scriptstyle p}\searrow & \downarrow \tilde{p} \\
& & B.
\end{array}
$$

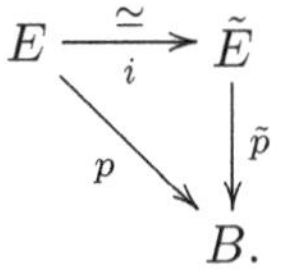

We need to prove that this is independent of the factorization. We consider two replacements of p by a fibration: $p_1 : E_1 \to B$ and $p_2 : E_2 \to B$. Note that by construction we have trivial cofibrations $E_1 \xrightarrow{\cong} E$ and $E_2 \xrightarrow{\cong} E$. We then consider the lift in the following diagram:

$$
\begin{array}{ccc}
E & \xrightarrow{\;\simeq\;} & E_2 \\
\simeq \downarrow & & \downarrow p_2 \\
E_1 & \xrightarrow{\;p_1\;} & B.
\end{array}
$$

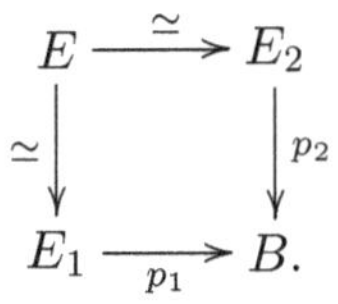

This is a weak equivalence by 2-out-of-3. As in the proof of Proposition 3.3 we obtain a weak equivalence between the fibres F_1 and F_2. By (i), this morphism is equivariant for the ΩB-action, so $[p_1] = [p_2]$.

(iii) Consider the homotopy pullback square in the statement of the proposition. By [**GJ99**, Lemma II.8.16], there exists a factorization of q into a trivial cofibration $i : E_2 \to \tilde{E}_2$ and fibration $\tilde{q} : \tilde{E}_2 \to B_2$ such that the induced morphism $E_1 \to B_1 \times_{B_2} \tilde{E}_2$ is a weak equivalence. Note that $f^*[q]$ is given by $B_1 \times_{B_2} \tilde{E}_2$. By (ii) we can take any fibration $\tilde{E}_1 \to B_1$ with $E_1 \to \tilde{E}_1$ a weak equivalence, and by (i), the homotopy fibres of $\tilde{E}_1$ and $B_1 \times_{B_2} \tilde{E}_2$ are weakly equivalent and the weak equivalence is ΩB-equivariant. Therefore $f^*[q] = [p]$. $\square$

4.2. Brown Representability

The Brown representability theorem is not really a single theorem, but rather a class of results stating conditions under which a set-valued functor on a model or homotopy category is representable. The first appearance is in the article of Brown

[**Bro62**] in which it is proven that a contravariant homotopy-continuous functor on the category of topological spaces is representable, with main application to the construction of spaces representing generalized cohomology theories. A more detailed analysis of why contravariant functors and pointed model categories are necessary assumptions was done in [**Hel81**]. Nowadays any reasonable textbook on algebraic topology contains a section on Brown representability for topological spaces.

There are not so many results on Brown representability for general, in particular unstable model categories. For a general model category, Brown representability usually fails, and at least some smallness assumptions are necessary. In this paper, we use the representability theorem by Jardine for compactly generated model categories, which is proven in [**Jar11**].

There are several names for the condition on the functors. Functors that satisfy the conditions for the representability were called half-exact in [**Bro62**], but we use the term homotopy-continuous. The terminology homotopy-continuous is reminiscent of Mac Lane's usage of the term *continuous* for a functor which preserves limits [**Mac98**, Section V.4]. Homotopy-continuous functors are the model category analogues of such continuous functors, as the Brown representability is a version of the adjoint functor theorem for model categories.

Definition 4.7 (Homotopy-Continuous Functor). *A functor $F : \mathcal{C}^{op} \to \mathcal{S}et_*$ on a pointed model category $\mathcal{C}$ is called* homotopy-continuous *if it satisfies the following assumptions:*

(HC i) *F takes weak equivalences to bijections.*

(HC ii) $F(*) = \{*\}$.

(HC iii) *For any coproduct $\bigvee_\alpha X_\alpha$ of a set $\{X_\alpha\}$ of objects of $\mathcal{C}$ the following* wedge axiom *is satisfied:*

$$F(\bigvee_\alpha X_\alpha) = \prod_\alpha F(X_\alpha).$$

(HC iv) *For any homotopy pushout*

$$\begin{array}{ccc} A & \longrightarrow & X \\ \downarrow & & \downarrow \\ B & \longrightarrow & Y, \end{array}$$

the induced morphism is surjective:

$$F(Y) \twoheadrightarrow F(B) \times_{F(A)} F(X).$$

This is called the Mayer-Vietoris axiom.

Now we recall Jardine's version of the Brown representability theorem [**Jar11**, Theorem 19]. In this version, we need the following definition, cf. [**Jar11**, Section 3]:

Definition 4.8. *A model category $\mathcal{C}$ is called* compactly generated, *if there is a set of compact cofibrant objects $\{K_i\}$ such that a map $f : X \to Y$ is a weak equivalence* *if and only if it induces a bijection*

$$[K_i, X] \xrightarrow{\cong} [K_i, Y]$$

for all objects K_i in the generating set.

This is a size condition that does not hold for all model categories of simplicial sheaves. It is explained in [**Jar11**, Section 3, p.88] that the injective model structure on the category simplicial sheaves on the Zariski resp. Nisnevich site is compactly generated.

Theorem 4.9 (Brown Representability (after Jardine)). *For a pointed, left proper, compactly generated model category $\mathcal{C}$ and a homotopy-continuous functor $F : \mathcal{C}^{op} \to \mathcal{S}et_*$, there exists an object Y of $\mathcal{C}$, a universal element $u \in F(Y)$, and a natural isomorphism*

$$T_u : \mathrm{Hom}_{\mathrm{Ho}(\mathcal{C})}(X, Y) \xrightarrow{\cong} F(X) : f \mapsto f^*(u)$$

for any object X of $\mathcal{C}$.

Corollary 4.10. *Let $\mathcal{C}$ be any left proper, compactly generated model category, let $F, G : \mathcal{C}^{op} \to \mathcal{S}et_*$ be homotopy-continuous functors with classifying spaces Y_F resp. Y_G and universal elements u_F resp. u_G. For any natural transformation $T : F \to G$ there exists a morphism $f : Y_F \to Y_G$, unique up to homotopy, such that the following diagram commutes for all $X \in \mathcal{C}$:*

$$\begin{array}{ccc}
\mathrm{Hom}_{\mathrm{Ho}(\mathcal{C})}(X, Y_F) & \xrightarrow{f_*} & \mathrm{Hom}_{\mathrm{Ho}(\mathcal{C})}(X, Y_G) \\
{\scriptstyle T_{u_F}(X)}\downarrow & & \downarrow{\scriptstyle T_{u_G}(X)} \\
F(X) & \xrightarrow[T(X)]{} & G(X)
\end{array}$$

Proof. We set $X = Y_F$. Then $T(Y_F) \circ T_{u_F}(Y_F)(\mathrm{id})$ yields an element of $G(X)$. By representability, we have that $T_{u_G}(Y_F)$ is an isomorphism and hence the element above is of the form $T_{u_G}(Y_F)(f)$ for a morphism $Y_F \to Y_G$, which is unique up to homotopy. □

4.3. Proof of Homotopy-Continuity

In this paragraph we will prove that the functor $\mathcal{H}^{\mathrm{pt}}(-, F)$ from Definition 4.1 is homotopy-continuous. Applying the Brown representability theorem discussed above, we get classifying spaces for fibre sequences and universal fibrations.

First note that $\mathcal{H}^{\mathrm{pt}}(*, F)$ is the singleton set consisting of the fibre sequence $F \to F \to *$, settling (HC ii).

The next serious thing to do is to show (HC i), i.e. that the functor $\mathcal{H}^{\mathrm{pt}}(-, F)$ is homotopically meaningful, in the sense that it carries weak equivalences between simplicial sheaves to isomorphisms of (pointed) sets. This implies in particular that there is a right derived functor $\mathbf{R}\mathcal{H}^{\mathrm{pt}}(-, F) : \mathrm{Ho}\,\mathcal{C}^{op} \to \mathcal{S}et_*$.

Proposition 4.11. *The functor $\mathcal{H}^{\mathrm{pt}}(-, F)$ sends weak equivalences of simplicial sheaves to bijections of pointed sets.*

Proof. Let $f : X \to Y$ be a weak equivalence, and consider $f^* : \mathcal{H}^{\mathrm{pt}}(Y, F) \to \mathcal{H}^{\mathrm{pt}}(X, F)$.

To show f^* is surjective, let $F \to E \xrightarrow{p} X$ be a fibre sequence with p a fibration. Consider the following diagram:

$$\begin{array}{ccc} E & \xrightarrow[\simeq]{i} & \tilde{E} \\ {\scriptstyle p}\downarrow & & \downarrow{\scriptstyle \tilde{p}} \\ X & \xrightarrow[f]{\simeq} & Y. \end{array}$$

Therein, $\tilde{E}$ is obtained by factoring $f \circ p$ into a trivial cofibration i and a fibration $\tilde{p}$. Since both i and f are weak equivalences, this square is a homotopy pullback. Hence by applying Proposition 4.6 we get $f^*\tilde{p} = p$.

To see that f^* is injective, let $p_1 : F \to E_1 \to Y$ and $p_2 : F \to E_2 \to Y$ be fibre sequences whose pullbacks are equivalent, i.e. $f^*p_1 = f^*p_2 \in \mathcal{H}^{\mathrm{pt}}(X, F)$. We assume that p_1 and p_2 are actually fibrations. By properness and the fact that f is a weak equivalence, we obtain weak equivalences $E_i \times_Y X \simeq f^*E_i \to E_i$. Therefore, the fibre sequences $F \to f^*E_i \to X$ and $F \to E_i \to Y$ are isomorphic in the homotopy category. Since we also assumed that $f^*p_1 = f^*p_2$, we have isomorphisms of fibre sequences in the homotopy category, which are also equivariant:

$$E_1 \xrightarrow{\simeq} f^*E_1 \xrightarrow{\simeq} f^*E_2 \xleftarrow{\simeq} E_2.$$

Thus p_1 and p_2 are equivalent fibre sequences. □

The following propositions will prove the two main parts of homotopy-continuity of the functor $\mathcal{H}^{\mathrm{pt}}(-, F)$, namely the Mayer-Vietoris and the wedge property. This is the point where we make essential use of the homotopy distributivity. This remarkable property allows to glue together fibrations defined on a covering of the base. The outcome will not be a fibration, but we still can determine the homotopy fibre, and therefore by Proposition 4.6, we know what the associated fibre sequence looks like. This is also the key argument in the work of Allaud [**All66**], although there is a lot more to do if one wants to work with homotopy equivalences of CW-complexes.

Proposition 4.12 (Mayer-Vietoris Axiom). *Let $B_1 \xleftarrow{\iota_1} B_0 \xrightarrow{\iota_2} B_2$ be a diagram of simplicial sheaves. We assume without loss of generality that $\iota_1 : B_0 \hookrightarrow B_1$ is in fact a cofibration, so that the homotopy pushout is given by the point-set pushout: $B := B_1 \cup^h_{B_0} B_2 = B_1 \cup_{B_0} B_2$.*

Then the induced morphism

$$\mathcal{H}^{\mathrm{pt}}(B_1 \cup_{B_0} B_2, F) \twoheadrightarrow \mathcal{H}^{\mathrm{pt}}(B_1, F) \times_{\mathcal{H}^{\mathrm{pt}}(B_0, F)} \mathcal{H}^{\mathrm{pt}}(B_2, F)$$

is surjective.

Proof. What we have to show is the following: assume given two fibre sequences $F \to E_1 \to B_1$ and $F \to E_2 \to B_2$, such that the corresponding pullbacks $\iota_1^*E_1$ and $\iota_2^*E_2$ are isomorphic fibre sequences via a given isomorphism $\rho : \iota_1^*E_1 \xrightarrow{\cong} \iota_2^*E_2$. Then we have to show that there is a fibre sequence over B inducing them compatibly.

We will apply Mather's cube theorem from Corollary 2.16. Since we know that the pullbacks of the fibre sequences are equivalent, the squares in the diagram are homotopy pullback squares:

$$\begin{array}{ccccc} E_1 & \longleftarrow & \iota_1^* E_1 \cong \iota_2^* E_2 & \longrightarrow & E_2 \\ {\scriptstyle p_1}\big\downarrow & & \big\downarrow & & \big\downarrow{\scriptstyle p_2} \\ B_1 & \underset{\iota_1}{\longleftarrow} & B_0 & \underset{\iota_2}{\longrightarrow} & B_2. \end{array}$$

The homotopy fibres of the vertical arrows are all weakly equivalent to F. Then Mather's cube theorem produces the following fibre sequence:

$$F \to E_1 \cup^h_{E_0} E_2 \xrightarrow{p} B \cong B_1 \cup^h_{B_0} B_2.$$

Actually, we only get the morphism p, and know that its homotopy fibre is F. Then we still have to do a fibrant replacement to really get a fibre sequence with total space $E := E_1 \cup^h_{E_0} E_2 \in \mathcal{H}^{\mathrm{pt}}(B, F)$.

What is left to show is that pulling back E to B_i yields equivalences $\phi : E_1 \xrightarrow{\cong} E$ and $\psi : E_2 \xrightarrow{\cong} E$ such that over B_0 we have $\phi = \psi \circ \rho$. This also follows from homotopy distributivity: We assume p has been rectified to a fibration. Since the squares

$$\begin{array}{ccc} E_i & \longrightarrow & E \\ \big\downarrow{\scriptstyle p_i} & & \big\downarrow{\scriptstyle p} \\ B_i & \longrightarrow & B \end{array}$$

are homotopy pullback squares, the map $E_i \to E$ factors through a unique weak equivalence from E_i to the point-set pullback of E along $B_i \to B$. This provides the equivalences ϕ and ψ. All we need to show is that the following diagram is commutative:

$$\begin{array}{ccccc} & \swarrow & \iota_1^* E_1 & \searrow & \\ & & \big\downarrow{\scriptstyle \rho} & & \\ B_0 & \longleftarrow & \iota_2^* E_2 & \longrightarrow & E \\ & \nwarrow & \big\downarrow{\scriptstyle \psi} & \nearrow & \\ & & E \times_B B_0 & & \end{array}$$

By the universal property of the pullback, this implies that $\psi \circ \rho = \phi$, since the morphism $\iota_1^* E_1 \to E \times_B B_0$ is by definition ϕ.

The upper left triangle commutes, since ρ was defined over B_0. The lower triangles commute because ψ was defined using the universal property of the pullback $E \times_B B_0$. The upper right triangle commutes because E was defined as the glueing of $\iota_1^* E_1$ and $\iota_2^* E_2$ along ρ. Therefore, we get that $\phi = \psi \circ \rho$. In particular, the image of $E \in \mathcal{H}^{\mathrm{pt}}(B, F)$ in $\mathcal{H}^{\mathrm{pt}}(B_1, F) \times_{\mathcal{H}^{\mathrm{pt}}(B_0, F)} \mathcal{H}^{\mathrm{pt}}(B_2, F)$ is exactly the class of (E_1, E_2, ρ) we started with.

Finally, note that the result of glueing two locally trivial fibre sequences in a homotopy colimit produces again a locally trivial fibre sequence. The canonical homotopy colimit decomposition reduces this assertion to the case of homotopy colimits of simplicial dimension zero representable sheaves, where it is obvious. □

Proposition 4.13 (Wedge Axiom). *Let B_α with $\alpha \in I$ be a set of pointed simplicial sheaves. Then*

$$\mathcal{H}^{\mathrm{pt}}(\bigvee_\alpha B_\alpha, F) \to \prod_\alpha \mathcal{H}^{\mathrm{pt}}(B_\alpha, F) : E \mapsto \iota_\alpha^*(E)$$

is a bijection, where $\iota_\alpha : B_\alpha \hookrightarrow \bigvee_\alpha B_\alpha$ denotes the inclusion. In particular, the collection $\mathcal{H}^{\mathrm{pt}}(\bigvee_\alpha B_\alpha, F)$ is a set if $\mathcal{H}^{\mathrm{pt}}(B_\alpha, F)$ is a set for every $\alpha \in I$.

Proof. Define E via the following homotopy pushout, where the maps $F \to E_\alpha$ are the ones from the definition of fibre sequence:

$$\begin{array}{ccc} \bigvee_\alpha F & \longrightarrow & \bigvee_\alpha E_\alpha \\ \downarrow & & \downarrow \\ F & \longrightarrow & E \end{array}$$

By homotopy distributivity we find that the following squares are homotopy pullbacks for all α:

$$\begin{array}{ccc} E_\alpha & \longrightarrow & E \\ \downarrow & & \downarrow \\ B_\alpha & \longrightarrow & \bigvee_\alpha B_\alpha. \end{array}$$

Therefore, the homotopy fibre of $\bigvee_\alpha p_\alpha$ is also F. Rectifying it to a fibre sequence, we get $\bigvee_\alpha E_\alpha \in \mathcal{H}^{\mathrm{pt}}(\bigvee_\alpha B_\alpha, F)$. This proves surjectivity.

Now assume given two fibre sequences E_1 and E_2 over $\bigvee_\alpha B_\alpha$. We first prove the weak equivalence $\bigvee_\alpha \iota_\alpha^* E \simeq E$ for any fibre sequence $E \to \bigvee_\alpha B_\alpha$. This follows from distributivity in categories of simplicial sheaves, since E is isomorphic to the colimit of $\iota_\alpha^* E_\alpha$. Then one can either use that the wedge is already the homotopy direct product, or again appeal to the homotopy distributivity. Then we have the sequence of weak equivalences:

$$E_1 \simeq \bigvee_\alpha \iota_\alpha^* E_1 \simeq \bigvee_\alpha \iota_\alpha^* E_2 \simeq E_2.$$

The middle weak equivalence follows from the fact [**Jar96**, Lemma 13.(3)] that for a set-indexed collection of weak equivalences $f_\alpha : B_\alpha \to Y_\alpha$, the morphism $\bigvee_\alpha f_\alpha$ is also a weak equivalence and the assumption that $\prod \iota_\alpha^* E_1 = \prod \iota_\alpha^* E_2$ in $\prod_\alpha \mathcal{H}^{\mathrm{pt}}(B_\alpha, F)$.

The set theory statement is then clear, since a set-indexed product of sets is again a set. □

Theorem 4.14. *Assume the model category $\Delta^{op}Shv(T)$ is compactly generated. The functor $\mathcal{H}^{\mathrm{pt}}(-,F)$ which associates to each simplicial sheaf X the set of fibre sequences over X with fibre F is homotopy continuous and therefore representable by a space B^fF. The space B^fF is unique up to weak equivalence.*

The universal element $u_F \in \mathcal{H}^{\mathrm{pt}}(B^fF,F)$ corresponds to the universal fibre sequence of simplicial sheaves with fibre F:

$$F \to E^fF \to B^fF$$

Remark 4.15. *(i) We use the notation B^fF to distinguish from other possible classifying spaces we will discuss later on.*

*(ii) There should be a simplicial functor associating to each space B the nerve of the category of fibre sequences over this space – one has to circumvent the obvious set-theoretical difficulty in the construction. It seems likely that the other Brown representability theorem [**Jar11**, Theorem 12] can then be applied to this functor. This would allow to remove the compact generation hypothesis in the above theorem. Anyway, the next section will show that classification of non-rooted fibrations is always possible.*

Using homotopy distributivity once again, we can construct change-of-fibre natural transformations, which via Brown representability for morphisms give rise to morphisms between the corresponding classifying spaces. We state the result without giving the obvious proof.

Proposition 4.16. *Assume the model category $\Delta^{op}Shv(T)$ is compactly generated. For any morphism $f: F \to F'$ of simplicial sheaves, there is a natural transformation $\mathcal{H}^{\mathrm{pt}}(-,F) \to \mathcal{H}^{\mathrm{pt}}(-,F')$. By Brown representability this is representable by a morphism $B^fF \to B^fF'$.*

Similarly, it is possible to generalize operations on fibrations, cf. [**Rud98**, Proposition 1.43] or [**May80**], to the simplicial sheaf setting.

Corollary 4.17. *Assume the model category $\Delta^{op}Shv(T)$ is compactly generated. There are morphisms of classifying spaces associated to fibrewise smash*

$$B^f(\wedge) : B^fF_1 \times B^fF_2 \to B^f(F_1 \wedge F_2),$$

and fibrewise suspension $B^fF \to B^f\Sigma F$.

5. Second Variant: Bar Construction

In this section, we explain the second approach to the construction of classifying spaces of fibre sequences. Again, this approach is a direct generalization of results that are known for topological spaces resp. simplicial sets. The first result in this direction is the work of Stasheff [**Sta63**] which proves that fibrations over CW-complexes with a given finite CW-complex as fibre can be classified by homotopy classes of maps into some CW-complex. In fact the classifying space is the classifying space of the topological monoid of homotopy self-equivalences of the fibre. The main idea in this approach is the construction of an associated principal bundle for a fibration. This associates to a fibration $p : E \to B$ a new fibration Prin(p) : Prin(E) $\to B$, whose fibre have the homotopy type of the topological monoid of homotopy self-equivalences.

A vast generalization of this can be found in [**May75**]. There, the double bar construction is used to construct the classifying spaces for fibrations with given fibre. Moreover, the notion of a category of fibres allows to classify fibrations with global structures. Again the main point in proving the classification theorem is a principalization construction which associates to a fibration a principal bundle.

These results can be translated to simplicial sheaves. One problem that appears in this setting is that principalization does not work a priori. The way around this is again the restriction to fibre sequences which are locally trivial in a given topology. These trivializations indeed allow to translate the principalization construction.

In the case of CW-complexes, the local triviality condition is no restriction at all: every point has a contractible neighbourhood U. For a fibration over such a contractible neighbourhood, the inclusion of the fibre is a weak equivalence (in fact a homotopy equivalence if all the spaces in sight are CW-complexes). This means that there is a morphism (over U) $E \to F \times U$ which is a weak equivalence. This provides the local trivializations in the case of CW-complexes. In fact, it allows to construct a morphism from the associated Čech-complex of a fine enough covering to the classifying space of the monoid of self-equivalences – for each intersection $U_i \cap U_j$ of contractible neighbourhoods there is a morphism $U_i \cap U_j \to \mathrm{hAut}_\bullet(F)$ corresponding to the composition of the two trivializations over U_i resp. U_j. The cocycle condition is not satisfied on the nose, but up to homotopy. Therefore, one obtains a morphism $U_i \cap U_j \cap U_k \times I \to \mathrm{hAut}_\bullet(X)$ etc. Since the realization of the Čech complex is homotopy equivalent to the CW-complex we started with, we obtain a map in the homotopy category $B \to B\,\mathrm{hAut}_\bullet(F)$. This is a slightly souped up version of the principalization construction, which also works in the simplicial sheaf setting. Hopefully, it has become clear with the above discussion that the local triviality condition on the fibre sequences comes in rather naturally in the bar construction approach.

5.1. Fibre Sequence Functor

We now define the functor which will be represented. In the case of the bar construction, this functor is the unpointed analogue of the one defined in Definition 4.1. It associates to an unpointed simplicial sheaf B the set of all locally trivial fibre sequences over B with fibre F. Therefore, it does not fix an equivalence between F and $p^{-1}(*)$.

Definition 5.1. *Recall the definition of locally trivial morphism with fibre F. Two locally trivial morphisms $p_1 : E_1 \to B$ and $p_2 : E_2 \to B$ with fibre F are said to be* equivalent *if there is a diagram in the homotopy category*

$$\begin{array}{ccc} E_1 & \xrightarrow{\alpha} & E_2 \\ {\scriptstyle p_1}\downarrow & & \downarrow{\scriptstyle p_2} \\ B & \xrightarrow[\mathrm{id}]{} & B \end{array}$$

where α is an isomorphism. We denote by $\mathcal{H}(X, F)$ the set of locally trivial morphisms over X with fibre F modulo the above equivalence relation.

Remark 5.2. *(i) Assuming that the p_i are fibrations, we can use the homotopy lifting to obtain a morphism which respects fibres. So if p_1 and p_2 are equivalent, then there is an equivalence which respects the fibres.*

(ii) The analogue of Proposition 4.5 can be proved in complete analogy, we omit the proof.

(iii) In case X is actually pointed, we can obtain the set $\mathcal{H}(X, F)$ by taking fibre sequences over X modulo the equivalence relation given by ladder diagrams in the homotopy category

$$\begin{array}{ccccc} F & \longrightarrow & E_1 & \longrightarrow & X \\ \downarrow{\scriptstyle \beta} & & \downarrow{\scriptstyle \alpha} & & \downarrow{\scriptstyle \mathrm{id}} \\ F & \longrightarrow & E_2 & \longrightarrow & X, \end{array}$$

where α and β should be isomorphisms.

5.2. Remarks on Categories of Fibres

In the following, we will not work in the full generality of categories of fibres. Rather we will only consider the fibre sequences which are locally trivial from Definition 4.1. However, we want to make a few remarks on the possible definition of category of fibres for simplicial sheaves.

The original definition of categories of fibres can be found in [**May75**]. A definition of categories of fibres in equivariant topology has been given Waner [**Wan80**, Definition 1.1.1] resp. French [**Fre03**, Definition 3.1] for equivariant homotopy theory. These definitions readily generalize to simplicial sheaves. One should however note that equivariant topology is a presheaf situation without a Grothendieck topology (at least in the case of finite groups) – in the full generalization it is therefore necessary to include a localization condition.

Definition 5.3 (Category of Fibres). *Let T be a site. A* category of fibres *is a subcategory $\mathcal{F}_T$ of the following category:*

- *Objects are morphisms $p : X \to U$ of simplicial sheaves, where U is the constant simplicial sheaf for a representable $U \in T$.*
- *Morphisms are commutative diagrams*

$$\begin{array}{ccc} X & \longrightarrow & X' \\ \downarrow & & \downarrow \\ U & \longrightarrow & U' \end{array}$$

Additionally, we require that

(CFi) The map $X \to U$ is required to be locally trivial in the T-topology.

(CFii) For a morphism

$$\begin{array}{ccc} X & \longrightarrow & X' \\ \downarrow & & \downarrow \\ U & \longrightarrow & U' \end{array}$$

there is a T-covering $\bigsqcup U_i \to U'$ such that the induced morphisms $X \times_{U'} U_i \to X' \times_{U'} U_i$ are weak equivalences of simplicial sheaves.

As in the equivariant definitions of categories of fibres one wants to have a simplicial sheaf F which serves as a model for the fibres: a corresponding category should contain at least the obvious objects $p_2 : F \times U \to U$ for $U \in T$, together with the obvious morphisms

$$\begin{array}{ccc} F \times U_1 & \xrightarrow{\mathrm{id} \times f} & F \times U_2 \\ {\scriptstyle p_2}\downarrow & & \downarrow{\scriptstyle p_2} \\ U_1 & \xrightarrow[f]{} & U_2 \end{array}$$

induced from $f : U_1 \to U_2$ in T.

The notion of Γ-completeness which appears in the cited works on categories of fibres basically state that the category of fibres should be closed under fibrant replacements. This is needed since some constructions (like glueing) yield quasi-fibrations instead of fibrations, and one would like to replace them by fibrations without losing the property that the fibres are elements in the category of fibres.

The basic definitions and results concerning categories of fibres and their principalizations can then be translated from e.g. [**Fre03**]. As said before, we will only consider locally trivial fibre sequences with given fibre, as defined in Definition 4.1. It is easy to check that this definition can be formulated as a special case of a category of fibres.

5.3. Homotopy Self-Equivalences

Most important for our studies in the sequel will be the simplicial monoid of homotopy self-equivalences of a simplicial sheaf. This is the obvious generalization of the homotopy self-equivalences of a simplicial set.

We first recall the definition of homotopy self-equivalences of simplicial sets. For more details on function complexes of simplicial sets, see [**GJ99**, Section I.5]. Function complexes in general model categories are constructed in [**DK80**]. A general discussion about what is known for the monoids of homotopy self-equivalences can be found in [**Rut97**].

Definition 5.4. *Let X be a fibrant simplicial set. Then there is a simplicial set $\mathbf{Hom}(X, X)$ whose set of n-simplices is given by*

$$\mathbf{Hom}(X, X)_n = \hom_{\Delta^{op}\mathcal{S}et}(X \times \Delta^n, X).$$

*This is a special case of function complexes of simplicial sets, cf. [**GJ99**]. By standard facts on function complexes, there is a fibration*

$$\mathbf{Hom}(X, X) \to \mathbf{Hom}(*, X) \simeq X,$$

therefore $\mathbf{Hom}(X, X)$ *is also a fibrant simplicial set.*

The monoid structure can be described as follows: for two maps $f, g : \Delta^n \times X \to X$*, their composition* $f \circ g$ *in the monoid* $\mathbf{Hom}(X, X)_n$ *is given by*

$$f \circ g : \Delta^n \times X \xrightarrow{D \times \mathrm{id}} \Delta^n \times \Delta^n \times X \xrightarrow{\mathrm{id} \times g} \Delta^n \times X \xrightarrow{f} X,$$

where $D : \Delta^n \to \Delta^n \times \Delta^n$ *is the diagonal morphism on the standard* n*-simplex* Δ^n.

It is obvious that the simplicial subset of morphisms $X \to X$ *which are weak equivalences is in fact a simplicial submonoid. The resulting monoid of homotopy self-equivalences is denoted by* $\mathrm{hAut}_\bullet(X)$.

Note that this monoid is group-like since X *is cofibrant and fibrant. In this case, a weak equivalence* $f : X \to X$ *is a homotopy equivalence and therefore its class in* $\pi_0\mathbf{Hom}(X, X)$ *has an inverse.*

The general definition of homotopy self-equivalences in general model category was given by Dwyer and Kan in [**DK80**]. Their construction yields for an object X in a model category $\mathcal{C}$ a function complex $\mathrm{hom}(X, X)$ which is a simplicial set. For simplicial sheaves, we can additionally use the internal Hom to obtain a simplicial sheaf of monoids of homotopy self-equivalences. It is explained in [**MV99**, Remark 1.1.7, Lemma 1.1.8] that the category of simplicial sheaves has internal hom-objects.

Definition 5.5. *Let* T *be a site, and let* X *be a fibrant simplicial sheaf. We define the sheaf of self-homotopy equivalences, which is a simplicial sheaf of monoids. By Theorem 2.1, the simplicial sheaves on* T *form a simplicial model category, hence for any two simplicial sheaves* X, Y *there is a simplicial set, the function complex* $\mathrm{Hom}(X, Y)$*, whose* n*-simplices are given by*

$$\mathrm{Hom}_{\Delta^{op}\mathcal{S}hv(T)}(X \times \Delta^n, Y).$$

In particular, we have a contravariant functor

$$T^{\mathrm{op}} \to \Delta^{op}\mathcal{S}et : (U \in T) \mapsto \mathrm{Hom}_{\Delta^{op}\mathcal{S}hv(T)}(X \times U, X).$$

This functor is representable by a simplicial sheaf which we again denote by

$$\mathbf{Hom}_{\Delta^{op}\mathcal{S}hv(T)}(X, X).$$

We can define a subpresheaf by taking for $U \in T$ *the subset of those morphisms* $\mathrm{Hom}_{\Delta^{op}\mathcal{S}hv(T)}(X \times U, X \times U)$ *which are weak equivalences of simplicial sheaves in* $\Delta^{op}\mathcal{S}hv(T)$*. Note that this is indeed a sheaf because weak equivalences are defined locally: given a covering* $\bigsqcup U_i \to U$ *and weak equivalences* $f_i : X \times U_i \to X \times U_i$ *which agree on the intersections, there is morphism* $f : U \to U$ *which is a weak equivalence if all the* f_i *are weak equivalences.*

The resulting simplicial sheaf of monoids will be denoted by $\mathrm{hAut}_\bullet(X)$*. The monoid structure is again given by composition as in Definition 5.4.*

Note that the simplicial sheaf of monoids $\mathrm{hAut}_\bullet(X)$ *is fibrant if* X *is. This is a consequence of the simplicial model structure on simplicial sheaves, cf. [**GJ99**, Proposition II.3.2]: the morphisms* $\mathbf{Hom}(X, X) \to \mathbf{Hom}(*, X)$ *and* $\mathbf{Hom}(*, X) \to \mathbf{Hom}(*, *) \cong *$ *induced from the morphism* $X \to *$ *are fibrations if* X *is fibrant.*

Lemma 5.6. *Let X be a fibrant simplicial sheaf on the site T. Then X is a left* $\mathrm{hAut}_\bullet(X)$ *space, i.e. there is an action*

$$\mathrm{hAut}_\bullet(X) \times X \to X.$$

Note that if X is fibrant, then a morphism $X \to X$ is a weak equivalence if and only if the morphism induced on sections $f(U) : X(U) \to X(U)$ is a weak equivalence of simplicial sets for all $U \in T$, cf. [**MV99**, Lemma I.1.10]. Therefore, $\mathrm{hAut}_\bullet(X)(U)$ acts on $X(U)$ via homotopy self-equivalences of simplicial sets. Note also that the action is really an action in $\Delta^{op}\mathcal{S}hv(T)$, not just an action in the homotopy category.

5.4. The Bar Construction

We repeat the definition and basic properties of the bar construction following [**May75**]. Again the setting changes from topological spaces to simplicial sheaves without major complications, cf. also [**MV99**, Example 4.1.11].

Definition 5.7 (Two-sided geometric bar construction)**.** *Let G be a simplicial sheaf of monoids on the site T. We assume that the inclusion of the identity $e \to G$ is a cofibration. For the injective model structure, this is no problem because every monomorphism is a cofibration. Let X and Y be simplicial sheaves, such that X has a left G-action and Y has a right G-action.*

Then there is a bisimplicial sheaf

$$B_{n,m}(Y, G, X) = (Y \times G^n \times X)_m.$$

For an object U of the site T, we have

$$B_{n,m}(Y, G, X)(U) = B_{n,m}(Y(U), G(U), X(U)),$$

and functoriality of the bar construction for simplicial sets provides the restriction maps to turn this into a simplicial sheaf. Similarly, the face and degeneracy maps are functorial, and hence provide $B_{n,m}(Y, G, X)$ above with the structure of bisimplicial sheaf. The diagonal $B_{n,n}(Y, G, X)$ is a simplicial sheaf, which we will denote $B(Y, G, X)$.

The classifying spaces for simplicial sheaves of monoids can then be obtained as $BG = B(*, G, *)$, and the universal G-bundle is given by the obvious functoriality:

$$EG = B(*, G, G) \to B(*, G, *) = BG.$$

The topology enters via a fibrant replacement: for a simplicial sheaf X, any morphism $X \to BG$ in the homotopy category can be represented up to homotopy by a morphism $X' \to BG$ for some suitable trivial local fibration $X' \to X$. The notion of trivial local fibration depends on the topology, as a trivial local fibration is a morphism of simplicial sheaves which induces a trivial Kan fibration of simplicial sets on the stalks. The fibrant replacement may change the global sections of $B(Y, G, X)$, but it does not change the homotopy types of the stalks, which therefore can be described as the bar constructions for the simplicial sets p^*Y, p^*G and p^*X.

The following properties of the bar construction for simplicial sheaves are direct consequences of the corresponding properties for simplicial sets resp. topological spaces, cf. [**May75**, Section 7]:

Proposition 5.8. *(i) The space $B(Y, G, X)$ is n-connected provided G is $(n-1)$-connected and X and Y are n-connected.*

(ii) If $f_1 : Y \to Y'$, $f_2 : G \to G'$ and $f_3 : X \to X'$ are weak equivalences of simplicial sheaves, then the morphism $f : B(Y, G, X) \to B(Y', G', X')$ is a weak equivalence.

(iii) For (Y, G, X) and (Y', G', X') the projections define a natural weak equivalence

$$B(Y \times Y', G \times G', X \times X') \to B(Y, G, X) \times B(Y', G', X').$$

(iv) Let $f : H \to G$ be a morphism of simplicial sheaves of monoids, and let $k : Z \to Y$ be an equivariant morphism of right G-spaces. Then the following diagrams are pullbacks:

$$\begin{array}{ccc} B(Z, H, X) & \xrightarrow{B(k,f,\mathrm{id})} & B(Y, G, X) \\ {\scriptstyle p}\downarrow & & \downarrow{\scriptstyle p} \\ B(Z, H, *) & \xrightarrow[B(k,f,\mathrm{id})]{} & B(Y, G, *). \end{array}$$

$$\begin{array}{ccc} B(Y, G, X) & \xrightarrow{q} & B(*, G, X) \\ {\scriptstyle p}\downarrow & & \downarrow{\scriptstyle p} \\ B(Y, G, *) & \xrightarrow[q]{} & BG \end{array}$$

Proof. For (i), note that n-connectedness means that the homotopy group *sheaves* $\pi_i(B(Y, G, X))$ are trivial for $i \leqslant n$. In particular, this does not imply that the simplicial sets $B(Y, G, X)(U)$ are n-connected for any $U \in T$.

All four statements are of a local nature, i.e. can be checked on stalks. The corresponding statements for topological spaces are Propositions 7.1, 7.3, 7.4 and 7.8 of [**May75**]. □

The following result is a version of [**May75**, Theorem 7.6, Proposition 7.9] for simplicial sheaves. It provides necessary fibre sequences for the proof of the classification theorem. Note that for any simplicial sheaf of monoids M, the monoid operation induces a monoid operation on the sheaf $\pi_0 M$. We say that M is *grouplike* if this operation turns $\pi_0 M$ into a sheaf of groups.

Theorem 5.9. *If G is grouplike, there are fibre sequences of simplicial sheaves*

(i) $X \to B(Y, G, X) \to B(Y, G, *)$,

(ii) $Y \to B(Y, G, X) \to B(*, G, X)$, and

(iii) $G \to Y \to B(Y, G, *)$.

Proof. The corresponding statements for simplicial sets resp. topological spaces can be found as [**May75**, Theorem 7.6, Proposition 7.9]. The corresponding statements are true for simplicial sheaves by Proposition 2.11: everything that locally (i.e. on stalks) looks like a fibre sequence, really is a fibre sequence. □

5.5. The Classification Theorem

Now we come to the proof of the classification theorem. The classifying space is given by the bar construction $B(*, \mathrm{hAut}_\bullet(F), *)$ and the universal fibre sequence is

$$F \to B(*, \mathrm{hAut}_\bullet(F), F) \to B(*, \mathrm{hAut}_\bullet(F), *).$$

It follows from the previous Theorem 5.9 that this is indeed a fibre sequence of simplicial sheaves.

The following is a version of May's classification result [**May75**, Theorem 9.2] for simplicial sheaves. The argument in the topological case can be found in [**Sta74, SW06**].

Theorem 5.10. *Let T be a site, F be a fibrant simplicial sheaf on T. Then there is a natural isomorphism of functors*

$$\mathcal{H}(X, F) \cong [X, B(*, \mathrm{hAut}_\bullet(F), *)],$$

where the right-hand side denotes the set of morphisms

$$X \to B(*, \mathrm{hAut}_\bullet(F), *)$$

in the homotopy category.

Proof. The universal fibre sequence is

$$F \to B(*, \mathrm{hAut}_\bullet(F), F) \to B(*, \mathrm{hAut}_\bullet(F), *).$$

We can replace this by an honest fibration of fibrant simplicial sheaves whose fibre is weakly equivalent to F. This can be viewed as an element of

$$\mathcal{H}(B(*, \mathrm{hAut}_\bullet(F), *), F)$$

which we denote by π.

(i) Now we define a natural transformation

$$\Psi : [X, B(*, \mathrm{hAut}_\bullet(F), *)] \to \mathcal{H}(X, F) : f \mapsto f^*\pi$$

This is well-defined and natural by Proposition 4.6.

(ii) In the other direction, we define

$$\Phi : \mathcal{H}(X, F) \to [X, B(*, \mathrm{hAut}_\bullet(F), *)]$$

via the following principalization construction. Let $F \to E \to X$ be a fibre sequence in $\mathcal{H}(X, F)$. By assumption, this is locally trivial, i.e. there exists a covering $\bigsqcup U_i \to X$ such that $E \times_X U_i \simeq F \times U_i$.

By composition of the two trivializations for U_i, U_j, we obtain a weak equivalence over $U_i \times_X U_J$:

$$\phi_{ij} : F \times (U_i \times_X U_j) \to F \times (U_i \times_X U_j),$$

which corresponds to a morphism $U_i \times_X U_j \to \mathrm{hAut}_\bullet(F)$.

Then there is a diagram of weak equivalences

$$\begin{array}{ccc} F\times(U_i\times_X U_j\times_X U_k) & \xrightarrow{\phi_{ij}} & F\times(U_i\times_X U_j\times_X U_k) \\ & \searrow^{\phi_{ik}} & \downarrow{\scriptstyle \phi_{jk}} \\ & & F\times(U_i\times_X U_j\times_X U_k). \end{array}$$

This diagram is not commutative but commutative up to homotopy, hence gives rise to a morphism $U_i \times_X U_j \times_X U_k \times \Delta^1 \to \mathrm{hAut}_\bullet(F)$.

In the usual way, we obtain a T-hypercovering $U_\bullet \to X$ and a morphism of simplicial sheaves $U_\bullet \to B(*, \mathrm{hAut}_\bullet(F), *)$. This is indeed a morphism

$$X \to B(*, \mathrm{hAut}_\bullet(F), *)$$

in the homotopy category because hypercoverings are locally trivial fibrations.

This is well-defined, since the category of hypercoverings is filtered. For any two hypercoverings $U_\bullet$ and $U'_\bullet$ and maps $U_\bullet \to B(*, \mathrm{hAut}_\bullet(F), *)$, there is a refinement $V_\bullet$ of both $U_\bullet$ and $U'_\bullet$ and a homotopy between the two corresponding maps $V_\bullet \to U_\bullet \to B(*, \mathrm{hAut}_\bullet(F), *)$ and $V_\bullet \to U'_\bullet \to B(*, \mathrm{hAut}_\bullet(F), *)$. For the basic assertions concerning hypercovers, see [**Fri82**].

(iii) The composition $\Psi \circ \Phi$ is the identity on $\mathcal{H}(X, F)$. This means that a fibre sequence $F \to E \to X$ is equivalent to $f^*\pi$ for $f : X \to B(*, \mathrm{hAut}_\bullet(F), *)$ the morphism constructed in (ii). By Proposition 4.6, it suffices to check this for the hypercovering $U_\bullet$. But since the fibre sequence over $U_\bullet$ is explicitly trivialized, the principalization consists of replacing $F \times U_i$ with $\mathrm{hAut}_\bullet(F) \times U_i$. The pullback of the universal fibre sequence along F replaces $\mathrm{hAut}_\bullet(F)$ again by F. Hence $\Psi \circ \Phi$ is the identity.

(iv) The composition $\Phi \circ \Psi$ is the identity on $[-, B(*, \mathrm{hAut}_\bullet(F), *)]$. Any map in the homotopy category from X to $B(*, \mathrm{hAut}_\bullet(F), *)$ can be represented by a hypercovering $U_\bullet \to X$ and a morphism $U_\bullet \to B(*, \mathrm{hAut}_\bullet(F), *)$. This hypercovering trivializes the corresponding fibre sequence, and the associated principal $\mathrm{hAut}_\bullet(F)$-bundle is obtained by replacing F by $\mathrm{hAut}_\bullet(F)$ as in (ii). The resulting map $f : U_\bullet \to B(*, \mathrm{hAut}_\bullet(F), *)$ is the map we started with. □

Remark 5.11. *In case X is pointed, the relation between the classifying spaces constructed in Section 4 and Section 5 is as follows: the global sections of the sheaf $\pi_0\, \mathrm{hAut}_\bullet(F)$ act on the set $\mathcal{H}^{\mathrm{pt}}(X, F)$, and the quotient modulo this action is $\mathcal{H}(X, F)$. We can not state a more general result as there are simplicial presheaves which can not be pointed because they do not have global sections.*

References

[**All66**] G. Allaud. *On the Classification of Fiber Spaces.* Math. Z. 92 (1966), 110–125.

[**Bro62**] E.H. Brown. *Cohomology theories.* Ann. Math. 75 (1962), 467–484.

[**DF96**] E. Dror Farjoun. *Cellular spaces, null spaces and homotopy localization.* Lecture Notes in Mathematics 1622, Springer (1996).

[DK80] W.G. Dwyer and D.M. Kan. *Function Complexes in Homotopical Algebra.* Topology 19 (1980), 427–440.

[DK84] W.G. Dwyer and D.M. Kan. *A classification theorem for diagrams of simplicial sets.* Topology 23 (1984), no. 2, 139–155.

[DL59] A. Dold and R. Lashof. *Principal quasi-fibrations and fibre homotopy equivalence of bundles.* Illinois J. Math. 3 (1959), 285–305.

[Dol63] A. Dold. *Partitions of unity in the theory of fibrations.* Ann. Math. (2) 78 (1963), 223–255.

[Dol66] A. Dold. *Halbexakte Homotopiefunktoren.* Lecture Notes in Mathematics 12. Springer-Verlag (1966).

[DT58] A. Dold and R. Thom. *Quasifaserungen und unendliche symmetrische Produkte.* Ann. of Math. (2) 67 (1958), 239–281.

[Dug01] D. Dugger. *Combinatorial model categories have presentations.* Adv. Math. 164 (2001), 177–201.

[Fre03] C.P. French. *The equivariant J-homomorphism.* Homology Homotopy Appl. 5 (2003), 161–212.

[Fri82] E.M. Friedlander. *Étale homotopy of simplicial schemes.* Annals of Mathematics Studies, 104. Princeton University Press, 1982.

[Gan65] T. Ganea. *A generalization of homology and homotopy suspension.* Comment. Math. Helv. 39 (1965), 295–322.

[GJ99] P.G. Goerss and J.F. Jardine. *Simplicial Homotopy Theory.* Progress in Mathematics 174. Birkhäuser (1999).

[Hel81] A. Heller. *On the representability of homotopy functors.* J. London Math. Soc.(2) 23 (1981), 551–562.

[Hir03] P.S. Hirschhorn. *Model categories and their localizations.* Mathematical Surveys and Monographs 99, American Mathematical Society (2003).

[Hor06] J. Hornbostel. *Localizations in motivic homotopy theory.* Math. Proc. Cambridge Philos. Soc. 140 (2006), 95–114.

[Hov98] M.A. Hovey. *Model categories.* Mathematical Surveys and Monographs 63, American Mathematical Society (1998).

[Jar87] J.F. Jardine. *Simplicial presheaves.* J. Pure Appl. Algebra 47 (1987), 35–87.

[Jar96] J.F. Jardine. *Boolean localization, in practice.* Doc. Math. 1(1996), 245–275.

[Jar11] J.F. Jardine. *Representability theorems for presheaves of spectra.* J. Pure Appl. Algebra 215 (2011), 77–88.

[Mac98] S. Mac Lane. *Categories for the working mathematician.* Graduate Texts in Mathematics 5. Springer (1998).

[Mat76] M. Mather. Pullbacks in homotopy theory. Canad. J. Math. 28 (1976), 225–263.

[May75] J.P. May. *Classifying Spaces and Fibrations.* Mem. Amer. Math. Soc. 155 (1975).

[May80] J.P. May. *Fibrewise localization and completion.* Trans. Amer. Math. Soc. 258 (1980), 127–146.

[Mil56] J.W. Milnor. *Construction of universal bundles II.* Ann. Math. 63 (1956), 430–436.

[MM92] S. Mac Lane and I. Moerdijk. *Sheaves in geometry and logic: a first introduction to topos theory.* Universitext, Springer (1992).

[MV99] F. Morel and V. Voevodsky. *$\mathbb{A}^1$-homotopy theory of schemes.* Publ. Math. Inst. Hautes Études Sci. 90 (1999), 45–143.

[Pup74] V. Puppe. *A remark on homotopy fibrations.* Manuscripta Math. 12 (1974), 113–120.

[Rez98] C. Rezk. *Fibrations and homotopy colimits of simplicial sheaves.* Preprint (1998), arXiv:math.AT/9811038.

[Rud98] Y.B. Rudyak. *On Thom spectra, orientability and cobordism.* Springer Monographs in Mathematics, Springer (1998).

[Rut97] J.W. Rutter. *Spaces of homotopy self-equivalences. A survey.* Lecture Notes in Mathematics, 1662. Springer-Verlag, Berlin, 1997.

[Sch82] R. Schön. *The Brownian classification of fiber spaces.* Arch. Math. (Basel) 39 (1982), 359–365.

[Sta63] J.D. Stasheff. *A classification theorem for fibre spaces.* Topology 2 (1963), 239–246.

[Sta70] J.D. Stasheff. *H-spaces and classifying spaces: foundations and recent developments.* Algebraic topology (Proc. Sympos. Pure Math., Vol. XXII, Univ. Wisconsin, Madison, Wis., 1970), pp. 247–272. Amer. Math. Soc., Providence, R.I., 1971.

[Sta74] J.D. Stasheff. *Parallel transport and classification of fibrations.* Algebraic and geometrical methods in topology (Conf. Topological Methods in Algebraic Topology, State Univ. New York, Binghamton, N.Y., 1973), pp. 1–17. Lecture Notes in Math., Vol. 428, Springer, Berlin, 1974.

[SW06] J.D. Stasheff and J. Wirth. *Homotopy transition cocycles.* J. Homotopy Relat. Struct. 1 (2006), no. 1, 273–283.

[Wan80] S. Waner. *Equivariant classifying spaces and fibrations.* Trans. Amer. Math. Soc. 258 (1980), 385–405.

[Wen07] M. Wendt. *On fibre sequences in motivic homotopy theory.* PhD thesis, Universität Leipzig (2007).

[Wen09] M. Wendt. *Fibre sequences and localization of simplicial sheaves.* Preprint, 2009.

[Wen10] M. Wendt. *On the $\mathbb{A}^1$-fundamental groups of smooth toric varieties.* Adv. Math. 223 (2010), 352–378.

This article may be accessed via WWW at `http://tcms.org.ge/Journals/JHRS/`

Matthias Wendt
`matthias.wendt@math.uni-freiburg.de`

Mathematisches Institut
Albert-Ludwigs- Universität Freiburg
Eckerstraße 1
79104, Freiburg im Breisgau
Germany

Journal of Homotopy and Related Structures, vol. 6(1), 2011, pp.39–63

ON THE COHOMOLOGY COMPARISON THEOREM

ALIN STANCU

(*communicated by James Stasheff*)

Abstract

A relative derived category for the category of modules over a presheaf of algebras is constructed to identify the relative Yoneda and Hochschild cohomologies with its homomorphism groups. The properties of a functor between this category and the relative derived category of modules over the algebra associated to the presheaf are studied. We obtain a generalization of the *Special Cohomology Comparison Theorem* of M. Gerstenhaber and S. D. Schack.

1. Introduction

Hochschild cohomology of a k-algebra A, denoted here $\mathrm{H}^\bullet(A, -)$, plays an important role in the study of associative algebras, by serving as a tool in the deformation theory of this class of algebras where, broadly speaking, deformations of A are parameterizations A_t, of associative algebras, such that for $t = 0$ one obtains A. We mention here only two of its many other interesting properties: first, separable algebras A are characterized by $\mathrm{H}^1(A, -) = 0$ and second, as discovered by Gerstenhaber, $\mathrm{H}^\bullet(A, A)$ has a rich algebraic structure (of G- algebra). In fact, one need not to restrict to a single algebra and, as M. Gerstenhaber and S. D. Schack did, may consider deformations of presheaves of algebras, or more general of diagrams of algebras, where the naturally defined Hochschild cohomology plays a similar role. The Hochschild cohomology of presheaves is interesting as a step to subsuming the deformation theory of complex manifolds in the deformation theory of associative algebras. The authors mentioned above associated to each presheaf of algebras $\mathbb{A}$ a single algebra $\mathbb{A}!$ and proved the *Special Cohomology Comparison Theorem* which states that Yoneda and Hochschild cohomologies of the presheaf and the algebra associated to the presheaf are isomorphic.

Note that Yoneda and Hochschild cohomologies are **relative** theories since k is a commutative ring that is not necessarily a field.

In this paper we develop a relative derived category, $\mathcal{D}_k^-(\mathbb{A} - \mathrm{bimod})$, of the category of bimodules over a presheaf $\mathbb{A}$ of k-algebras, one where the relative Yoneda

---

This paper was inspired by [7] and it would have not been possible without the support of Samuel D. Schack.

Received January 08, 2009, revised November 23, 2010; published on January 18, 2011.

2000 Mathematics Subject Classification: Primary: 18, Secondary: 16.

Key words and phrases: Hochschild cohomology, derived category, cohomology comparison theorem.

cohomology, $\mathrm{Ext}^i_{\mathbb{A}-\mathbb{A},k}(\mathbb{M},\mathbb{N})$, so in particular Hochschild cohomology, can be regarded as homomorphism groups, $Mor_{\mathcal{D}_k^-(\mathbb{A}-\mathrm{bimod})}(\mathbb{M}_\bullet,\mathbb{N}_\bullet[i])$. The reader should be aware that the term 'presheaf of k-algebras' is used to describe functors $\mathbb{A}$, defined on posets $\mathcal{C}$, with images in the category of k-algebras. In this context, we also show that the functor !, induced between the relative derived categories of $\mathbb{A}$-bimod and $\mathbb{A}$!-bimod, is full and faithful and we obtain a generalization of the *Special Cohomology Comparison Theorem.*

This natural construction may be part of providing a more conceptual interpretation for the Hochschild cohomology of a presheaf of algebras together with its Gerstenhaber bracket, that of the Lie algebra of an algebraic group (*i.e* a group valued functor). In the case of a single algebra over a field B. Keller, in [6], identifies $\mathrm{H}^\bullet(A,A)$ with the Lie algebra of an algebraic group by regarding $\mathrm{H}^i(A,A)$ as a homomorphism group $Mor_{\mathcal{D}(A-\mathrm{bimod})}(A^\bullet,A^\bullet[i])$ in the derived category $\mathcal{D}(A-\mathrm{bimod})$ and then establishing a bijection between the latter groups and certain infinitesimal deformations of A which have a natural Lie bracket. Since the Gerstenhaber bracket exists on the Hochschild cohomology of presheaves of algebras presumably a similar interpretation exists for this situation too. To adapt Keller's technique to this case one needs to find the "correct" derived category that allows the interpretation of the relative Hochschild cohomology as Hom groups.

2. Resolutions, adjoint functors and the functor !

Let k be a commutative ring and $\mathcal{C}$ a poset viewed as a category in the usual way: for each $i \leqslant j$ there is a unique map $\varphi^{ij} : i \longrightarrow j$. When A is a k-algebra and M any A bimodule we assume M to be symmetric over k. (*i.e.* $ax = xa$ for all $x \in M$ and $a \in k$.) A presheaf of k-algebras over $\mathcal{C}$ is a functor $\mathbb{A} : \mathcal{C}^{op} \longrightarrow k$-**alg**. We will denote $\mathbb{A}(i)$ by $\mathbb{A}^i$. A presheaf as above is a special case of functor defined from a small category to the category of k algebras. In [1] these functors are called a "diagrams".

The category $\mathbb{A}$-bimod is the category whose objects are $\mathbb{A}$-bimodules and the maps are maps of bimodules. An $\mathbb{A}$-bimodule $\mathbb{M}$ is a presheaf of abelian groups such that $\mathbb{M}^i$ is an $\mathbb{A}^i$-bimodule $(\forall) i \in \mathcal{C}$ and for all $i \leqslant j$ the map $T^{ij}_{\mathbb{M}} : \mathbb{M}^j \longrightarrow \mathbb{M}^i$ is an $\mathbb{A}^j$-bimodule map. An $\mathbb{A}$-bimodule map $\eta : \mathbb{M} \longrightarrow \mathbb{N}$ is a natural transformation in which η^i is an $\mathbb{A}^i$-bimodule map $(\forall) i \in \mathcal{C}$.

In defining Yoneda cohomology of the category $\mathbb{A}$-bimod 'allowable' maps play a vital role. A map $\eta : \mathbb{M} \longrightarrow \mathbb{N}$ is **allowable** if $(\forall) i \in \mathcal{C}$ the map $\eta^i : \mathbb{M}^i \longrightarrow \mathbb{N}^i$ admits a k-bimodule splitting map $k^i : \mathbb{N}^i \longrightarrow \mathbb{M}^i$ satisfying $\eta^i k^i \eta^i = \eta^i$. We do not require the splitting maps k^i to be natural. An $\mathbb{A}$-bimodule $\mathbb{P}$ is a **relative projective** if for every allowable epimorphism $\mathbb{M} \longrightarrow \mathbb{N}$ the induced map $Hom_{\mathbb{A}-\mathbb{A}}(\mathbb{P},\mathbb{M}) \longrightarrow Hom_{\mathbb{A}-\mathbb{A}}(\mathbb{P},\mathbb{N})$ is an epimorphism of sets.

A **relative projective allowable resolution** of an $\mathbb{A}$-bimodule $\mathbb{M}$ is an exact sequence $\cdots \longrightarrow \mathbb{P}_n \cdots \longrightarrow \mathbb{P}_1 \longrightarrow \mathbb{P}_0 \longrightarrow \mathbb{M} \longrightarrow 0$ in which all $\mathbb{P}_n$ are relative projective $\mathbb{A}$- bimodules and all maps are allowable. The category $\mathbb{A}$-bimod has enough relative projective bimodules and each bimodule has a relative projective allowable resolution. Moreover, there is a functorial way of getting this type of resolutions. The construction of such a resolution is due to M. Gerstenhaber and S. D. Schack (see [1]) and is based on two facts: First, the 'forgetful' functor $\mathbb{A}$-bimod$\longrightarrow\mathbb{K}$-bimod has a left adjoint $\mathbb{A}\otimes_{\mathbb{K}} - \otimes_{\mathbb{K}}\mathbb{A}$, where $\mathbb{K}$ is the constant presheaf $\mathbb{K}^i = k$, $(\forall)i \in \mathcal{C}$. For each $\mathbb{N} \in \mathbb{A}$-bimod we set $(\mathbb{A}\otimes_{\mathbb{K}}\mathbb{N}\otimes_{\mathbb{K}}\mathbb{A})^i = \mathbb{A}^i \otimes_k \mathbb{N}^i \otimes_k \mathbb{A}^i$ and the map $\mathbb{A}^j \otimes_k \mathbb{N}^j \otimes_k \mathbb{A}^j \longrightarrow \mathbb{A}^i \otimes_k \mathbb{N}^i \otimes_k \mathbb{A}^i$ corresponding to $i \leqslant j$ in $\mathcal{C}$ is just $\varphi^{ij}\otimes T^{ij}_{\mathbb{N}}\otimes\varphi^{ij}$.

The corresponding categorical bar resolution, of [2], of an $\mathbb{A}$-bimodule $\mathbb{N}$, denoted $\mathcal{B}_\bullet(\mathbb{N})$, is allowable and since $\mathcal{B}_q(\mathbb{N}) = \mathbb{A}\otimes_{\mathbb{K}}\mathcal{B}_{q-1}(\mathbb{N})\otimes_{\mathbb{K}}\mathbb{A}$ we have that $\mathcal{B}_q(\mathbb{N})^i$ is a relative projective $\mathbb{A}^i$-bimodule $(\forall)i \in \mathcal{C}$. In addition, the resolution has a functorial contracting homotopy $x_q : \mathcal{B}_q(\mathbb{N}) \longrightarrow \mathcal{B}_{q+1}(\mathbb{N})$, $x_q(a) = 1\otimes a\otimes 1$.

Second, observe that $(\forall)i \in \mathcal{C}$ the functor $(i)^* : \mathbb{A}$-bimod $\longrightarrow \mathbb{A}^i$-bimod defined by $(i)^*\mathbb{M} = \mathbb{M}^i$ admits a left adjoint $(i)_! : \mathbb{A}^i$-bimod$\longrightarrow \mathbb{A}$-bimod, where $(i_!M)^h = \mathbb{A}^h \otimes_{\mathbb{A}^i} M \otimes_{\mathbb{A}^i} \mathbb{A}^h$ if $h \leqslant i$ and $(i_!M)^h = 0$ otherwise. If $h \leqslant j \leqslant i$ the map $(i_!M)^j \longrightarrow (i_!M)^h$ is $\varphi^{hj}\otimes Id_M\otimes\varphi^{hj}$ and it is zero otherwise.

Combining the functors $(i)^*$ we obtain a single exact functor $\mathcal{R} : \mathbb{A}$-bimod$\longrightarrow \prod_{i\in\mathcal{C}}(\mathbb{A}^i$-bimod), defined on objects by $\mathcal{R}\mathbb{M} = \prod_{i\in\mathcal{C}}\mathbb{M}^i$ and whose left adjoint $\mathcal{L}$ is defined on objects by $\mathcal{L}M_i = \coprod_{i\in\mathcal{C}}(i)_!M_i$. Applying again the categorical bar resolution of [2] we obtain an allowable resolution with a functorial contracting homotopy. We denote this resolution by $\mathcal{S}_\bullet$. Thus $\mathcal{S}_p = (\mathcal{L}\mathcal{R})^{p+1} = \mathcal{L}\mathcal{R}\mathcal{S}_{p-1}$ and the boundary maps $d_p : \mathcal{S}_{p+1} \longrightarrow \mathcal{S}_p$ are defined inductively by $d_p = \varepsilon_{\mathcal{S}_p} - \mathcal{L}\mathcal{R}d_{p-1}$, where $d_{-1} = \varepsilon$ is the counit of the adjunction. The contracting homotopy is the unit $\eta_{\mathcal{R}\mathcal{S}_p} : \mathcal{R}\mathcal{S}_p \longrightarrow \mathcal{R}\mathcal{S}_{p+1}$.

Here is a more direct description of $\mathcal{S}_\bullet$. Let [p] be the linearly ordered set $\{0 < 1 < \cdots < p\}$. A covariant functor $\sigma : [p] \to \mathcal{C}$ is called a p-simplex. Thus p-simplices are objects of the functor category $\mathcal{C}^{[p]}$. The domain of σ is defined as $\sigma(0)$ and is denoted by $d\sigma$. Similarly, the codomain of σ is defined as $\sigma(p)$ and is denoted by $c\sigma$. For each p-simplex σ we write $\sigma = (\sigma^{01}, \ldots, \sigma^{p-1,p})$ and define

$$\sigma_r = \begin{cases} (\sigma^{12}, \ldots, \sigma^{p-1,p}) & \text{if} \quad r=0 \\ (\sigma^{01}, \ldots, \sigma^{r-1,r+1}, \ldots, \sigma^{p-1,p}) & \text{if} \quad 0 < r < p \\ (\sigma^{01}, \ldots, \sigma^{p-2,p-1}) & \text{if} \quad r=p \end{cases}$$

Note that $d\sigma_r = d\sigma = \sigma(0)$ if $r \neq 0$ and $d\sigma_0 = \sigma(1)$. Similarly, $c\sigma_r = c\sigma = \sigma(p)$ if $r \neq p$ and $c\sigma_p = \sigma(p-1)$. Also, note that $d\sigma \leqslant d\sigma_r$ and $c\sigma_r \leqslant c\sigma$ and recall that the structure maps defining presheaves and bimodules are contravariant.

For $\mathbb{N} \in \mathbb{A}$-bimod and $p \geqslant 0$ we have $\mathcal{S}_p\mathbb{N} = \coprod_{\sigma\in\mathcal{C}^{[p]}}\mathcal{S}^\sigma_p\mathbb{N}$, where $\mathcal{S}^\sigma_p\mathbb{N} = (d\sigma)_!(\mathbb{A}^{d\sigma}\otimes_{\mathbb{A}^{c\sigma}}\mathbb{N}^{c\sigma}\otimes_{\mathbb{A}^{c\sigma}}\mathbb{A}^{d\sigma})$ and $\mathbb{A}^{d\sigma}$ is an $\mathbb{A}^{c\sigma}$-bimodule via the map $\varphi^{d\sigma,c\sigma} : \mathbb{A}^{c\sigma} \longrightarrow \mathbb{A}^{d\sigma}$.

For $p \geqslant 0$, the boundary $\partial : \mathcal{S}_p\mathbb{N} \longrightarrow \mathcal{S}_{p-1}\mathbb{N}$ is a sum $\partial = \sum_{r=0}^{p}(-1)^r\partial_r$ where the restriction of ∂_r to $\mathcal{S}^\sigma_p$ is denoted $\partial^\sigma_r : \mathcal{S}^\sigma_p\mathbb{N} = (d\sigma)_!(\mathbb{A}^{d\sigma}\otimes_{\mathbb{A}^{c\sigma}}\mathbb{N}^{c\sigma}\otimes_{\mathbb{A}^{c\sigma}}\mathbb{A}^{d\sigma}) \longrightarrow (d\sigma_r)_!(\mathbb{A}^{d\sigma_r}\otimes_{\mathbb{A}^{c\sigma_r}}\mathbb{N}^{c\sigma_r}\otimes_{\mathbb{A}^{c\sigma_r}}\mathbb{A}^{d\sigma_r}) = \mathcal{S}^{\sigma_r}_{p-1}\mathbb{N}$.

We obtain that for $h \leqslant d\sigma$ and $a\otimes n\otimes a' \in (\mathcal{S}^\sigma_p\mathbb{N})^h = \mathbb{A}^h\otimes_{\mathbb{A}^{c\sigma}}\mathbb{N}^{c\sigma}\otimes_{\mathbb{A}^{c\sigma}}\mathbb{A}^h$, $\partial^\sigma_r(a\otimes n\otimes a') = a\otimes T^{c\sigma_r,c\sigma}_{\mathbb{N}}(n)\otimes a' \in (\mathcal{S}^{\sigma_r}_{p-1}\mathbb{N})^h$. Here $T^{c\sigma_r,c\sigma}_{\mathbb{N}}$ is the structure map of the bimodule $\mathbb{N}$ corresponding to $c\sigma_r \leqslant c\sigma$. In particular, when $r = 0$ we get $\partial^\sigma_0(a\otimes n\otimes a') = a\otimes n\otimes a'$, and when $r = p$ we get $\partial^\sigma_p(a\otimes n\otimes a') = a\otimes T^{c\sigma_p,c\sigma}_{\mathbb{N}}(n)\otimes a'$.

The augmentation map $\varepsilon : \mathcal{S}_0\mathbb{N} = \coprod_{i\in\mathcal{C}} \mathcal{S}_0^i = \coprod_{i\in\mathcal{C}} (i)_!(\mathbb{A}^i \otimes_{\mathbb{A}^i} \mathbb{N}^i \otimes_{\mathbb{A}^i} \mathbb{A}^i) \longrightarrow \mathbb{N}$ is defined on the components $(i)_!(\mathbb{A}^i \otimes_{\mathbb{A}^i} \mathbb{N}^i \otimes_{\mathbb{A}^i} \mathbb{A}^i)$. For $h \leqslant i$, $(i)_!(\mathbb{A}^i \otimes_{\mathbb{A}^i} \mathbb{N}^i \otimes_{\mathbb{A}^i} \mathbb{A}^i)^h = \mathbb{A}^h \otimes_{\mathbb{A}^i} \mathbb{N}^i \otimes_{\mathbb{A}^i} \mathbb{A}^h \longrightarrow \mathbb{N}^h$ is given by $1 \otimes n \otimes 1 \longrightarrow T_{\mathbb{N}}^{hi}(n)$.

For $i \in \mathcal{C}$ the contracting homotopy $\kappa_p^i : (\mathcal{S}_p\mathbb{N})^i \longrightarrow (\mathcal{S}_{p+1}\mathbb{N})^i$ is given componentwise by $(\mathcal{S}_p^\sigma\mathbb{N})^i \longrightarrow (\mathcal{S}_{p+1}^{(i,\sigma)}\mathbb{N})^i = identity$, where $(\mathcal{S}_{p+1}^{(i,\sigma)}\mathbb{N}) = 0$ if $i \not\leqslant d\sigma$. If $i \leqslant d\sigma$, then (i,σ) is the simplex $(i, d\sigma = \sigma(0), \ldots, \sigma(p))$.

In general the above resolution is not a relative projective resolution, but it is when each $\mathbb{N}^i$ is a relative projective $\mathbb{A}^i$-bimodule. Thus, to construct a relative projective allowable resolution of an $\mathbb{A}$-bimodule $\mathbb{N}$ we take the resolution $\mathcal{B}_\bullet(\mathbb{N}) \longrightarrow \mathbb{N}$, determined by the forgetful functor and its left adjoint, and then apply $\mathcal{S}_\bullet$ to it to obtain a double complex $\mathcal{S}_\bullet\mathcal{B}_\bullet(\mathbb{N})$. Take now the total complex of this double complex to get the desired resolution.

The Hochschild cohomology of a presheaf $\mathbb{A}$ is defined to be the relative Yoneda cohomology of $\mathbb{A}$.

That is,

$$\mathbf{H}^\bullet(\mathbb{A}, -) = \mathbf{Ext}^\bullet_{\mathbb{A}-\mathbb{A}}(\mathbb{A}, -).$$

It plays a crucial role in the study of deformations of diagrams of algebras and it has the same rich structure as the Hochschild cohomology of a single algebra.

If $\mathbb{P}_\bullet \to \mathbb{A}$ is a relative projective allowable resolution of $\mathbb{A}$ then $\mathbf{H}^\bullet(\mathbb{A}, -)$ is the homology of the complex $\mathbf{Hom}_{\mathbb{A}-\mathbb{A}}(\mathbb{P}_\bullet, -)$.

To each presheaf of algebras $\mathbb{A}$ over $\mathcal{C}$ we can associate a single algebra $\mathbb{A}! =$ row-finite $\mathcal{C} \times \mathcal{C}$ matrices (a_{ij}) with $a_{ij} \in \mathbb{A}^i$ if $i \leqslant j$ and $a_{ij} = 0$ otherwise. The addition is componentwise and the multiplication $(a_{ij})(b_{ij}) = (c_{ij})$ is induced by the matrix multiplication with the understanding that, for $h \leqslant i \leqslant j$, the summand $a_{hi}b_{ij}$ of c_{hj} is regarded as $a_{hi}b_{ij} = a_{hi}\varphi^{hi}(b_{ij})$. For our purpose it is convenient to use the equivalent representation $\mathbb{A}! = \prod_{i\in\mathcal{C}} \coprod_{i\leqslant j} \mathbb{A}^i\varphi^{ij}$, as k-bimodule. Here φ^{ij} serve to distinguish distinct copies of $\mathbb{A}^i$ from one another. The general element of $\mathbb{A}^i\varphi^{ij}$ will be denoted $a^i\varphi^{ij}$. The multiplication is defined componentwise and subject to the rule: $(a^h\varphi^{hi})(a^j\varphi^{jl}) = a^h\varphi^{hi}(a^j)\varphi^{hl}$ if $i = j$ and 0 otherwise.

Let 1_i the unit element of $\mathbb{A}^i$. Since $(a^h\varphi^{hi})(1_i\varphi^{ij}) = a^h\varphi^{hj}$ and $(1_i\varphi^{hi})(a^i\varphi^{ij}) = \varphi^{hi}(a^i)\varphi^{hj}$ we may abbreviate $1_i\varphi^{ij}$ to φ^{ij}. The maps φ^{ij} are then elements of $\mathbb{A}!$ and $\varphi^{hi}\varphi^{ij} = \varphi^{hj}$; $\varphi^{hi}\varphi^{jl} = 0$ if $i \neq j$.

We define the functor $! : \mathbb{A}\text{-bimod} \longrightarrow \mathbb{A}!\text{-bimod}$, such that $\mathbb{A} \longrightarrow \mathbb{A}!$, by setting for any $\mathbb{A}$-bimodule $\mathbb{M}$, $\mathbb{M}! = \prod_{i\in\mathcal{C}} \coprod_{i\leqslant j} \mathbb{M}^i\varphi^{ij}$ as a k-bimodule. The actions of $\mathbb{A}!$ are defined by:

$$(a^h\varphi^{hi})(m^i\varphi^{ij}) = a^h T_{\mathbb{M}}^{hi}(m^i)\varphi^{hj}$$

$$(m^h\varphi^{hi})(a^i\varphi^{ij}) = m^h\varphi^{hi}(a^i)\varphi^{hj}$$

$$(a^h\varphi^{hi})(m^j\varphi^{jl}) = 0 = (m^h\varphi^{hi})(a^j\varphi^{jl}), \text{if} i \neq j.$$

For $\eta \in Hom_{\mathbb{A}-\mathbb{A}}(\mathbb{N}, \mathbb{M})$ define $\eta! \in Hom_{\mathbb{A}!-\mathbb{A}!}(\mathbb{N}!, \mathbb{M}!)$ by $\eta!(n^i\varphi^{ij}) = \eta^i(n^i)\varphi^{ij}$.

We will use the following proposition due to M. Gerstenhaber and S. D. Schack.

Proposition 2.1. *The functor* $! : \mathbb{A}$*-bimod*$\longrightarrow \mathbb{A}!$*-bimod is exact, preserves allowability and is full and faithful.*

Proof. see [2] □

In fact, M. Gerstenhaber and S. D. Schack proved in [2] the "Special Cohomology Comparison Theorem" (SCCT).

Theorem(SCCT). *Let $\mathcal{C}$ be an arbitrary poset and $\mathbb{A}$ a presheaf over $\mathcal{C}$. The functor ! induces an isomorphism of relative Yoneda cohomologies*

$$Ext^{\bullet}_{\mathbb{A}-\mathbb{A}}((-),(-)) \cong Ext^{\bullet}_{\mathbb{A}!-\mathbb{A}!}((-)!,(-)!).$$

In particular, we have an isomorphism of relative Hochschild cohomologies $H^{\bullet}(\mathbb{A},(-)) \cong H^{\bullet}(\mathbb{A}!,(-)!)$.

An important consequence of this theorem is that the deformation theories of $\mathbb{A}$ and of $\mathbb{A}!$ are equivalent, if the poset $\mathcal{C}$ has a terminator. Another is that $H^{\bullet}(\mathbb{A}!,\mathbb{A}!)$ has a G-algebra structure. These results can be found in their full generalization to diagrams in [2], but we will not deal with them here. We will however generalize the SCCT to derived categories and prove theorems 3.9 and 4.1. The SCCT follows as a corollary from these theorems. To do this we need to introduce a subcategory of the category of $\mathbb{A}!$-bimod. The image of ! lies in a full subcategory of $\mathbb{A}!$-bimod. This is the category of **aligned** bimodules, $\mathbb{A}!$-albimod. The main reason to consider it here is that the functor ! has a left adjoint when restricted to ! : $\mathbb{A}$-bimod $\longrightarrow$ $\mathbb{A}!$-albimod. Thus, for every $\mathbb{A}!$ bimodule X we set $X_{al} = \prod_{i\in\mathcal{C}} \coprod_{i\leqslant j} \varphi^{ii}X\varphi^{jj}$ with the obvious $\mathbb{A}!$ bimodule structure.

Definition 2.2. An $\mathbb{A}!$ bimodule X is said to be **aligned** if the k linear map $X \longrightarrow \prod_{i\in\mathcal{C}}\prod_{j\in\mathcal{C}} \varphi^{ii}X\varphi^{jj}$, $x \longrightarrow <\varphi^{ii}x\varphi^{jj}>$ induces an $\mathbb{A}!$ bimodule isomorphism $\alpha_X : X \longrightarrow X_{al} = \prod_{i\in\mathcal{C}}\coprod_{i\leqslant j} \varphi^{ii}X\varphi^{jj}$.

For each $\mathbb{A}!$-bimodule map $f : X \longrightarrow Y$, the restriction of f to $\varphi^{ii}X\varphi^{jj}$ is a k linear, even a $\mathbb{A}^i$-$\mathbb{A}^j$-bimodule map $f^{ij} : \varphi^{ii}X\varphi^{jj} \longrightarrow \varphi^{ii}Y\varphi^{jj}$ since $f(\varphi^{ii}x\varphi^{jj}) = \varphi^{ii}f(x)\varphi^{jj}$ lies in $\varphi^{ii}Y\varphi^{jj}$. Thus, f gives rise to a family of k linear maps $f^{ij} : \varphi^{ii}X\varphi^{jj} \longrightarrow \varphi^{ii}Y\varphi^{jj}$ such that $f^{hj}(a^h\varphi^{hi}\cdot x) = a^h\varphi^{hi}\cdot f^{ij}(x)$ and $f^{iq}(x\cdot a^j\varphi^{jq}) = f^{ij}(x)\cdot a^j\varphi^{jq}$ $\forall x \in \varphi^{ii}X\varphi^{jj}$, $a^h \in \mathbb{A}^h$, $a^j \in \mathbb{A}^j$ and $h \leqslant i \leqslant j \leqslant q$ in $\mathcal{C}$.

In fact these are exactly the conditions necessary on such a collection of maps for $f_{al} = \prod_{i\in\mathcal{C}}\coprod_{i\leqslant j} f^{ij}$ to be an $\mathbb{A}!$-bimodule map $X_{al} \longrightarrow Y_{al}$. One can easily see that $\mathbb{A}!$-albimod is abelian, and that both the inclusion functor inc : $\mathbb{A}!$-albimod$\longrightarrow$ $\mathbb{A}!$-bimod and the alignment functor $(-)_{al}$: $\mathbb{A}!$-bimod$\longrightarrow$ $\mathbb{A}!$-albimod, $X \longrightarrow X_{al}$ are exact and preserve allowability and that $\alpha : Id_{\mathbb{A}!-albimod} \longrightarrow (-)_{al} \circ inc$ is a natural isomorphism.

Now, we describe a method of producing relative projective allowable resolutions of aligned bimodules of the form $\mathbb{N}!$ that we will use to replace complexes of aligned bimodules with relative projective ones in a suitable derived category. We begin with a result due to M. Gerstenhaber and S. D. Schack.

Proposition 2.3. *1. For each $i \leqslant j$ in $\mathcal{C}$ the restriction functor $(-)^{ij}$: $\mathbb{A}!$-albimod$\longrightarrow$ $\mathbb{A}^i$-mod-$\mathbb{A}^j$, $X \longrightarrow \varphi^{ii}X\varphi^{jj}$ is exact and preserves allowability.*
2. The functor $(-)^{ij}$ has a left adjoint L_{ij} that preserves relative projectivity.

Proof. Part 1 is obvious. For 2, define $L_{ij} : \mathbb{A}^i\text{-mod-}\mathbb{A}^j \longrightarrow \mathbb{A}!\text{-albimod}$ as follows:
$L_{ij}(N)^{hl} = \begin{cases} \mathbb{A}^h \otimes_{\mathbb{A}^i} |N|_{jl} & \text{if } h \leqslant i \leqslant j \leqslant l \\ 0 & \text{otherwise} \end{cases}$

Here, $|N|_{jl}$ is N viewed as a left $\mathbb{A}^i$-module and a right $\mathbb{A}^l$-module via the map φ^{jl}. The actions of $\mathbb{A}!$ are given by

$$a^r\varphi^{rh}(a^h \otimes n) = a^r\varphi^{rh}(a^h) \otimes n \in L_{ij}(N)^{rl}$$

$$(a^h \otimes n)a^l\varphi^{lm} = a^h \otimes n\varphi^{jl}(a^l) \in L_{ij}(N)^{hm},$$

for $a^h \otimes n \in L_{ij}(N)^{hl}$ and $a^r\varphi^{rh}, a^l\varphi^{lm} \in \mathbb{A}!$.

One can check now that we have a natural isomorphism

$$Hom_{\mathbb{A}!-\text{albimod}}(L_{ij}(N), X) \leftrightarrows Hom_{\mathbb{A}^i-\mathbb{A}^j}(N, X^{ij})$$

for all $X \in \mathbb{A}!$-albimod and $N \in \mathbb{A}^i$-mod-$\mathbb{A}^j$.

If $P \in \mathbb{A}^i$-mod-$\mathbb{A}^j$ is relative projective then the natural isomorphism $Hom_{\mathbb{A}!-\text{albimod}}(L_{ij}(P), -) \cong Hom_{\mathbb{A}^i\mathbb{A}^j}(P, (-)^{ij}) = Hom_{\mathbb{A}^i\mathbb{A}^j}(P, -) \circ (-)^{ij}$ is a composite of functors which preserve allowable epimorphisms, so $L_{ij}(P)$ is relative projective. (for more details see [2]) □

Modeled on the M. Gerstenhaber - S. D. Schack resolution $\mathcal{S}_\bullet$, C. B. Kullmann obtained in [3] an allowable resolution $\mathcal{T}_\bullet\mathbb{N} \longrightarrow \mathbb{N}!$ in $\mathbb{A}!$-albimod as follows. For $p \geqslant 0$ let $\mathcal{T}_p\mathbb{N} = \coprod_{\sigma\in C[p]} \mathcal{T}_p^\sigma\mathbb{N}$, where the coproduct is taken in $\mathbb{A}!$-albimod (constructed by applying $(-)_{al}$ to that in $\mathbb{A}$-bimod), where $\mathcal{T}_p^\sigma\mathbb{N} = L_{d\sigma,c\sigma}(\mathbb{A}^{d\sigma} \otimes_{\mathbb{A}^{c\sigma}} \mathbb{N}^{c\sigma})$.

For $h \leqslant d\sigma \leqslant c\sigma \leqslant l$ we have a natural isomorphism $(\mathcal{T}_p^\sigma\mathbb{N})^{hl} = \mathbb{A}^h \otimes_{\mathbb{A}^{d\sigma}} |\mathbb{A}^{d\sigma} \otimes_{\mathbb{A}^{c\sigma}} \mathbb{N}^{c\sigma}|_{c\sigma,l} \cong \mathbb{A}^h \otimes_{\mathbb{A}^{c\sigma}} |\mathbb{N}^{c\sigma}|_{c\sigma.l}$ and we use this identification to define the differentials. If $p \geqslant 1$ we define $d : \mathcal{T}_p\mathbb{N} \longrightarrow \mathcal{T}_{p-1}\mathbb{N}$ as a sum $d = \sum_{r=0}^p (-1)^r d_r$, where each d_r is determined by its restriction to $\mathcal{T}_p^\sigma\mathbb{N}$ and for $h \leqslant d\sigma \leqslant c\sigma \leqslant l$ and $a \otimes n \in (\mathcal{T}_p^\sigma\mathbb{N})^{hl} = \mathbb{A}^h \otimes_{\mathbb{A}^{c\sigma}} |N^{c\sigma}|_{c\sigma,l}$, we have $d_r^\sigma(a \otimes n) = a \otimes T_{\mathbb{N}}^{c\sigma_r,c\sigma}(n) \in \mathbb{A}^h \otimes_{\mathbb{A}^{c\sigma_r}} |\mathbb{N}^{c\sigma_r}|_{c\sigma_r,l} = (\mathcal{T}_{p-1}^\sigma\mathbb{N})^{hl}$. If $p = 0$ the map $\varepsilon^{\mathcal{T}} : \mathcal{T}_0\mathbb{N} = \coprod_i \mathcal{T}_0^{(i)}\mathbb{N} \longrightarrow \mathbb{N}!$ is determined by $(\mathcal{T}_0^{(i)}\mathbb{N})^{hl} = \mathbb{A}^h \otimes_{\mathbb{A}^i} |\mathbb{N}^i|_{il} \longrightarrow \mathbb{N}!^{hl} = \mathbb{N}^h\varphi^{hl}$, $a \otimes n \longrightarrow aT_{\mathbb{N}}^{hi}(n)\varphi^{hl}$, for $h \leqslant i \leqslant l$.

It is easy to check that $\mathcal{T}_\bullet\mathbb{N} \longrightarrow \mathbb{N}!$ is a chain complex and it is in fact an allowable resolution since it has a contracting homotopy induced by $\kappa_p : (\mathcal{T}_p^\sigma\mathbb{N})^{hl} \longrightarrow (\mathcal{T}_{p+1}^{(h,\sigma)}\mathbb{N})^{hl}$, $\kappa_p = identity$, where (h, σ) is the simplex $(h, \sigma(0), \ldots, \sigma(p))$ if $h \leqslant \sigma(0)$ and $\mathcal{T}_{p+1}^{(h,\sigma)}\mathbb{N} = 0$ if $h \nleq d\sigma$.

In general $\mathcal{T}_p\mathbb{N}$ is not a relative projective aligned $\mathbb{A}!$-bimodule, but it is when each $\mathbb{N}^i$ is relative projective $\mathbb{A}^i$-bimodule. To obtain a relative projective aligned resolution, for each $\mathbb{A}$-bimodule $\mathbb{N}!$, take the relative projective resolution $\mathcal{B}_\bullet(\mathbb{N})$, apply ! and then $\mathcal{T}_\bullet$ to obtain a double complex. Now, take the total complex to obtain the desired resolution.

We conclude this section with a result which connects $\mathcal{T}_\bullet$ and $\mathcal{S}_\bullet$ via a left adjoint of !. Because the only source for the following theorem is [3] the proof is included in the Appendix A.

Theorem 2.4. *1. The functor* $! : \mathbb{A}$*-bimod*$\longrightarrow \mathbb{A}!$*-albimod admits a left adjoint* $\mathfrak{i} : \mathbb{A}!$*-albimod*$\longrightarrow \mathbb{A}$*-bimod.*
2. There are natural isomorphisms $\mathcal{T}_p\mathbb{N}\mathfrak{i} \longrightarrow \mathcal{S}_p\mathbb{N}$ *which induce a natural isomorphism of complexes* $(\mathcal{T}_\bullet\mathbb{N} \longrightarrow \mathbb{N}!)\mathfrak{i}$ *and* $(\mathcal{S}_\bullet\mathbb{N} \longrightarrow \mathbb{N})$.

Proof. see Appendix A. □

3. Derived categories and Hochschild cohomology

Let $Kom^-(\mathbb{A} - \text{bimod})$ the category of bounded to the right complexes of $\mathbb{A}$-bimodules

$$\mathbb{M}_\bullet := \cdots \mathbb{M}_n \longrightarrow \cdots \qquad \cdots \longrightarrow \mathbb{M}_1 \longrightarrow \mathbb{M}_0 \longrightarrow 0$$

A map between two complexes $\mathbb{M}_\bullet$ and $\mathbb{N}_\bullet$ is a collection of maps $f = (f_i) : \mathbb{M}_i \to \mathbb{N}_i$, one for each positive integer i, which commute with the differentials of $\mathbb{M}_\bullet$ and $\mathbb{N}_\bullet$. We do not require the maps defining the complexes or the maps between complexes to be k-split.
Similarly, we define $Kom^-(\mathbb{A}! - \text{albimod})$ and $Kom^-(\mathbb{A}! - \text{bimod})$.

Definition 3.1. 1) A map $f : M_\bullet \longrightarrow N_\bullet$ in $Kom^-(\mathbb{A}!-\text{bimod})$ (or $Kom^-(\mathbb{A}!-\text{albimod})$) is a **relative quasi-isomorphism** if its cone $\mathcal{C}(f)_\bullet$ is contractible when considered as a complex of k-bimodules.
2) A map $f : \mathbb{M}_\bullet \longrightarrow \mathbb{N}_\bullet$ in $Kom^-(\mathbb{A}-\text{bimod})$ is a **relative quasi-isomorphism** if the maps of complexes $f^i : \mathbb{M}^i_\bullet \longrightarrow \mathbb{N}^i_\bullet$ have contractible cones, when considered as complexes of k-bimodules, for all $i \in \mathcal{C}$.

The word "**relative**" in the above definition is used as a reminder to the reader that Yoneda and Hochschild cohomologies are relative theories, since k is a commutative ring that is not necessarily a field. It is the relative Yoneda groups that we want to view as homomorphism groups in a suitable category.

Proposition 3.2. *Let A be any k-algebra and $f : M_\bullet \longrightarrow N_\bullet$ a map of complexes of A bimodules in $Kom^-(A - bimod)$. Then, f is a relative quasi-isomorphism if and only if there exists $\gamma : N_\bullet \longrightarrow M_\bullet$ a map of complexes of k-bimodules such that $f\gamma \sim id_{N_\bullet}$ and $\gamma f \sim id_{M_\bullet}$ in $Kom^-(k - bimod)$, where '$\sim$' stands for homotopy equivalence.*

Proof. $'\Rightarrow'$
Assume that f is a relative quasi-isomorphism. Thus $\mathcal{C}(f)_\bullet$ is contractible when regarded as a complex of k-bimodules, so there exist $s = (s_n) : \mathcal{C}(f)^{n-1}_\bullet \longrightarrow \mathcal{C}(f)^n_\bullet$ maps of k-bimodules such that $sd_{\mathcal{C}(f)_\bullet} + d_{\mathcal{C}(f)_\bullet}s = id$. We may assume that

$$s = \begin{pmatrix} \alpha & \gamma \\ \beta & \delta \end{pmatrix} \text{ and}$$

$$d_{\mathcal{C}(f)_\bullet} = \begin{pmatrix} -d_{M_\bullet} & 0 \\ f & d_{N_\bullet} \end{pmatrix},$$

where $\alpha : M_{\bullet-1} \longrightarrow M_\bullet$, $\beta : M_{\bullet-1} \longrightarrow N_{\bullet+1}$, $\gamma : N_\bullet \longrightarrow M_\bullet$ and $\delta : N_\bullet \longrightarrow N_{\bullet+1}$ are k linear maps. Since $sd_{\mathcal{C}(f)_\bullet} + d_{\mathcal{C}(f)_\bullet}s = id$, we obtain $-\alpha d_{M_\bullet} + \gamma f - d_{M_\bullet}\alpha = id_{M_\bullet}$, $-\beta d_{M_\bullet} + \delta f + f\alpha + d_{N_\bullet}\beta = 0$, $\gamma d_{N_\bullet} - d_{M_\bullet}\gamma = 0$ and $\delta d_{N_\bullet} + f\gamma + d_{N_\bullet}\delta = id_{N_\bullet}$.

Thus, γ is a map of complexes of k-bimodules and since $\delta d_{N_\bullet} + d_{N_\bullet}\delta = id_{N_\bullet} - f\gamma$ and $\alpha d_{M_\bullet} + d_{M_\bullet}\alpha = \gamma f - id_{M_\bullet}$, we have $f\gamma \sim id_{N_\bullet}$ and $\gamma f \sim id_{M_\bullet}$ in $Kom^-(k - bimod)$.

$'\Leftarrow'$

Assume $f\gamma \sim id_{N_\bullet}$ and $\gamma f \sim id_{M_\bullet}$ in $Kom^-(k - bimod)$, so there are maps $s^{N_\bullet}$ and $s^{M_\bullet}$ such that $f\gamma - id_{N_\bullet} = s^{N_\bullet}d_{N_\bullet} + d_{N_\bullet}s^{N_\bullet}$ and $\gamma f - id_{M_\bullet} = s^{M_\bullet}d_{M_\bullet} + d_{M_\bullet}s^{M_\bullet}$.

The map $s_\bullet^{\mathcal{C}(f)} = \begin{pmatrix} s^{M_\bullet} + \gamma(s^{N_\bullet}f - fs^{M_\bullet}) & \gamma \\ s^{N_\bullet}(fs^{M_\bullet} - s^{N_\bullet}f) & -s^{N_\bullet} \end{pmatrix}$ is a homotopy. Indeed,

$$s_\bullet^{\mathcal{C}(f)} d_{\mathcal{C}(f)_\bullet} + d_{\mathcal{C}(f)_\bullet} s_\bullet^{\mathcal{C}(f)} =$$

$$= \begin{pmatrix} id_{M_\bullet} - \gamma s^{N_\bullet} f d_{M_\bullet} + \gamma f s^{M_\bullet} d_{M_\bullet} - d_{M_\bullet}\gamma s^{N_\bullet} f + d_{M_\bullet}\gamma f s^{M_\bullet} & 0 \\ s^{N_\bullet}s^{N_\bullet} f d_{M_\bullet} + f\gamma s^{N_\bullet} f - d_{N_\bullet} s^{N_\bullet}s^{N_\bullet} f - s^{N_\bullet} f\gamma f & id_{N_\bullet} \end{pmatrix}$$

$$= \begin{pmatrix} id_{M_\bullet} & 0 \\ 0 & id_{N_\bullet} \end{pmatrix} = id_{\mathcal{C}(f)_\bullet}.$$ Thus $\mathcal{C}(f)_\bullet$ is contractible in $Kom^-(k - bimod)$.

□

Proposition 3.2. allows us to conclude that if any two of f, g or fg are relative quasi-isomorphisms then so is the third. We prove now the following

Proposition 3.3. *The class of relative quasi-isomorphisms in the homotopic category* $\mathcal{K}^-(\mathbb{A} - \mathrm{bimod})$ *is localizing.*

Proof. We showed already that the class of relative quasi-isomorphisms is closed under the composition of maps. To conclude this class is localizing we need to justify two facts:

1) The extension conditions: For every $f \in Mor_{\mathcal{K}^-(\mathbb{A}-\mathrm{bimod})}$ and s relative quasi-isomorphism there exist $g \in Mor_{\mathcal{K}^-(\mathbb{A}-\mathrm{bimod})}$ and t relative quasi-isomorphism such that the following square

$$\begin{array}{ccc} \mathbb{N}_\bullet & \xrightarrow{f} & \mathbb{M}_\bullet \\ {\scriptstyle t}\downarrow & & \downarrow{\scriptstyle s} \\ \mathbb{K}_\bullet & \xrightarrow{g} & \mathbb{L}_\bullet \end{array}$$

(resp.

$$\begin{array}{ccc} \mathbb{L}_\bullet & \xrightarrow{g} & \mathbb{K}_\bullet \\ {\scriptstyle s}\downarrow & & \downarrow{\scriptstyle t} \\ \mathbb{M}_\bullet & \xrightarrow{f} & \mathbb{N}_\bullet \end{array}$$

is commutative.

2) Given f, g two morphisms from $\mathbb{N}_\bullet$ to $\mathbb{M}_\bullet$, the existence of s relative quasi-isomorphism with $sf = sg$ is equivalent to the existence of t relative quasi-isomorphism with $ft = gt$.

The proof of theorem 4, chapter 3 in [5], which states that the class of quasi-isomorphisms (not relative) in the homotopic category of an abelian category is localizing, can be used entirely so we will not reproduce it here. One needs to note for 1) that the cone of the map t constructed there is the same, in $\mathcal{K}^-(\mathbb{A}-\text{bimod})$, as the cone of s; and for 2) that the cone of the map t constructed there is the cone of s shifted by 1. Thus in both cases t is a relative quasi-isomorphism. □

Remark that the same result is true for $\mathcal{K}^-(\mathbb{A}!-\text{bimod})$ and $\mathcal{K}^-(\mathbb{A}!-\text{albimod})$.

We now define the relative derived categories by formally inverting all relative quasi-isomorphisms.

Definition 3.4. Let $\mathcal{A}$ be any of the categories $\mathbb{A}$-bimod, $\mathbb{A}!$-bimod or $\mathbb{A}!$-albimod and $\sum$ the appropriate class of relative quasi-isomorphisms.

$$\mathcal{D}_k^-(\mathcal{A}) := \mathcal{K}^-(\mathcal{A})(\Sigma^{-1}),$$

where $\mathcal{K}^-$ is the corresponding homotopy category.

Because $\sum$ is localizing we may regard the morphisms, in any of the relative derived categories defined above, as equivalence classes of diagrams

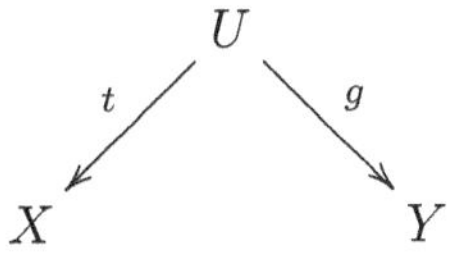

The maps t and g are morphisms in the homotopy category with $t \in \sum$. These diagrams are usually called roofs and we adopt this terminology. In addition, because $\sum$ is a localizing class the relative derived categories defined above are triangulated.

We begin studying the objects of $\mathcal{D}_k^-(\mathbb{A}-\text{bimod})$ with the complexes of relative projective bimodules.

Lemma 3.5. *Let* $\mathbb{P}_\bullet$ *be a complex of relative projective* $\mathbb{A}$-bimodules *and* $\mathbb{R}_\bullet \xrightarrow{f} \mathbb{P}_\bullet$ *a relative quasi-isomorphism. We have*

$$Mor_{\mathcal{K}^-(\mathbb{A}-\text{bimod})}(\mathbb{P}_\bullet, \mathcal{C}(f)_\bullet) = 0.$$

Proof. Because f is a relative quasi-isomorphism the cone $\mathcal{C}(f)_i$ is acyclic and allowable $(\forall) i \in \mathcal{C}$. Given $g \in Mor_{\mathcal{K}^-(\mathbb{A}-\text{bimod})}(\mathbb{P}_\bullet, \mathcal{C}(f)_\bullet)$ we show that $g = (g)_i : \mathbb{P}_i \longrightarrow \mathcal{C}(f)_i, i \geqslant 0$ is homotopic to 0 inductively. Since $\mathbb{P}_0$ is a complex of relative projective $\mathbb{A}$-bimodules we obtain that the map g_0 from $\mathbb{P}_0$ to $\mathcal{C}(f)_0$ can be lifted to a map $\delta_0 : \mathbb{P}_0 \longrightarrow \mathcal{C}(f)_1$ such that $d_{\mathcal{C}(f)_1}\delta_0 = g_0$. The image of $g_1 - \delta_0 d_{\mathbb{P}_1}$ is contained in the image of $d_{\mathcal{C}(f)_1}$ so it has a lifting $\delta_1 : \mathbb{P}_1 \longrightarrow \mathcal{C}(f)_2$ such that $d_{\mathcal{C}(f)_2}\delta_1 = g_1 - \delta_0 d_{\mathbb{P}_1}$. Now, the image of $g_2 - \delta_1 d_{\mathbb{P}_2}$ is contained in the image of $d_{\mathcal{C}(f)_2}$ and the conclusion follows inductively. □

Proposition 3.6. *Let* $\mathbb{P}_\bullet$ *be a complex of relative projective bimodules in* $Kom^-(\mathbb{A}-$bimod). *The canonical map*

$$Mor_{\mathcal{K}^-(\mathbb{A}-\text{bimod})}(\mathbb{P}_\bullet, \mathbb{M}_\bullet) \xrightarrow{can} Mor_{\mathcal{D}_k^-(\mathbb{A}-\text{bimod})}(\mathbb{P}_\bullet, \mathbb{M}_\bullet)$$

is an isomorphism for all $\mathbb{M}_\bullet \in Kom^-(\mathbb{A} - \text{bimod})$.

Proof. To prove the injectivity let $\mathbb{P}_\bullet \xrightarrow{\alpha} \mathbb{M}_\bullet$ and $\mathbb{P}_\bullet \xrightarrow{\beta} \mathbb{M}_\bullet$ such that their corresponding roofs:

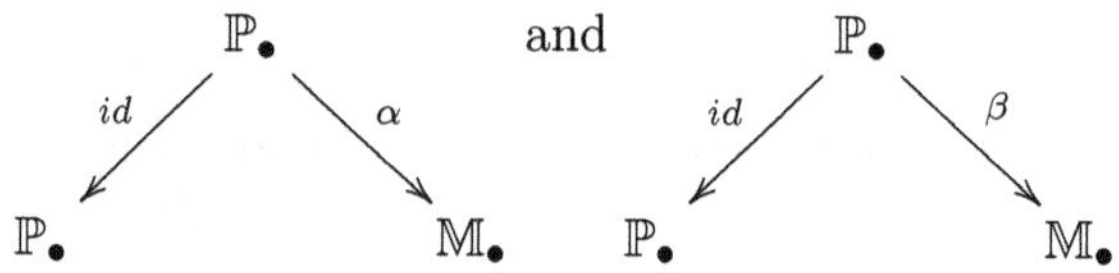

are equivalent in $\mathcal{D}_k^-(\mathbb{A} - \text{bimod})$. Thus, we have the commutative diagram in $\mathcal{K}^-(\mathbb{A} - \text{bimod})$

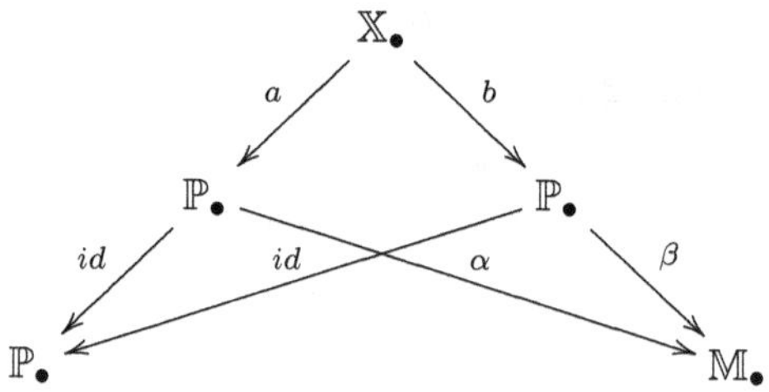

We obtain $a = b$ and $\alpha a = \beta b$.

To check that $\alpha = \beta$, apply $Mor_{\mathcal{K}^-(\mathbb{A}-bimod)}(\mathbb{P}_\bullet, -)$ to the distinguished triangle $\mathbb{X}_\bullet \xrightarrow{a} \mathbb{P}_\bullet \longrightarrow \mathcal{C}(a)_\bullet \longrightarrow \mathbb{X}_\bullet[1]$ and use previous lemma to see that $Mor_{\mathcal{K}^-(\mathbb{A}-bimod)}(\mathbb{P}_\bullet, \mathcal{C}(a)_\bullet) = 0$. This implies the existence of a map c such that $ac = id_{\mathbb{P}_\bullet}$ in $\mathcal{K}^-(\mathbb{A} - \text{bimod})$ and the injectivity follows from here.

For a morphism in $Mor_{\mathcal{D}_k^-(\mathbb{A}-\text{bimod})}(\mathbb{P}_\bullet, \mathbb{M}_\bullet)$ represented by the roof

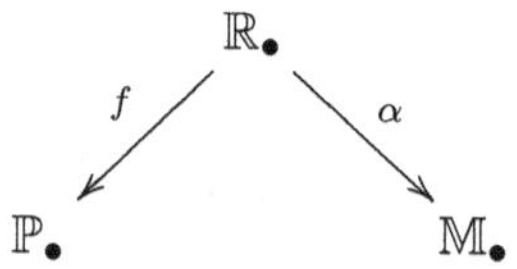

the distinguished triangle $\mathbb{R}_\bullet \xrightarrow{f} \mathbb{P}_\bullet \longrightarrow \mathcal{C}(f)_\bullet \longrightarrow \mathbb{R}_\bullet[1]$ induces a long exact sequence by applying $Mor_{\mathcal{K}^-(\mathbb{A}-\text{bimod})}(\mathbb{P}_\bullet, (-))$ to it. Again, by the previous lemma $Mor_{\mathcal{K}^-(\mathbb{A}-\text{bimod})}(\mathbb{P}_\bullet, \mathcal{C}(f)_\bullet) = 0$, thus the map

$$Mor_{\mathcal{K}^-(\mathbb{A}-\text{bimod})}(\mathbb{P}_\bullet, \mathbb{R}_\bullet) \xrightarrow{f} Mor_{\mathcal{K}^-(\mathbb{A}-\text{bimod})}(\mathbb{P}_\bullet, \mathbb{P}_\bullet)$$

is onto, so $(\exists)$ a map $\mathbb{P}_\bullet \xrightarrow{s} \mathbb{R}_\bullet$ such that $fs = id_{\mathbb{P}_\bullet}$ in $\mathcal{K}^-(\mathbb{A} - \text{bimod})$. Since f is a relative quasi-isomorphism s is a relative quasi-isomorphism, so we have the commutative diagram:

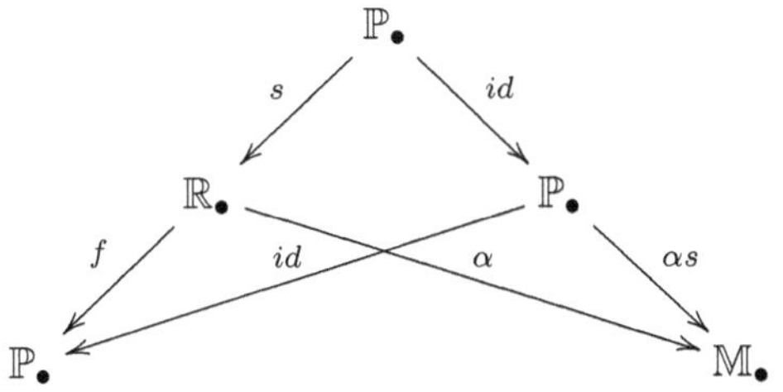

Thus, the roofs

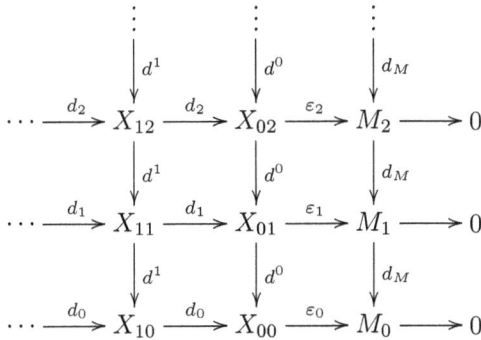

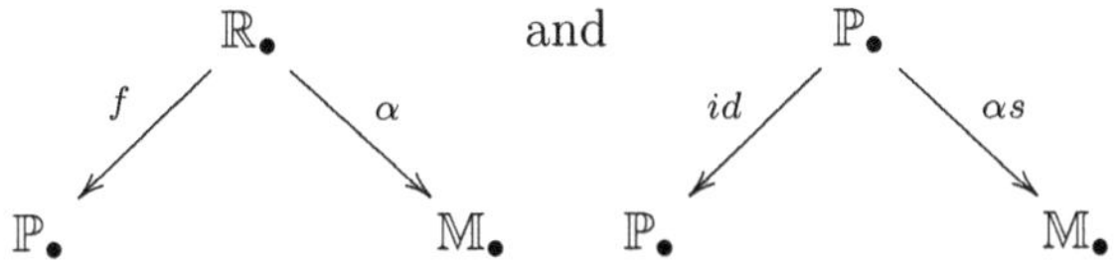

are equivalent and since the second is the image of αs the surjectivity is proved. □

Note that relative projective complexes in $Kom^-(\mathbb{A}! - \text{bimod})$ and $Kom^-(\mathbb{A}! - \text{albimod})$ satisfy the same property.

We prove now that each complex of $\mathbb{A}$-bimodules is relative quasi-isomorphic to a complex of relative projective bimodules. For this we need the following

Proposition 3.7. *Let A be a k algebra and assume that we have a double complex of A bimodules*

$$\begin{array}{ccccccccc}
 & & \vdots & & \vdots & & \vdots & & \\
 & & \downarrow{\scriptstyle d^1} & & \downarrow{\scriptstyle d^0} & & \downarrow{\scriptstyle d_M} & & \\
\cdots & \xrightarrow{d_2} & X_{12} & \xrightarrow{d_2} & X_{02} & \xrightarrow{\varepsilon_2} & M_2 & \longrightarrow & 0 \\
 & & \downarrow{\scriptstyle d^1} & & \downarrow{\scriptstyle d^0} & & \downarrow{\scriptstyle d_M} & & \\
\cdots & \xrightarrow{d_1} & X_{11} & \xrightarrow{d_1} & X_{01} & \xrightarrow{\varepsilon_1} & M_1 & \longrightarrow & 0 \\
 & & \downarrow{\scriptstyle d^1} & & \downarrow{\scriptstyle d^0} & & \downarrow{\scriptstyle d_M} & & \\
\cdots & \xrightarrow{d_0} & X_{10} & \xrightarrow{d_0} & X_{00} & \xrightarrow{\varepsilon_0} & M_0 & \longrightarrow & 0
\end{array}$$

such that:

a) Each row is k contractible. (i.e. There exist k-bimodule maps

$X_{(k-1)i} \xrightarrow{t_i^k} X_{ki}$ such that $d_i t_i^{k+1} + t_i^k d_i = id_{X_{ki}}$.)

b) The following diagrams are commutative:

$$\begin{array}{ccccccc}
X_{ki} & \xleftarrow{t_i^k} & X_{(k-1)i} & \quad & X_{0i} & \xleftarrow{t_i^0} & M_i \\
{\scriptstyle d^k}\downarrow & & \downarrow{\scriptstyle d^{k-1}} & & {\scriptstyle d^0}\downarrow & & \downarrow{\scriptstyle d_M} \\
X_{k(i-1)} & \xleftarrow{t_{i-1}^k} & X_{(k-1)(i-1)} & & X_{0(i-1)} & \xleftarrow{t_{i-1}^0} & M_{i-1}
\end{array}$$

for all $k, i \geqslant 0$, *Then*

1. $M_\bullet \xrightarrow{t_\bullet^0} (TotX_{\bullet\bullet})$ *and* $(TotX_{\bullet\bullet}) \xrightarrow{\varepsilon_\bullet} M_\bullet$ *are maps of complexes of* k*-bimodules, where* $\varepsilon_i = 0$ *on* $X_{jk}, j + k = i$ *if* $j > 0$.

2. $\varepsilon_\bullet t_\bullet^0 = id_{M_\bullet}$ *and* $t_\bullet^0 \varepsilon_\bullet \sim id_{TotX_{\bullet\bullet}}$ *in* $Kom^-(k - bimod)$, *where* $\sim$*=homotopy equivalence.*

Proof. 1. The map $t_\bullet^0$ is a map of complexes by b) and $\varepsilon_\bullet$ is a map of complexes because $d_M \varepsilon_{i+1} = d^0 \varepsilon_i$ and $\varepsilon_i d_i = 0$.

2. The only thing to prove here is $t_\bullet^0 \varepsilon_\bullet \sim id_{TotX_{\bullet\bullet}}$ in $Kom^-(k - bimod)$. For $n \geqslant 0$ we define the map $(TotX_{\bullet\bullet})^n \xrightarrow{h^n} (TotX_{\bullet\bullet})^{n+1}$ by $h^n := (t_0^{n+1}, t_1^n, \ldots, t_n^1, 0)$. It is a simple exercise to check that $h^\bullet d_{TotX_{\bullet\bullet}} + d_{TotX_{\bullet\bullet}} h^\bullet = id - t_\bullet^0 \varepsilon_\bullet$. □

Theorem 3.8. *For each* $\mathbb{M}_\bullet \in \mathcal{D}_k^-(\mathbb{A} - \text{bimod})$ *there exist* $\mathcal{U}\mathbb{M}_\bullet \in \mathcal{D}_k^-(\mathbb{A} - \text{bimod})$ *and* $\mathcal{U}\mathbb{M}_\bullet \xrightarrow{\varepsilon} \mathbb{M}_\bullet$ *a relative quasi-isomorphism such that* $\mathcal{U}\mathbb{M}_\bullet$ *is a complex of relative projective* $\mathbb{A}$*-bimodules.*

Proof. We described in section 2 a method of constructing a relative projective allowable resolution $Tot\mathcal{S}_\bullet\mathcal{B}_\bullet(\mathbb{M}) \longrightarrow \mathbb{M}$, for each $\mathbb{M} \in \mathbb{A}$-bimod. We use this for each term $\mathbb{M}_i$ of the complex $\mathbb{M}_\bullet$, $i \geqslant 0$. We obtain a double complex with augmented column $\mathbb{M}_\bullet$. In addition, each row is contractible and for all $p \in \mathcal{C}$ we obtain a double complex of $\mathbb{A}^p$-bimodules which satisfies the conditions of the previous proposition. Thus, by taking the total complex of the double complex with augmented column $\mathbb{M}_\bullet$ we obtain the desired complex of relative projective $\mathbb{A}$-bimodules, $\mathcal{U}\mathbb{M}_\bullet$, together with a relative quasi-isomorphism $\mathcal{U}\mathbb{M}_\bullet \xrightarrow{\varepsilon} \mathbb{M}_\bullet$. □

Note that the same argument shows that for each complex $\mathbb{M}_\bullet! \in \mathcal{D}_k^-(\mathbb{A}! - \text{bimod})$ the total complex, $Tot\mathcal{T}_\bullet\mathbb{M}_\bullet$, of the double complex $\mathcal{T}_\bullet\mathbb{M}_\bullet$ obtained by taking the allowable resolution of each $\mathbb{M}_i!$ described in section 2, gives a relative quasi-isomorphism $(Tot\mathcal{T}_\bullet\mathbb{M}_\bullet) \xrightarrow{\varepsilon} \mathbb{M}_\bullet!$. In addition, by theorem 2.4., the left adjoint ¡ to ! has the property that $(Tot\mathcal{T}_\bullet\mathbb{M}_\bullet \xrightarrow{\varepsilon} \mathbb{M}_\bullet!)$¡ is isomorphic to $Tot\mathcal{S}_\bullet\mathbb{M}_\bullet \xrightarrow{\varepsilon ¡} \mathbb{M}_\bullet$, so ε¡ is a relative quasi-isomorphism.

To see how the relative derived categories defined earlier relate to Hochschild cohomology recall that given a presheaf of k-algebras $\mathbb{A}$ the relative Hochschild cohomology of $\mathbb{A}$, denoted $\mathbf{H}^\bullet(\mathbb{A}, (-))$, is the same as the relative Yoneda cohomology $\mathbf{Ext}^\bullet_{\mathbb{A}-\mathbb{A}}(\mathbb{A}, (-))$ of the category of $\mathbb{A}$-bimodules. The word **relative** appears as an indication that k is not necessarily a field, in general only a commutative ring.

Thus, the relative Hochschild cohomology of a presheaf of algebras, with coefficients in an arbitrary $\mathbb{A}$-bimodule $\mathbb{M}$, is computed by taking any relative projective allowable resolution of $\mathbb{A}$, applying $Hom_{\mathbb{A}-\mathbb{A}}((-), \mathbb{M})$ to it and then taking the homology of the resulting complex.

Theorem 3.9. $\mathbf{Ext}^i_{\mathbb{A}-\mathbb{A}}(\mathbb{M}, \mathbb{N}) \simeq Mor_{\mathcal{D}^-_k(\mathbb{A}-\mathrm{bimod})}(\mathbb{M}_\bullet, \mathbb{N}_\bullet[i])$. *In particular,* $\mathbf{H}^i(\mathbb{A}, \mathbb{N}) \simeq Mor_{\mathcal{D}^-_k(\mathbb{A}-\mathrm{bimod})}(\mathbb{A}_\bullet, \mathbb{N}_\bullet[i])$.

Proof. Let $Tot\mathcal{B}_\bullet\mathcal{S}_\bullet\mathbb{M}$ the relative allowable projective resolution described in section 2. (same as $\mathcal{U}\mathbb{M}_\bullet$ in this case since $\mathbb{M}_i = 0$, $(\forall)i \neq 0$.) Using proposition 3.6. and theorem 3.8. we obtain the isomorphisms
$\mathbf{Ext}^i_{\mathbb{A}-\mathbb{A}}(\mathbb{M}, \mathbb{N}) = H^i(Hom_{\mathbb{A}-\mathbb{A}}(\mathcal{U}\mathbb{M}_\bullet, \mathbb{N})) = Mor_{\mathcal{K}^-(\mathbb{A}-\mathrm{bimod})}(\mathcal{U}\mathbb{M}_\bullet, \mathbb{N}_\bullet[i])$
$\cong Mor_{\mathcal{D}^-_k(\mathbb{A}-\mathrm{bimod})}(\mathcal{U}\mathbb{M}_\bullet, \mathbb{N}_\bullet[i]) \cong Mor_{\mathcal{D}^-_k(\mathbb{A}-\mathrm{bimod})}(\mathbb{M}_\bullet, \mathbb{N}_\bullet[i])$. □

4. Functors between derived categories

The functor $\mathbb{A} - \mathrm{bimod} \xrightarrow{!} \mathbb{A}! - \mathrm{bimod}$ is exact and preserves allowability so it induces a functor between the corresponding relative derived categories. In this section we prove the following property of the induced functor.

Theorem 4.1. *The functor* $\mathcal{D}^-_k(\mathbb{A} - \mathrm{bimod}) \xrightarrow{!} \mathcal{D}^-_k(\mathbb{A}! - \mathrm{bimod})$ *is full and faithful. That is,*

$$Mor_{\mathcal{D}^-_k(\mathbb{A}-\mathrm{bimod})}(\mathbb{M}_\bullet, \mathbb{N}_\bullet) \xrightarrow{!} Mor_{\mathcal{D}^-_k(\mathbb{A}!-\mathrm{bimod})}(\mathbb{M}_\bullet!, \mathbb{N}_\bullet!)$$

is an isomorphism of sets for all $\mathbb{M}_\bullet, \mathbb{N}_\bullet \in \mathcal{D}^-_k(\mathbb{A} - \mathrm{bimod})$.

The difficulties in proving the theorem reside in two places. First, since the morphisms in $\mathcal{D}^-_k(\mathbb{A} - \mathrm{bimod})$ and $\mathcal{D}^-_k(\mathbb{A}! - \mathrm{bimod})$ are equivalence classes of roofs, it is not clear how one can find ancestors in $\mathcal{D}^-_k(\mathbb{A} - \mathrm{bimod})$ for arbitrary roofs in $\mathcal{D}^-_k(\mathbb{A}! - \mathrm{bimod})$.

A good sign for that would be the existence of a left adjoint for !, but there is none. Fortunately, a left adjoint exists between $\mathbb{A}$-bimod and the full subcategory of $\mathbb{A}$!-bimod of aligned bimodules. Second, left adjoints do not necessarily preserve all relative quasi-isomorphisms. However, this left adjoint preserves some that can be used to trace back ancestors for any roof in $Mor_{\mathcal{D}^-_k(\mathbb{A}!-\mathrm{bimod})}(\mathbb{M}_\bullet!, \mathbb{N}_\bullet!)$.

We will prove that $\mathcal{D}^-_k(\mathbb{A} - \mathrm{bimod}) \xrightarrow{!} \mathcal{D}^-_k(\mathbb{A}! - \mathrm{albimod})$ and the inclusion $\mathcal{D}^-_k(\mathbb{A}! - \mathrm{albimod}) \xrightarrow{inc} \mathcal{D}^-_k(\mathbb{A}! - \mathrm{bimod})$ are full and faithful.

Proposition 4.2. *The functor*

$$\mathcal{D}^-_k(\mathbb{A} - \mathrm{bimod}) \xrightarrow{!} \mathcal{D}^-_k(\mathbb{A}! - \mathrm{albimod})$$

is full and faithful.

Proof. To prove the proposition we need to show that

$$Mor_{\mathcal{D}^-_k(\mathbb{A}-\mathrm{bimod})}(\mathbb{M}_\bullet, \mathbb{N}_\bullet) \xrightarrow{!} Mor_{\mathcal{D}^-_k(\mathbb{A}!-\mathrm{albimod})}(\mathbb{M}_\bullet!, \mathbb{N}_\bullet!)$$

is an isomorphism for all $\mathbb{M}_\bullet$ and $\mathbb{N}_\bullet \in \mathcal{D}_k^-(\mathbb{A}-\text{bimod})$.

Since for all $\mathbb{M}_\bullet \in \mathcal{D}_k^-(\mathbb{A}-\text{bimod})$ there exist $\mathcal{U}\mathbb{M}_\bullet \xrightarrow{\varepsilon} \mathbb{M}_\bullet$ relative quasi-isomorphism in $\mathcal{D}_k^-(\mathbb{A}-\text{bimod})$ such that $\mathcal{U}\mathbb{M}_i$ is relative projective for all i, we may assume that $\mathbb{M}_\bullet$ is a complex of relative projective $\mathbb{A}$ bimodules. This is because of the commutative diagram

$$\begin{array}{ccc} Mor_{\mathcal{D}_k^-(\mathbb{A}-\text{bimod})}(\mathbb{M}_\bullet,\mathbb{N}_\bullet) & \xrightarrow{!} & Mor_{\mathcal{D}_k^-(\mathbb{A}!-\text{albimod})}(\mathbb{M}_\bullet!,\mathbb{N}_\bullet!) \\ \varepsilon\downarrow & & \downarrow\varepsilon! \\ Mor_{\mathcal{D}_k^-(\mathbb{A}-\text{bimod})}(\mathcal{U}\mathbb{M}_\bullet,\mathbb{N}_\bullet) & \xrightarrow{!} & Mor_{\mathcal{D}_k^-(\mathbb{A}!-\text{albimod})}(\mathcal{U}\mathbb{M}_\bullet!,\mathbb{N}_\bullet!) \end{array}$$

where ε and $\varepsilon!$ are isomorphisms.

Because $(\mathbb{M}_i)^p$ is a relative projective $\mathbb{A}^p$-bimodule, $(\forall)p \in \mathcal{C}$, each $\mathbb{M}_i!$ admits a resolution of relative projective aligned $\mathbb{A}!$-bimodules obtained using $\mathcal{T}_\bullet$. The total complex of the double complex obtained by taking the resolution of each $\mathbb{M}_i!$ gives a relative quasi-isomorphism $Tot(\mathcal{T}_\bullet\mathbb{M}_\bullet) \xrightarrow{\varepsilon} \mathbb{M}_\bullet!$, where each $Tot(\mathcal{T}_\bullet\mathbb{M}_\bullet)_i$ is a relative projective aligned $\mathbb{A}!$ bimodule.

Moreover, the left adjoint ¡ has the property that $(Tot\mathcal{T}_\bullet\mathbb{M}_\bullet \xrightarrow{\varepsilon} \mathbb{M}_\bullet!)$¡ is isomorphic to $Tot\mathcal{S}_\bullet\mathbb{M}_\bullet \xrightarrow{\varepsilon ¡} \mathbb{M}_\bullet$ and ε¡ is a relative quasi-isomorphism. Now, given any roof

$$\begin{array}{ccc} & X_\bullet & \\ {}^{s}\swarrow & & \searrow^{f} \\ \mathbb{M}_\bullet! & & \mathbb{N}_\bullet! \end{array}$$

in $Mor_{\mathcal{D}_k^-(\mathbb{A}!-\text{albimod})}(\mathbb{M}_\bullet!,\mathbb{N}_\bullet!)$ take $Tot(\mathcal{T}_\bullet\mathbb{M}_\bullet) \xrightarrow{\varepsilon} \mathbb{M}_\bullet!$ as above.

By applying $Mor_{\mathcal{D}_k^-(\mathbb{A}!-\text{albimod})}(Tot(\mathcal{T}_\bullet\mathbb{M}_\bullet),(-))$ to the distinguished triangle

$$X_\bullet \xrightarrow{s} \mathbb{M}_\bullet! \longrightarrow \mathcal{C}(s)_\bullet \longrightarrow X_\bullet[1]$$

we obtain a long exact sequence.
In this sequence $Mor_{\mathcal{D}_k^-}(Tot(\mathcal{T}_\bullet\mathbb{M}_\bullet),\mathcal{C}(s)_\bullet) = 0$ because $\mathcal{C}(s)_\bullet$ is contractible, as a complex of k-bimodules, and $Tot(\mathcal{T}_\bullet\mathbb{M}_\bullet)$ is a complex of relative projective aligned $\mathbb{A}!$ bimodules, so the map

$$Mor_{\mathcal{D}_k^-}(Tot(\mathcal{T}_\bullet\mathbb{M}_\bullet),X) \xrightarrow{s} Mor_{\mathcal{D}_k^-}(Tot(\mathcal{T}_\bullet\mathbb{M}_\bullet),\mathbb{M}_\bullet!)$$

is onto. Because $\varepsilon \in Mor_{\mathcal{D}_k^-(\mathbb{A}!-\text{albimod})}(Tot(\mathcal{T}_\bullet\mathbb{M}_\bullet),\mathbb{M}_\bullet!)$, there exist $q \in Mor_{\mathcal{D}_k^-(\mathbb{A}!-\text{albimod})}(Tot(\mathcal{T}_\bullet\mathbb{M}_\bullet),X_\bullet)$ such that the diagram

$$\begin{array}{ccc} & & Tot(\mathcal{T}_\bullet\mathbb{M}_\bullet) \\ & {}^{q}\swarrow & \downarrow\varepsilon \\ X_\bullet & \xrightarrow{s} & \mathbb{M}_\bullet! \end{array}$$

commutes. The map q is a relative quasi-isomorphism because both s and ε are and we have the equivalence of roofs

$$\begin{array}{ccc} & X_\bullet & \\ {}^{s}\swarrow & & \searrow^{f} \\ \mathbb{M}_\bullet ! & & \mathbb{N}_\bullet ! \end{array} \qquad \text{and} \qquad \begin{array}{ccc} & Tot(\mathcal{T}_\bullet \mathbb{M}_\bullet) & \\ {}^{\varepsilon}\swarrow & & \searrow^{fq} \\ \mathbb{M}_\bullet ! & & \mathbb{N}_\bullet ! \end{array}$$

because the diagram

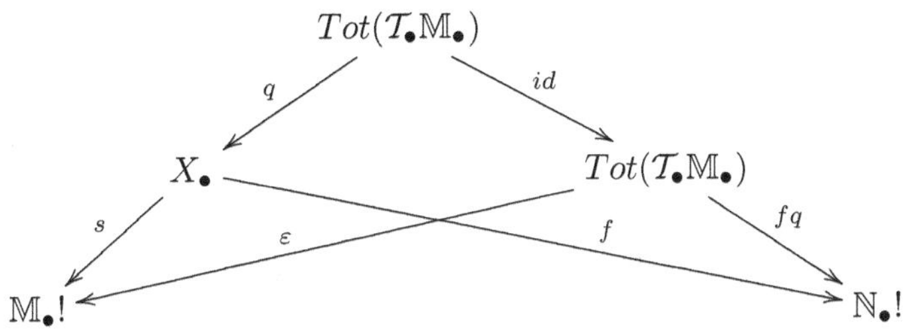

is commutative.

Since ($Tot\mathcal{T}_\bullet \mathbb{M}_\bullet \xrightarrow{\varepsilon} \mathbb{M}_\bullet !$)¡ is isomorphic to $Tot\mathcal{S}_\bullet \mathbb{M}_\bullet \xrightarrow{\varepsilon\text{¡}} \mathbb{M}_\bullet$ and ε¡ is a relative quasi-isomorphism, the roof

$$\begin{array}{ccc} & (Tot(\mathcal{T}_\bullet \mathbb{M}_\bullet))\text{¡} & \\ {}^{\varepsilon_{\mathbb{M}_\bullet} \varepsilon \text{¡}}\swarrow & & \searrow^{\varepsilon_{\mathbb{N}_\bullet} f \text{¡} q \text{¡}} \\ \mathbb{M}_\bullet & & \mathbb{N}_\bullet \end{array}$$

exists in $\mathcal{D}_k^-(\mathbb{A} - \text{bimod})$.
Here, $\varepsilon_{\mathbb{M}_\bullet}$ and $\varepsilon_{\mathbb{N}_\bullet}$ are the maps of complexes induced by the counit of the adjunction

$\mathbb{A} - \text{bimod} \underset{\text{¡}}{\overset{!}{\rightleftarrows}} \mathbb{A}! - \text{albimod}$. The image of this roof via ! is

$$\begin{array}{ccc} & [(Tot(\mathcal{T}_\bullet \mathbb{M}_\bullet)\text{¡}])! & \\ {}^{\varepsilon_{\mathbb{M}_\bullet} !\varepsilon \text{¡}!}\swarrow & & \searrow^{\varepsilon_{\mathbb{N}_\bullet} !f \text{¡} !q \text{¡}!} \\ \mathbb{M}_\bullet ! & & \mathbb{N}_\bullet ! \end{array}$$

and is equivalent to

$$\begin{array}{ccc} & Tot(\mathcal{T}_\bullet \mathbb{M}_\bullet) & \\ {}^{\varepsilon}\swarrow & & \searrow^{fq} \\ \mathbb{M}_\bullet ! & & \mathbb{N}_\bullet ! \end{array} \quad .$$

This results from the commutative diagram

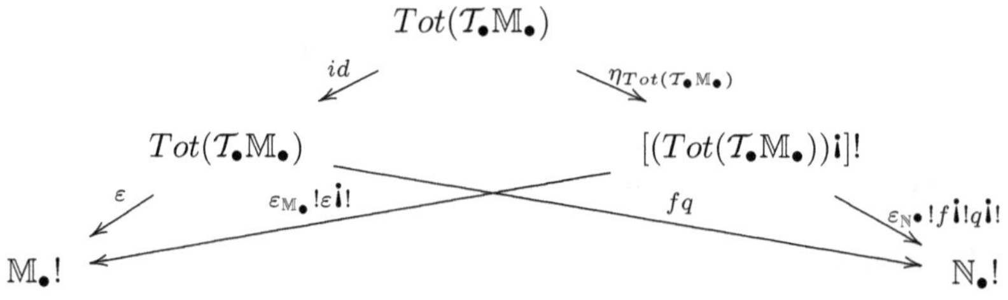

(1) $\varepsilon_{M_\bullet}!(\varepsilon\mathbf{i})!\eta_{Tot(\mathcal{T}_\bullet M_\bullet)} = \varepsilon$ and

(2) $\varepsilon_{N_\bullet}![f\mathbf{i}qi]!\eta_{Tot(\mathcal{T}_\bullet M_\bullet)} = fq$

To check (1) observe that we have $\varepsilon_{M_\bullet}!\eta_{M_\bullet}! = id_{M_\bullet}!$ by the adjunction. In addition, the functoriality of η induces the commutative square

$$
\begin{array}{ccc}
Tot(\mathcal{T}_\bullet M_\bullet) & \xrightarrow{\ \varepsilon\ } & M_\bullet! \\
{\scriptstyle \eta_{Tot(\mathcal{T}_\bullet M_\bullet)}}\downarrow & & \downarrow{\scriptstyle \eta_{M_\bullet}!} \\
[(Tot(\mathcal{T}_\bullet M_\bullet))\mathbf{i}]! & \xrightarrow{\ (\varepsilon\mathbf{i})!\ } & [(M_\bullet!)\mathbf{i}]!
\end{array}
$$

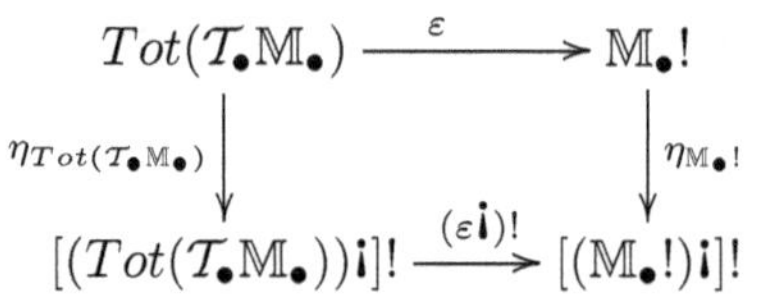

Thus, we have $(\varepsilon\mathbf{i})!\eta_{Tot(\mathcal{T}_\bullet M_\bullet)} = \eta_{M_\bullet}!\varepsilon$ and by composing with $\varepsilon_{M_\bullet}!$ we obtain (1). Similarly one may check (2).

To prove injectivity, let

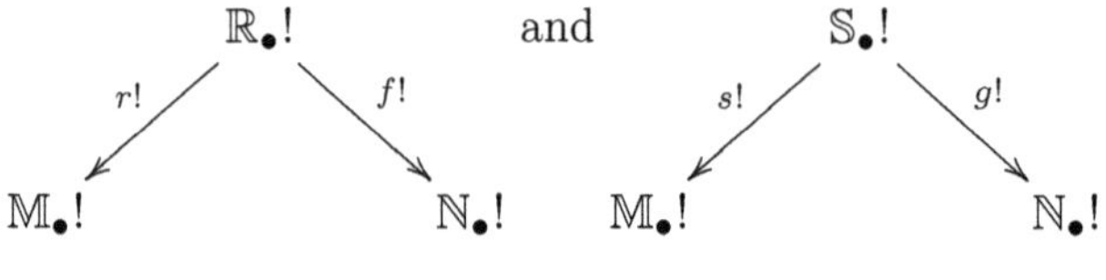

be equivalent roofs in $\mathcal{D}_k^-(\mathbb{A}! - \text{albimod})$. One may assume that $\mathbb{R}_\bullet$ is a complex or relative projective $\mathbb{A}$ bimodules. To see this, let $\mathcal{U}M_\bullet \xrightarrow{\ \varepsilon\ } M_\bullet$ the relative quasi-isomorphism with $\mathcal{U}M_i$ relative projective $\mathbb{A}$-bimodules.

Again, applying $Mor_{\mathcal{D}_k^-(\mathbb{A}-\text{bimod})}(\mathcal{U}M_\bullet, (-))$ to the distinguished triangle $\mathbb{R}_\bullet \xrightarrow{\ r\ } M_\bullet \longrightarrow \mathcal{C}(r)_\bullet \longrightarrow \mathbb{R}[1]_\bullet$, in $\mathcal{D}_k^-(\mathbb{A}-\text{bimod})$, we obtain a long exact sequence where $Mor_{\mathcal{D}_k^-}(\mathcal{U}M_\bullet, \mathcal{C}(r)_\bullet) = 0$.

This implies the existence of a map t such that the following diagram

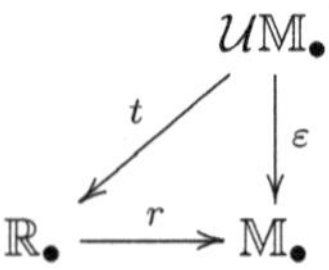

commutes. In addition, t is a relative quasi-isomorphism, since r and ε are and we

have the equivalent roofs

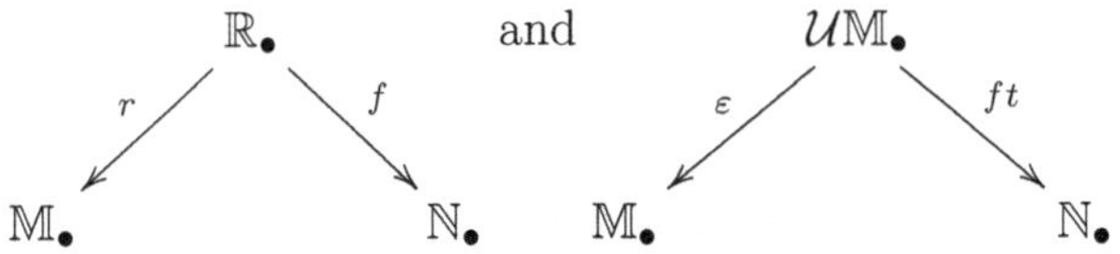

in $\mathcal{D}_k^-(\mathbb{A} - \text{bimod})$ because of the following commutative diagram

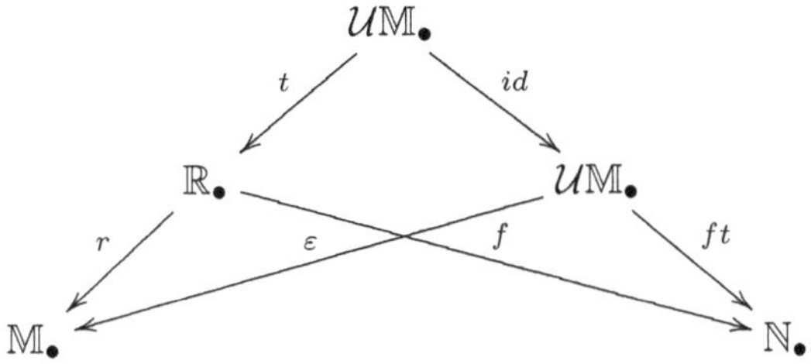

This implies the the equivalence of

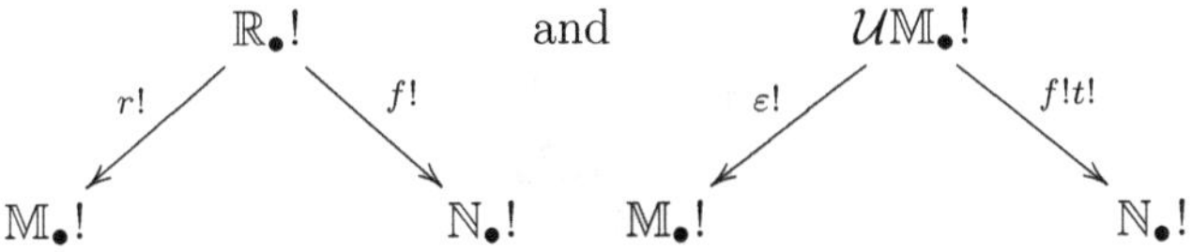

in $\mathcal{D}_k^-(\mathbb{A}! - \text{albimod})$. So, we may assume that $\mathbb{R}_\bullet$ is a complex of relative projective $\mathbb{A}$-bimodules. The equivalence of

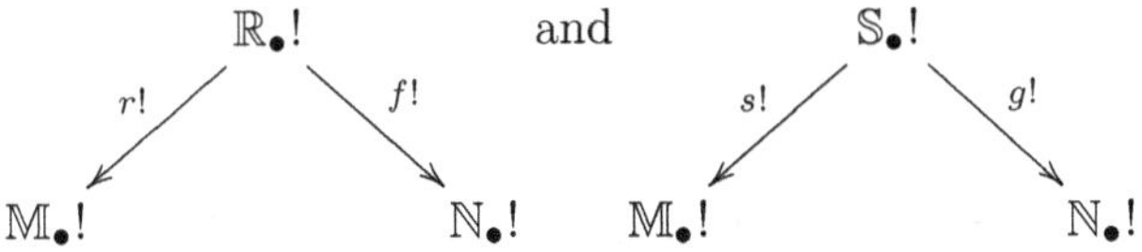

translates into the existence of a commutative diagram

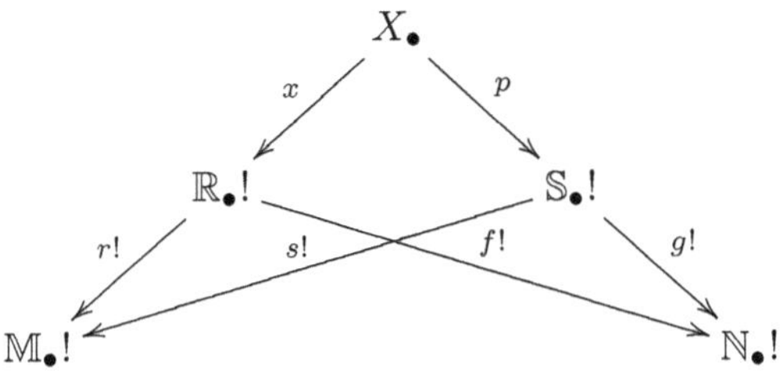

Here, $X_\bullet \in \mathcal{D}_k^-(\mathbb{A}! - \text{albimod})$ and x is a relative quasi-isomorphism such that $f!x = g!p$, (1) and $s!p = r!x$, (2). Since x is a relative quasi-isomorphism and $TotT_\bullet\mathbb{R}_\bullet$ is a complex of aligned relative projective $\mathbb{A}!$-bimodules there exist j such that the diagram

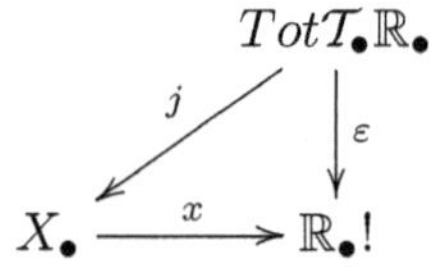

TotT•R•
j
ε
X•
x
R•!

is commutative. Moreover, j is a relative quasi-isomorphism because ε and x are. We obtain the commutative diagram

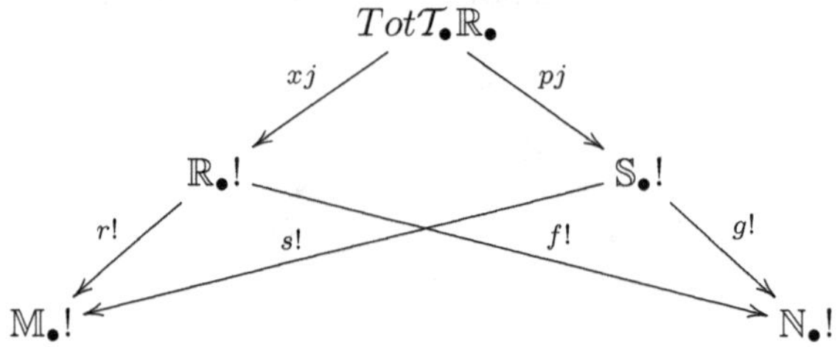

TotT•R•
xj
pj
R•!
S•!
r!
s!
f!
g!
M•!
N•!

because $f!xj = g!pj$, by (1) and $s!pj = r!xj$, by (2). Because ! is full and faithful we have the isomorphism $(\mathbb{T}_\bullet !)\text{¡} \xrightarrow{\varepsilon_{\mathbb{T}_\bullet}} \mathbb{T}_\bullet$ for all $\mathbb{T}_\bullet$ in $\mathcal{D}_k^-(\mathbb{A}-\text{bimod})$ and so $(r!)\text{¡}$ and $(s!)\text{¡}$ are relative quasi-isomorphisms in $\mathcal{D}_k^-(\mathbb{A}-\text{bimod})$. In addition, $\varepsilon\text{¡}$ is a relative quasi-isomorphism and we get the commutative diagram

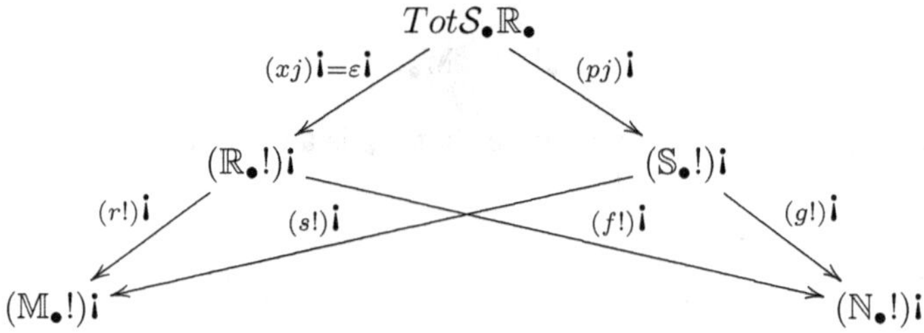

TotS•R•
(xj)¡=ε¡
(pj)¡
(R•!)¡
(S•!)¡
(r!)¡
(s!)¡
(f!)¡
(g!)¡
(M•!)¡
(N•!)¡

Finally, we obtain the equivalence of

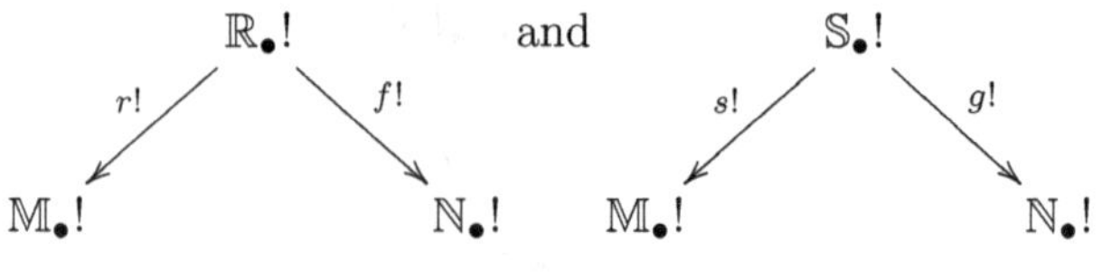

R•!
and
S•!
r!
f!
s!
g!
M•!
N•!
M•!
N•!

by constructing

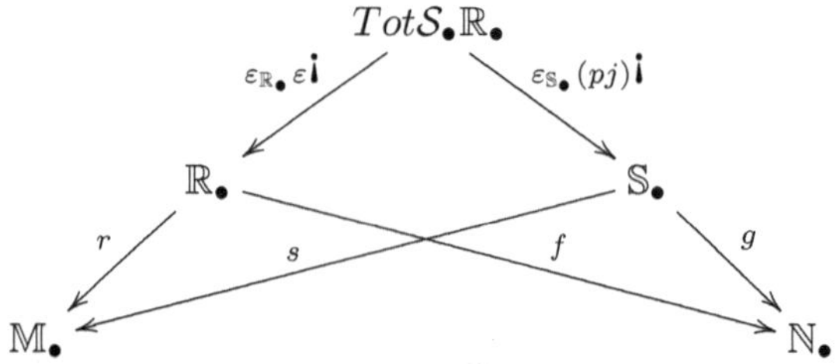

TotS•R•
ε_R• ε¡
ε_S• (pj)¡
R•
S•
r
s
f
g
M•
N•

This is because $f\varepsilon_{\mathbb{R}_\bullet}\varepsilon\text{¡} = \varepsilon_{\mathbb{N}_\bullet}(f!)\text{¡}\varepsilon\text{¡} = \varepsilon_{\mathbb{N}_\bullet}(g!)\text{¡}(pj)\text{¡} = g\varepsilon_{\mathbb{S}_\bullet}(pj)\text{¡}$ and $s\varepsilon_{\mathbb{S}_\bullet}(pj)\text{¡} = \varepsilon_{\mathbb{M}_\bullet}(s!)\text{¡}(pj)\text{¡} = \varepsilon_{\mathbb{M}_\bullet}(r!)\text{¡}\varepsilon\text{¡} = r\varepsilon_{\mathbb{R}_\bullet}\varepsilon\text{¡}$. □

We show now that the inclusion

$$\mathcal{D}_k^-(\mathbb{A}! - \text{albimod}) \xrightarrow{inc} \mathcal{D}_k^-(\mathbb{A}! - \text{bimod})$$

is full and faithful. The lack of an adjoint in this case requires a two step process of replacing the top of each roof by a complex of aligned bimodules. For $X \in \mathbb{A}! - \text{bimod}$, let $X^+ := \prod_{i \in C} \varphi^{ii} X$. This defines an exact functor $\mathbb{A}! - \text{bimod} \xrightarrow{+} \mathbb{A}! - \text{bimod}$ that preserves allowability, so also relative quasi-isomorphisms.

We also have the natural maps $X \xrightarrow{\beta_X} X^+$, $x \longrightarrow < \varphi^{ii} x >$ and $X_{al} \xrightarrow{\gamma_X} X^+$, $< x_{ij} > \longrightarrow < \sum_{j \geqslant i} x_{ij} >$. Also, if X is aligned both β_X and γ_X are isomorphisms and $\beta_X = \gamma_X \alpha_X$, where α is the natural isomorphism $\alpha : Id_{\mathbb{A}! - albimod} \longrightarrow (-)_{al} \circ inc$.

Proposition 4.3. *The functor*

$$\mathcal{D}_k^-(\mathbb{A}! - \text{albimod}) \xrightarrow{inc} \mathcal{D}_k^-(\mathbb{A}! - \text{bimod})$$

is full and faithful.

Proof. We have to prove that

$$Mor_{\mathcal{D}_k^-(\mathbb{A}! - \text{albimod})}(M_\bullet, N_\bullet) \xrightarrow{inc} Mor_{\mathcal{D}_k^-(\mathbb{A}! - \text{bimod})}(M_\bullet, N_\bullet)$$

is an isomorphism of sets for all $M_\bullet$ and $N_\bullet \in \mathcal{D}_k^-(\mathbb{A}! - \text{albimod})$. First, we prove that the map is onto. For any roof

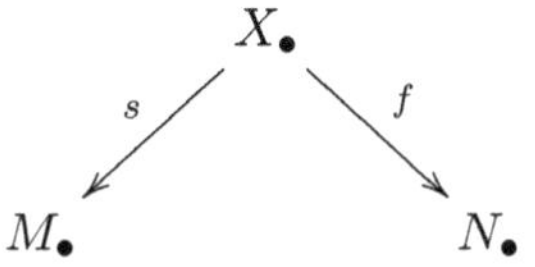

in $Mor_{\mathcal{D}_k^-(\mathbb{A}! - \text{bimod})}(M_\bullet, N_\bullet)$ we have the equivalences

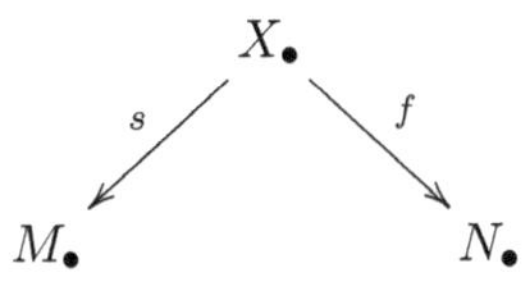

and

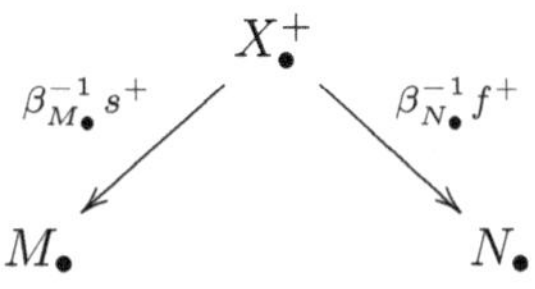

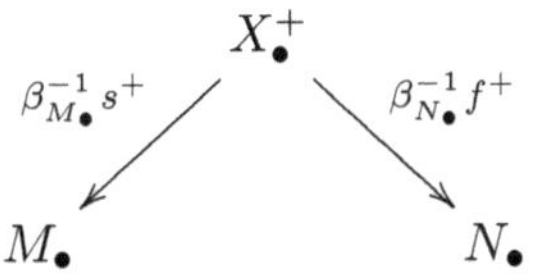

and

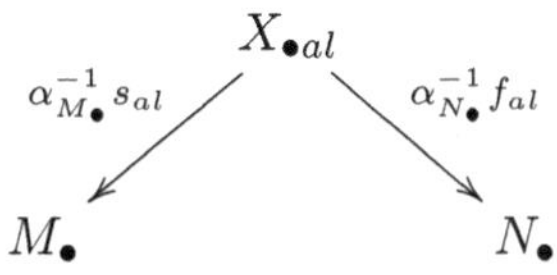

To see this, observe that since β is a natural transformation we have $s^+ \beta_{X_\bullet} = \beta_{M_\bullet} s$ and $f^+ \beta_{X_\bullet} = \beta_{N_\bullet} f$.

In addition, because $M_\bullet$ and $N_\bullet$ are aligned $\beta_{M_\bullet}$ and $\beta_{N_\bullet}$ are isomorphisms and we obtain $\beta_{M_\bullet}^{-1} s^+ \beta_{X_\bullet} = s$ and $\beta_{N_\bullet}^{-1} f^+ \beta_{X_\bullet} = f$.

This implies the first equivalence because the diagram

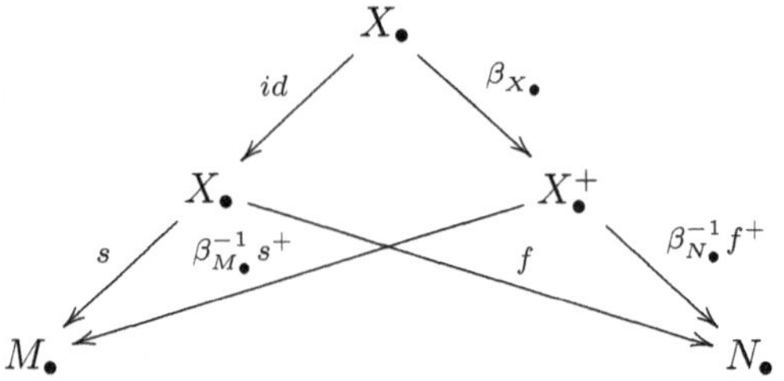

is commutative.

For the second equivalence, since γ is natural we have $s^+\gamma_{X_\bullet} = \gamma_{M_\bullet} s_{al}$ and $f^+\gamma_{X_\bullet} = \gamma_{N_\bullet} f_{al}$.

Because $M_\bullet$ and $N_\bullet$ are aligned $\gamma_{M_\bullet}, \gamma_{N_\bullet}, \alpha_{M_\bullet}, \alpha_{N_\bullet}, \beta_{M_\bullet}$ and $\beta_{N_\bullet}$ are isomorphisms, so we get $\beta_{M_\bullet}^{-1} s^+\gamma_{X_\bullet} = \beta_{M_\bullet}^{-1}\gamma_{M_\bullet} s_{al} = \alpha_{M_\bullet}^{-1}\gamma_{M_\bullet}^{-1}\gamma_{M_\bullet} s_{al} = \alpha_{M_\bullet}^{-1} s_{al}$ and $\beta_{N_\bullet}^{-1} f^+\gamma_{X_\bullet} = \beta_{N_\bullet}^{-1}\gamma_{N_\bullet} f_{al} = \alpha_{N_\bullet}^{-1}\gamma_{N_\bullet}^{-1}\gamma_{N_\bullet} f_{al} = \alpha_{N_\bullet}^{-1} f_{al}$.

The diagram

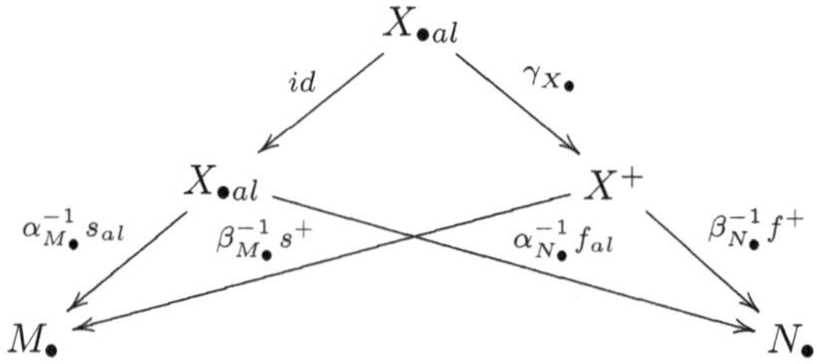

is commutative and implies the second equivalence. Now, the surjectivity follows since the roof

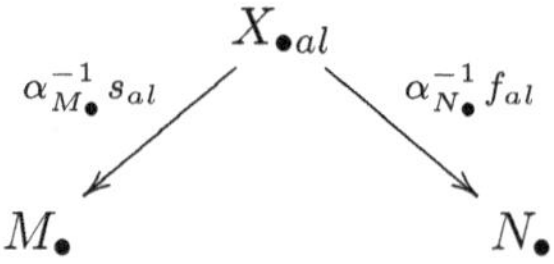

exists in $Mor_{\mathcal{D}_k^-(\mathbb{A}!-\text{albimod})}(M_\bullet, N_\bullet)$ and its image is equivalent to

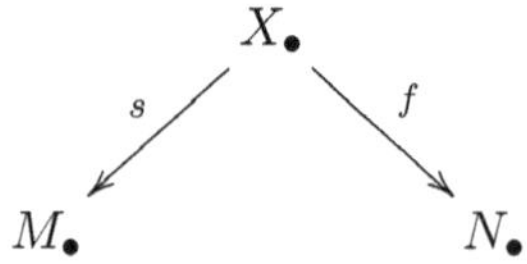

in $Mor_{\mathcal{D}_k^-(\mathbb{A}!-\text{bimod})}(M_\bullet, N_\bullet)$.

To prove the injectivity, let

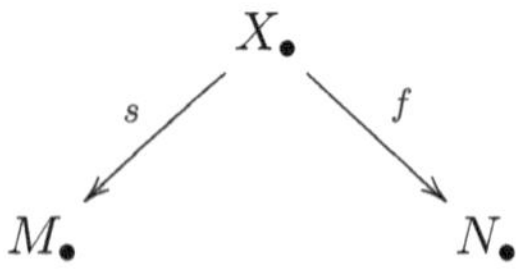

and

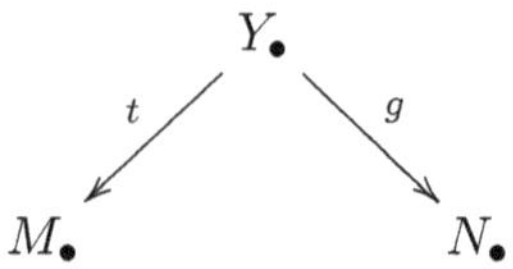

in $Mor_{\mathcal{D}_k^-(\mathbb{A}!-\text{albimod})}(M_\bullet, N_\bullet)$ equivalent in $Mor_{\mathcal{D}_k^-(\mathbb{A}!-\text{bimod})}(M_\bullet, N_\bullet)$.
Thus, we have a commutative diagram

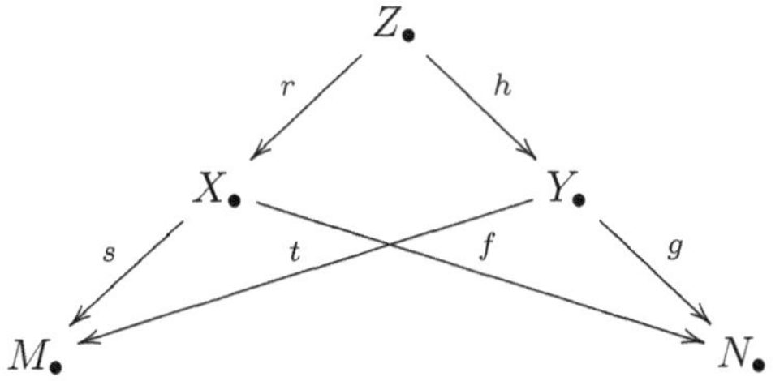

where r and s are relative quasi-isomorphisms. Since the alignment functor preserves relative quasi-isomorphisms and $M_\bullet$, $N_\bullet$, $X_\bullet$ and $Y_\bullet$ are complexes of aligned $\mathbb{A}!$ bimodules we have the commutative diagram

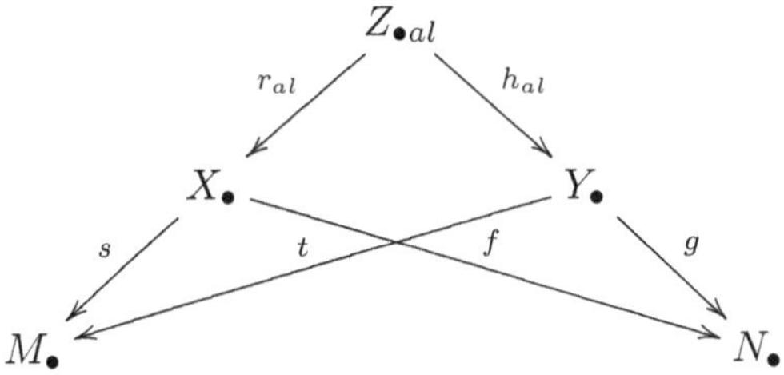

which implies the equivalence of roofs

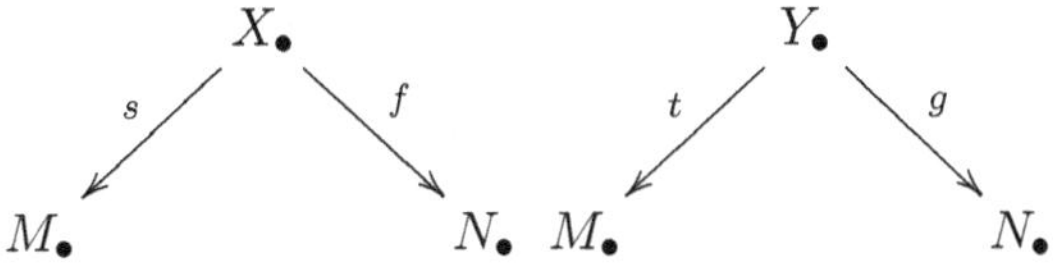

in $Mor_{\mathcal{D}_k^-(\mathbb{A}!-\text{albimod})}(M_\bullet, N_\bullet)$, and so the injectivity of *inc*. □

The proof of theorem 4.1. follows now easily combining propositions 4.2. and 4.3. In particular, we obtain the following theorem of [2], due to M. Gerstenhaber and S. D. Schack.

Corollary 4.4. *(Special Cohomology Comparison Theorem)*
The functor ! induces an isomorphism of relative Yoneda cohomologies

$$Ext^\bullet_{\mathbb{A}-\mathbb{A}}((-),(-)) \cong Ext^\bullet_{\mathbb{A}!-\mathbb{A}!}((-)!,(-)!).$$

In particular, we have an isomorphism of relative Hochschild cohomologies

$$H^\bullet(\mathbb{A},(-)) \cong H^\bullet(\mathbb{A}!,(-)!).$$

Proof.

$$Ext^i_{\mathbb{A}-\mathbb{A}}(\mathbb{M},\mathbb{N}) \cong Mor_{\mathcal{D}_k^-(\mathbb{A}-\text{bimod})}(\mathbb{M}_\bullet, \mathbb{N}_\bullet[i]) \cong$$

$$\cong Mor_{\mathcal{D}_k^-(\mathbb{A}!-\text{bimod})}(\mathbb{M}_\bullet!, \mathbb{N}_\bullet[i]!) \cong Ext^i_{\mathbb{A}!-\mathbb{A}!}(\mathbb{M}!, \mathbb{N}!).$$

□

Note : By taking a very different approach Wendy Lowen and Michel Van Den Bergh also proved in [8] that the functor ! is full and faithful.

AppendixA. theorem 2.4.

Theorem AppendixA.1. *1. The functor* $! : \mathbb{A}$*-bimod*$\longrightarrow \mathbb{A}!$*-albimod admits a left adjoint* $\text{¡} : \mathbb{A}!$*-albimod*$\longrightarrow \mathbb{A}$*-bimod.*
2. There are natural isomorphisms $\mathcal{T}_p\mathbb{N}\text{¡} \longrightarrow \mathcal{S}_p\mathbb{N}$ *which induce a natural isomorphism of complexes* $(\mathcal{T}_\bullet\mathbb{N} \longrightarrow \mathbb{N}!)\text{¡}$ *and* $(\mathcal{S}_\bullet\mathbb{N} \longrightarrow \mathbb{N})$*.*

Proof. 1. The left adjoint is the restriction of a functor $\text{¡} : \mathbb{A}!$-bimod$\longrightarrow \mathbb{A}$-bimod. For any $\mathbb{A}!$-bimodule X and $i \in \mathcal{C}$ we define $X\text{¡}^i$ as the colimit of a particular functor over the poset $\mathcal{C}_i^1$ whose elements are the 1-simplices of $\mathcal{C}_i = \{j | j \geqslant i\}$. The ordering is $\sigma \ll \tau \Leftrightarrow \sigma = \tau$ or σ is degenerate ($d\sigma = c\sigma$), τ is not, and either $d\sigma = d\tau$ or $d\sigma = c\tau$. We denote by $X^{pq} = \varphi^{pp} X \varphi^{qq}$. Define $F_X^i : \mathcal{C}_i^1 \longrightarrow \mathbb{A}^i$-bimod on each object σ to be the coequalizer of the $\mathbb{A}^i$-bimodule maps $X^{i,d\sigma} \otimes_k \mathbb{A}^{d\sigma} \rightrightarrows X^{i,c\sigma} \otimes_{\mathbb{A}^{c\sigma}} \mathbb{A}^i$ given by $x \otimes a \to x\varphi^{d\sigma,c\sigma} \otimes \varphi^{i,d\sigma}(a)$ and $x \otimes a \to xa\varphi^{d\sigma,c\sigma} \otimes 1$. For $\sigma \ll \tau$ in $\mathcal{C}_i^1$, the map $F_X^i(\sigma\tau) : F_X^i(\sigma) \to F_X^i(\tau)$ is defined by $\overline{x \otimes a} \to \overline{x\varphi^{c\sigma,c\tau} \otimes a}$.

Let $X\text{¡}^i := colim F_X^i$, $\forall i \in \mathcal{C}$ and ι_σ^i the canonical map $F_X^i(\sigma) \to X\text{¡}^i$. To show that $X\text{¡}$ is an $\mathbb{A}$-bimodule, for each $h \leqslant i$ in $\mathcal{C}$, we have to define a map $T_{X\text{¡}}^{hi} : X\text{¡}^i \to X\text{¡}^h$ such that $T_{X\text{¡}}^{hj} = T_{X\text{¡}}^{hi} T_{X\text{¡}}^{ij}$ if $h \leqslant i \leqslant j$.

First, we have a natural transformation $\Gamma_X^{hi} : F_X^i \to_{hi} | - |_{hi} \circ F_X^h \circ inc_{hi}$, where $inc_{hi} : \mathcal{C}_i^1 \to \mathcal{C}_h^1$ is the inclusion functor induced by $\mathcal{C}_i \subset \mathcal{C}_h$ and ${}_{hi}| - |_{hi} : \mathbb{A}^h$-bimod$\to \mathbb{A}^i$-bimod is the forgetful functor. To define Γ_X^{hi} observe that, for each $\sigma \in \mathcal{C}^1$, left multiplication by φ^{hi} is an $\mathbb{A}^i$-$\mathbb{A}^{c\sigma}$ bimodule map : $X^{i,c\sigma} \to X^{h,c\sigma}$, while $\varphi^{hi} : \mathbb{A}^i \to \mathbb{A}^h$ is an $\mathbb{A}^{c\sigma}$-$\mathbb{A}^i$ bimodule map. The map of $\mathbb{A}^i$-bimodules $(\varphi^{hi} \cdot -) \otimes \varphi^{hi} : X^{i,c\sigma} \otimes_{\mathbb{A}^{c\sigma}} \mathbb{A}^i \to_{hi} |X^{h,c\sigma} \otimes_{\mathbb{A}^{c\sigma}} \mathbb{A}^i|_{hi}$ induces the $\mathbb{A}^i$-bimodule map $(\Gamma_X^{hi})_\sigma : F_X^i(\sigma) \to_{hi} |F_X^h(\sigma)|_{hi}$ given by $\overline{x \otimes a} \to \overline{\varphi^{hi}x \otimes \varphi^{hi}(a)}$.

Second, let $T_{X\text{¡}}^{hi}$ be the composite of maps $X\text{¡}^i = colim F_X^i \to colim({}_{hi}| - |_{hi} \circ F_X^h \circ inc_{hi}) =_{hi} |colim(F_X^h \circ inc_{hi})|_{hi} \to_{hi} |colim F_X^h|_{hi} =_{hi} |X\text{¡}^h|_{hi}$. Thus we have $T_{X\text{¡}}^{hi}(\iota_\sigma^i(\overline{x \otimes a})) = \iota_\sigma^h(\overline{\varphi^{hi}x \otimes \varphi^{hi}(a)})$.

One may easily check now the identity $T_{X\text{¡}}^{hj} = T_{X\text{¡}}^{hi} T_{X\text{¡}}^{ij}$ for $h \leqslant i \leqslant j$, so $X\text{¡}$ is an $\mathbb{A}$-bimodule.

So far we have defined ¡ on the objects of $\mathbb{A}!$-bimod so we need to define it on maps. Let $g : X \to Y$ be an $\mathbb{A}!$-bimodule map. The restriction of g to X^{ij} is an $\mathbb{A}^i$-$\mathbb{A}^j$ bimodule map, $g : X^{ij} \to Y^{ij}$, and for $\sigma \in \mathcal{C}^1$, the $\mathbb{A}^i$-bimodule map $g \otimes id : X^{i,c\sigma} \otimes_{\mathbb{A}^{c\sigma}} \mathbb{A}^i \to Y^{i,c\sigma} \otimes_{\mathbb{A}^{c\sigma}} \mathbb{A}^i$ induces the map $\widetilde{g}_\sigma^i : F_X^i(\sigma) \to F_Y^i(\sigma)$ defined by $\overline{x \otimes a} \to \overline{g(x) \otimes a}$. Its easy to check the naturality of $\widetilde{g}_\sigma^i$ since g is a $\mathbb{A}!$-bimodule map and by taking the colimits we obtain an $\mathbb{A}^i$-bimodule map $g\text{¡}^i : X\text{¡}^i \to Y\text{¡}^i$, given by $g\text{¡}^i(\iota_\sigma^i(\overline{x \otimes a})) = \iota_\sigma^i(\widetilde{g}_\sigma^i(\overline{x \otimes a}))$. These are the components of an $\mathbb{A}$-bimodule map, *i.e.*$T^{hi}_{Y\text{¡}} \circ g\text{¡}^i = g\text{¡}^h \circ T^{hi}_{X\text{¡}}$ for $h \leqslant i$ because of the commutative diagram

$$\begin{array}{ccc} F^i_X & \xrightarrow{\Gamma^{hi}_X} & {}_{hi}| - |_{hi} \circ F^h_X \circ inc_{hi} \\ \Big\downarrow \widetilde{g}^i & & \Big\downarrow id \circ \widetilde{g}^h \circ id \\ F^i_Y & \xrightarrow{\Gamma^{hi}_Y} & {}_{hi}| - |_{hi} \circ F^h_Y \circ inc_{hi} \end{array}$$

In addition, one has $(g_1 g_2)\text{¡} = (g_1)\text{¡} \circ (g_2)\text{¡}$ and $(id)\text{¡} = id$ since $\widetilde{g_1 g_2}^i = \widetilde{g_1}^i \widetilde{g_2}^i$ and $\widetilde{id}^i = id$, so ¡ is a functor.

We now prove that the functor constructed above is a left adjoint to !, when restricted to $\mathbb{A}$-albimod. Let X be an aligned $\mathbb{A}$!-bimodule. For $i \leqslant j$ we define $\eta^{ij}_X : X^{ij} \to (X\text{¡})!^{ij}$ to be the $\mathbb{A}^i$-$\mathbb{A}^j$ bimodule map $\eta^{ij}_X = \iota^i_{(j \leqslant j)}(x \otimes 1)\varphi^{ij}$. One may check that for $h \leqslant i \leqslant j \leqslant q$, $a^h \in \mathbb{A}^h$, $a^j \in \mathbb{A}^j$ and $x \in X^{ij}$ we have $\eta^{hj}_X(a^h \varphi^{hi} \cdot x) = a^h \varphi^{hi} \cdot \eta^{ij}_X(x)$ and $\eta^{iq}_X(x \cdot a^j \varphi^{jq}) = \eta^{ij}_X(x) \cdot a^j \varphi^{jq}$ so the family of maps η^{ij}_X determine an $\mathbb{A}$!-bimodule natural map $\eta_X : X \to (X\text{¡})!$.

Let $\mathbb{N} \in \mathbb{A}$-bimod. To define the components of the counit $\varepsilon^i_{\mathbb{N}} : (\mathbb{N}!)\text{¡}^i = colim F^i_{\mathbb{N}!} \to \mathbb{N}^i$, we define a family of $\mathbb{A}^i$-bimodule maps $\varepsilon^{i,\sigma}_{\mathbb{N}} : F^i_{\mathbb{N}!}(\sigma) \to \mathbb{N}^i$ such that $\varepsilon^{i,\tau}_{\mathbb{N}} \circ F^i_{\mathbb{N}!}(\sigma\tau) = \varepsilon^{i,\sigma}_{\mathbb{N}}$, for $\sigma \ll \tau$ in $\mathcal{C}^1_i$ and use the universal property of colimits. The $\mathbb{A}^i$-bilinear function $\mathbb{N}^i \varphi^{i,c\sigma} \times \mathbb{A}^i \to \mathbb{N}^i$, $(n\varphi^{i.c\sigma}, a) \to na$ is $\mathbb{A}^{c\sigma}$-balanced and the induced $\mathbb{A}^i$-bimodule map $\mathbb{N}^i \varphi^{i,c\sigma} \otimes_{\mathbb{A}^{c\sigma}} \mathbb{A}^i \to \mathbb{N}^i$ vanishes on $\{n\varphi^{d\sigma,c\sigma} \otimes \varphi^{i.d\sigma}(a) - na\varphi^{d\sigma.c\sigma} \otimes 1 \mid n \in \mathbb{N}^i \varphi^{i,d\sigma}, a \in \mathbb{A}^{d\sigma}\}$ so for each $\sigma \in \mathcal{C}^1_i$, we obtain the $\mathbb{A}^i$-bimodule map $\varepsilon^{i,\sigma}_{\mathbb{N}} : F^i_{\mathbb{N}!}(\sigma) \to \mathbb{N}^i$, $\overline{n\varphi^{i,c\sigma} \otimes a} \to na$. We have that $\varepsilon^{i,\tau}_{\mathbb{N}} \circ F^i_{\mathbb{N}!}(\sigma\tau)(n\varphi^{i,c\sigma} \otimes a) = \varepsilon^{i,\tau}_{\mathbb{N}}(n\varphi^{i,c\sigma} \cdot \varphi^{c\sigma,c\tau} \otimes a) = na = \varepsilon^{i,\sigma}_{\mathbb{N}}(n\varphi^{i,c\sigma} \otimes a)$ and thus the map $\varepsilon^i_{\mathbb{N}}$ is given by $\varepsilon^i_{\mathbb{N}}(\iota^i_\sigma(\overline{n\varphi^{i,c\sigma} \otimes a})) = na$. The maps $\varepsilon^i_{\mathbb{N}}$ determine a natural map $\varepsilon_{\mathbb{N}} : (\mathbb{N}!)\text{¡} \to \mathbb{N}$ of $\mathbb{A}$-bimodules since for $h \leqslant i$ and $\sigma \in \mathcal{C}^1_i$ we have $T^{hi}_{\mathbb{N}} \varepsilon^{i,c\sigma}_{\mathbb{N}}(\overline{n\varphi^{i,c\sigma} \otimes a}) = T^{hi}_{\mathbb{N}}(na) = T^{hi}_{\mathbb{N}}(n) \cdot a = T^{hi}_{\mathbb{N}}(n)\varphi^{hi}(a)$ while $\varepsilon^{h,\sigma}_{\mathbb{N}}(\Gamma^{hi}_{\mathbb{N}!})_\sigma(\overline{n\varphi^{i,c\sigma} \otimes a}) = \varepsilon^{h,\sigma}_{\mathbb{N}}(\overline{\varphi^{hi} \cdot n\varphi^{i,c\sigma} \otimes \varphi^{hi}(a)}) = \varepsilon^{h,\sigma}_{\mathbb{N}}(\overline{T^{hi}_{\mathbb{N}}(n)\varphi^{h,c\sigma} \otimes \varphi^{hi}(a)}) = T^{hi}_{\mathbb{N}}(n)\varphi^{hi}(a)$.

To finish the proof we show that η and ε form an adjoint pair. To see that $\varepsilon_{\mathbb{N}}! \circ \eta_{\mathbb{N}!} = id_{\mathbb{N}!}$ it is enough to check this on each $\mathbb{N}^i \varphi^{ij}$ and $\varepsilon_{\mathbb{N}}!(\eta_{\mathbb{N}!}(n\varphi^{ij}) = \varepsilon_{\mathbb{N}}!(\iota^i_{(j \leqslant j)}(n\varphi^{ij} \otimes 1)\varphi^{ij}) = \varepsilon^i_{\mathbb{N}}(\iota^i_{(j \leqslant j)}(n\varphi^{ij} \otimes 1))\varphi^{ij} = (n \cdot 1)\varphi^{ij} = n\varphi^{ij}$, as required.

Last, for each $X \in \mathbb{A}$-albimod we need to verify that $\varepsilon_{X\text{¡}} \circ \eta_X \text{¡} = id_{X\text{¡}}$. This can be checked on a set of $\mathbb{A}^i$-bimodule generators for each component $X\text{¡}^i$ and the set $\{\iota^i_{(j \leqslant j)}(x \otimes 1) \mid j \geqslant i, x \in X^{ij}\}$ has this property. Since $(\varepsilon^i_{X\text{¡}} \circ \eta_X \text{¡}^i)(\iota^i_{(j \leqslant j)}(x \otimes 1))) = \varepsilon^i_{X\text{¡}}(\iota^i_{(j \leqslant j)}(\eta_X(x) \otimes 1)) = \varepsilon^i_{X\text{¡}}(\iota^i_{(j \leqslant j)}(\iota^i_{(j \leqslant j)}(x \otimes 1)\varphi^{ij} \otimes 1)) = \iota^i_{(j \leqslant j)}(x \otimes 1) \cdot 1$ we obtain the required identity.

2. Because both $\mathcal{T}_p\mathbb{N}$ and $\mathcal{S}_p\mathbb{N}$ are coproducts and ¡, as a left adjoint, preserves colimits it is enough to find natural isomorphisms $\gamma^\sigma : (\mathcal{T}^\sigma_p)\text{¡} \longrightarrow \mathcal{S}^\sigma_p$ such that, for

$0 \leqslant r \leqslant p$, the following square

$$\begin{array}{ccc}
(\mathcal{T}_p^{\sigma}\mathbb{N})\text{¡} & \xrightarrow{(d_r^{\mathcal{T},\sigma})\text{¡}} & (\mathcal{T}_{p-1}^{\sigma_r}\mathbb{N})\text{¡} \\
\gamma_{\mathbb{N}}^{\sigma}\downarrow & & \downarrow\gamma_{\mathbb{N}}^{\sigma_r} \\
\mathcal{S}_p^{\sigma}\mathbb{N} & \xrightarrow{d_r^{\mathcal{S},\sigma}} & \mathcal{S}_{p-1}^{\sigma_r}\mathbb{N}
\end{array}$$

commutes, where, when $p = 0$, we interpret the right column as the counit $\varepsilon_{\mathbb{N}} : (\mathbb{N}!)\text{¡} \longrightarrow \mathbb{N}$ and $d_0^{\mathcal{T}}$ and $d_0^{\mathcal{S}}$ as the augmentations. To construct the isomorphisms, for $p > 0$, observe that, for each $\sigma \in \mathcal{C}^{[p]}$, the diagram

$$\begin{array}{ccc}
\mathbb{A}-\text{bimod} & \xrightarrow{\;!\;} & \mathbb{A}!-\text{albimod} \\
(d\sigma)^*\downarrow & & \downarrow(-)^{d\sigma,c\sigma} \\
\mathbb{A}^{d\sigma}-\text{bimod} & & \mathbb{A}^{d\sigma}-\text{mod}-\mathbb{A}^{c\sigma} \\
{}_{d\sigma,c\sigma}|-|_{d\sigma,c\sigma}\downarrow & & \downarrow{}_{d\sigma,c\sigma}|-| \\
\mathbb{A}^{c\sigma}-\text{bimod} & \xrightarrow{id} & \mathbb{A}^{c\sigma}-\text{bimod}
\end{array}$$

is commutative. Since each functor in it admits a left adjoint and ${}_{d\sigma,c\sigma}| - |_{d\sigma,c\sigma} \circ (d\sigma)^* = {}_{d\sigma,c\sigma} | - | \circ (-)^{d\sigma,c\sigma} \circ !$ we have the isomorphisms, natural in N,

$$\gamma_N^{\sigma} : (L_{d\sigma,c\sigma}(\mathbb{A}^{d\sigma} \otimes_{\mathbb{A}^{c\sigma}} N))\text{¡} \longrightarrow (d\sigma)_!(\mathbb{A}^{d\sigma} \otimes_{\mathbb{A}^{c\sigma}} N \otimes_{\mathbb{A}^{c\sigma}} \mathbb{A}^{d\sigma}).$$

The $\mathbb{A}^i$ bimodule $((L_{d\sigma,c\sigma}(\mathbb{A}^{d\sigma} \otimes_{\mathbb{A}^{c\sigma}} N)))\text{¡}^i$ is generated by $\{\iota^i_{(j\leqslant j)}((1\otimes n)\otimes 1) \mid j \geqslant c\sigma, n \in N\}$ for $i \leqslant d\sigma$ and is 0 if $i \nleqslant d\sigma$. Tracing through the adjunction we obtain that, for each $i \in \mathcal{C}$, $\gamma_N^{\sigma,i}(\iota^i_{(j\leqslant j)}((1 \otimes n) \otimes 1)) = 1 \otimes n \otimes 1$.

For all $\mathbb{N}$ and σ we define $\gamma_{\mathbb{N}}^{\sigma} = \gamma_{\mathbb{N}^{c\sigma}}^{\sigma}$ and, for $p > 0$, we have that $(d_r^{\mathcal{T},\sigma})\text{¡}^i(\iota^i_{(j\leqslant j)}((1\otimes n)\otimes 1)) = \iota^i_{(j\leqslant j)}(d_r^{\mathcal{T},\sigma}(1\otimes n)\otimes 1) = \iota^i_{(j\leqslant j)}((1\otimes T_{\mathbb{N}}^{c\sigma_r,d\sigma}(n))\otimes 1)$, while $d_r^{\mathcal{S},\sigma}(1\otimes n\otimes 1) = 1\otimes T_{\mathbb{N}}^{c\sigma_r,c\sigma}(n)\otimes 1$, so the square is commutative. When $p = 0$, we obtain, for $\sigma \in \mathcal{C}^{[0]}$ and $i \leqslant d\sigma = c\sigma \leqslant j$ that $\varepsilon_{\mathbb{N}}^i \circ (\varepsilon^{\mathcal{T},\sigma})\text{¡}^i(\iota^i_{(j\leqslant j)}((1 \otimes n) \otimes 1)) = \varepsilon_{\mathbb{N}}^i(\iota^i_{(j\leqslant j)}(\varepsilon_{ij}^{\mathcal{T},\sigma}(1 \otimes n) \otimes 1)) = \varepsilon_{\mathbb{N}}^i(\iota^i_{(j\leqslant j)}(T_{\mathbb{N}}^{i,c\sigma}(n)\varphi^{ij} \otimes 1)) = T_{\mathbb{N}}^{i,c\sigma}(n) = \varepsilon^{\mathcal{S},\sigma,i}(1 \otimes n \otimes 1) = \varepsilon^{\mathcal{S},\sigma,i} \circ \gamma_{\mathbb{N}}^{\sigma,i}(\iota^i_{(j\leqslant j)}((1 \otimes n) \otimes 1))$, so the square commutes in this case too. □

References

[1] M. Gerstenhaber and S. D. Schack, *Algebraic Cohomology and Deformation Theory*, Deformation Theory of Algebras and Structures and Applications, Kluwer, Dordrecht (1988) 11-264.

[2] M. Gerstenhaber and S. D. Schack, *The Cohomology of Presheaves of Algebras: Presheaves over a Partially Ordered Set*, Trans. Amer. Math. Soc. 310 (1988) 135-165.

[3] C. B. Kullmann, *Adjoints and Cohomology for Presheaves of Algebras Over a Poset* , SUNY at Buffalo Ph.D. thesis, (1998).

[4] S. MacLane, *Homology* , Spinger-Verlag, Berlin, (1967).

[5] S. I. Gelfand and Yu. I. Manin, *Methods of Homological Algebra*, Springer Verlag (1996).

[6] B. Keller, *Hochschild Cohomology and the Derived Picard Group*, Journal of Pure and Applied Algebra 190 (2004), 177-196.

[7] A. Stancu, *Hochschild Cohomology and Derived Categories*, SUNY at Buffalo Ph.D. thesis, (2006).

[8] W. Lowen and M. Van Den Bergh, *A Hochschild Cohomology Comparison Theorem for Prestaks*, arXiv:0905.2354v1 [math.KT]

This article may be accessed via WWW at `http://tcms.org.ge/Journals/JHRS/`

Alin Stancu
`stancu_alin1@colstate.edu`

Department of Mathematics
Columbus State University
Columbus, GA 31907, USA

Journal of Homotopy and Related Structures, vol. 6(1), 2011, pp.65–69

MOD 2 MORAVA K-THEORY FOR FROBENIUS COMPLEMENTS OF EXPONENT DIVIDING $2^n \cdot 9$

MALKHAZ BAKURADZE

(*communicated by Vladimir Vershinin*)

Abstract

We determine the cohomology rings $K(s)^*(B\mathcal{G})$ at 2 for all finite Frobenius complements $\mathcal{G}$ of exponent dividing $2^n \cdot 9$.

Let V be an abelian group, and let $\mathcal{G}$ be a group of automorphisms of V. If $\mathcal{G}$ has exponent $2^n \cdot 3^k$ for $0 \leqslant n$ and $0 \leqslant k \leqslant 2$ and $\mathcal{G}$ acts freely on V, then $\mathcal{G}$ is finite (see [**6**] Theorem 1.1). Every finite group that acts freely on an abelian group is isomorphic to a Frobenius complement in some finite Frobenius group (see [**6**] Lemma 2.6). By the classification of finite Frobenius complements (see [**7**]) the quotient of $\mathcal{G}$ by its maximal normal 3-subgroup $\mathcal{H}$ is isomorphic to a cyclic 2-group $\mathcal{C}$, a generalized quaternion group Q, the binary tetrahedral group $2\mathcal{T}$ of order 24 (or SL(2,3)), or the binary octahedral group $2\mathcal{O}$ of order 48. Then Atiyah-Hirzebruch-Serre spectral sequence for $\mathcal{H} \lhd \mathcal{G}$ implies that at 2 the ring $K(s)^*(B\mathcal{G})$ is isomorphic to $K(s)^*(B\mathcal{K})$, for $\mathcal{K} = \mathcal{G}/\mathcal{H}$ is either $\mathcal{C}, Q, 2\mathcal{T}, 2\mathcal{O}$. For the cyclic group $\mathcal{C} = \mathbb{Z}/2^k$, $K(s)^*(B\mathbb{Z}/2^k) = \mathbb{F}_2[v_s, v_s^{-1}][u]/(u^{2^{ks}})$. For the generalized quaternion group $Q_{2^{m+2}}$ we have Theorem 1.1 of [**4**]. We deduce Morava K-theory rings at 2 for the groups $2\mathcal{T}$ and $2\mathcal{O}$ as certain subgroups in $K(s)^*(BQ_8)$ and $K(s)^*(BQ_{16})$ respectively (Proposition 5 and Proposition 6.)

In [**3**] we proved the following formula for the first Chern class of the transferred line complex bundle: Let $X \to Y$ be the regular two covering defined by free action of $\mathbb{Z}/2$ on X and let $\theta \to Y$ be the associated line complex bundle; Let $\xi \to X$ be a complex line bundle and let $\zeta \to Y$ be the plane bundle, transferred from ξ by Atiyah transfer [**2**]. Then for $Tr^* : K(s)(X) \to K(s)^*(Y)$, the transfer homomorphism [**1**] for our covering $X \to Y$, one has

$$Tr^*(c_1(\xi)) = c_1(\theta) + c_1(\zeta) + v_s \sum_{i=1}^{s-1} c_1(\theta)^{2^s - 2^i} c_2(\zeta)^{2^{i-1}}. \tag{1}$$

We show that formula 1 plays major role in the ring structure $K(s)^*(B\mathcal{G})$ at 2 for aforementioned groups and gives another derivations for some related rank one Lie groups.

Research was supported by Volkswagen Foundation, Ref.: I/84 328 and by GNSF, ST08/3-387
Received October 06, 2010, revised January 12, 2011; published on February 19, 2011.
2000 Mathematics Subject Classification: 55N20, 55R12, 55R40.
Key words and phrases: Morava K-theory, transfer homomorphism, Chern class.

Much of our note is written in terms of Theorem 1.1 of [**4**]. Let

$$G = \langle a, b \mid a^{2^{m+1}} = 1,\ b^2 = a^e, bab^{-1} = a^r\rangle, \quad m \geqslant 1$$

and either $e = 0$, $r = -1$ (the dihedral group $D_{2^{m+2}}$ of order 2^{m+2}), $e = 2^m$, $r = -1$ (the generalized quaternion group $Q_{2^{m+2}}$) or $m \geqslant 2$, $e = 0$, $r = 2^m - 1$ (the semidihedral group $SD_{2^{m+2}}$).

Spectral sequence consideration (see [**8**]) imply that $K(s)(BG)$ is generated by following Chern classes $|c| = |x| = 2$, $|c_2| = 4$:

$$c = c_1(\eta_1),\ \eta_1 : G/\langle a\rangle \cong \mathbf{Z}/2 \to \mathbb{C}^*,\ b \mapsto -1;$$
$$x = c_1(\eta_2),\ \eta_2 : G/\langle a^2, b\rangle \cong \mathbf{Z}/2 \to \mathbb{C}^*,\ a \mapsto -1;$$

and $c_2 = c_2(\xi_{\pi_1})$, where $\xi_{\pi_1} \to B\langle a, b\rangle$ is the plane bundle transferred from the canonical line bundle $\xi \to B\langle a\rangle$, for the double covering $\pi_1 : B\langle a\rangle \to B\langle a, b\rangle$ corresponding to η_1.

The ring structure is the result of the formula for transferred first Chern class 1. See [**4**].

Let N be the normalizer of $U(1)$ in S^3. The normalizes of the maximal torus in $SO(3)$ is $O(2) = U(1) \rtimes \mathbb{Z}/2$ and $\mathbb{Z}/2$ acts on $K(s)^*BU(1) = K(s)^*[[u]]$ by $[-1]_F(u)$ as above.

Since $BU(1)\hat{}_p = [colim_n B\mathbb{Z}/(p^n)]\hat{}_p$, we have

$$K(s)^*(BO(2)) = K(s)^*(lim_m(BD_{2^{m+2}})) = K(s)^*(lim_m(BSD_{2^{m+2}}))$$

and

$$K(s)^*(BN) = K(s)^*(lim_m(BQ_{2^{m+2}})).$$

Thus Theorem 1.1 of [**4**] implies

Corollary 1. $K(s)^*(BO(2)) = K(s)^*[[c, c_2]]/(c^{2^s}, v_s c \sum_{i=1}^{s} c^{2^s - 2^i} c_2^{2^{i-1}})$, where $c = c_1(det\eta)$ and $c_2 = c_2(\eta)$ are the Chern classes of the bundle $\eta \to BO(2)$, the complexification of canonical $O(2)$ bundle.

Corollary 2. $K(s)^*(BN) = K^*(s)[[c, c_2]]/(c^{2^s}, c^2 + v_s c \sum_{i=1}^{s} c^{2^s - 2^i} c_2^{2^{i-1}})$, where $c = c_1(\nu)$ is the Chern class of ν the pullback bundle of the canonical real line bundle by $N \to N/U(1) = \mathbb{Z}/2$ and $c_2 = c_2(p^*(\zeta))$ is the Euler class of the pullback bundle of the canonical quaternionic line bundle by the inclusion $N \subset S^3$.

Then $RP^2 \to BO(2) \to BO(3)$ is the projective bundle of the canonical $SO(3)$ bundle. Hence the pullback of the complexification of this canonical $SO(3)$ bundle splits over $BO(2)$ as $\eta \oplus det\eta$. Note that $c_1(det\eta) = c_1(\eta) + v_s c_2(\eta)^{2^{s-1}}$ modulo transfer for the covering $BU(1) \to BO(2)$. Thus $K(s)^*(BSO(3))$ is subring in $K(s)^*(BO(2))$ generated by $v = c^2 + v_s c c_2^{2^{s-1}} + c_2$ and $w = cc_2$. This implies

Corollary 3. $K(s)^*(BSO(3)) = K(s)^*[[v, w]](f_s(v, w), g_s(v, w))$, where $|v| = 4$, $|w| = 6$, and $f_s = f_s(v, w)$, $g_s = g_s(v, w)$ are determined by $f_2 = vw$, $g_2 = w^2$ and for $s > 2$

$$f_s = \begin{cases} f_{s-1}^2 & s \text{ even}, \\ \frac{f_{s-1}g_{s-1}}{v} + wv^{2^{s-1}-1} & s \text{ odd}, \end{cases}$$

$$g_s = \begin{cases} g_{s-1}^2 & s \text{ odd}, \\ \frac{f_{s-1}g_{s-1}}{v} + wv^{2^{s-1}-1} & s \text{ even}. \end{cases}$$

Our main result is the following.

Let $\mathcal{G}$ be a group acting freely on an abelian group. Let $\mathcal{G}$ be of exponent dividing $2^n \cdot 9$ (hence $\mathcal{G}$ is necessarily finite, as above) and let $\mathcal{H} \lhd \mathcal{G}$ be the maximal normal 3-subgroup.

Theorem 4. *As a ring $K(s)^*(B\mathcal{G})$ has one of the following forms*

(i) If $\mathcal{G}/\mathcal{H}=Q_8$, then $K(s)^*(B\mathcal{G}) = K(s)^*[c, x, c_2]/R$
and the relations R are determined by

$$c^{2^s} = x^{2^s} = 0,\ v_s c c_2^{2^{s-1}} = v_s \sum_{i=1}^{s-1} c^{2^s-2^i+1} c_2^{2^{i-1}} + c^2,\ v_s^2 c_2^{2^s} = c^2 + cx + x^2,$$
$$v_s x c_2^{2^{s-1}} = v_s \sum_{i=1}^{s-1} x^{2^s-2^i+1} c_2^{2^{i-1}} + x^2.$$

(ii) If $\mathcal{G}/\mathcal{H}=Q_{2^{m+2}}$, $m > 1$, then $K(s)^*(B\mathcal{G}) = K(s)^*[c, x, c_2]/R$,
and the relations R are determined by

$$c^{2^s} = x^{2^s} = 0,\ v_s c c_2^{2^{s-1}} = v_s \sum_{i=1}^{s-1} c^{2^s-2^i+1} c_2^{2^{i-1}} + c^2,\ v_s^{2\kappa(m)} c_2^{2^{ms}} = cx + x^2,$$
$$v_s x c_2^{2^{s-1}} = v_s x \sum_{i=1}^{s-1} c^{2^s-2^i} c_2^{2^{i-1}} + \sum_{i=1}^{ms} v_s^{1+\kappa(m)+2^{ms}-2^i} c_2^{(2^{ms}+1)2^{s-1}-(2^s-1)2^{i-1}}$$
$$+ cx,$$
where $\kappa(m) = \frac{2^{ms}-1}{2^s-1}$.

(iii) If $\mathcal{G}/\mathcal{H}=2\mathcal{T}$, then $K(s)^*(B\mathcal{G}) = K(s)^*[c_2]/c_2^{(2^s+1)2^{s-1}}$.

(iv) If $\mathcal{G}/\mathcal{H}=2\mathcal{O}$, then
$K(s)^*(B\mathcal{G}) = K(s)^*[c, c_2]/(c^{2^s}, c^2 + v_s c \sum_{i=1}^{s} c^{2^s-2^i} c_2^{2^{i-1}}, c_2^{(2^s+1)2^{s-1}})$.

(v) If $\mathcal{G}/\mathcal{H}=\mathbb{Z}/2^k$, then $K(s)^*(B\mathcal{G}) = K(s)^*[c]/c^{2^{ks}}$.

Here in all cases $|c| = |x| = 2$, $|c_2| = 4$.

The statement (v) is clear. (i) and (ii) follow from Theorem 1.1 of [**4**] for Q_8 and $Q_{2^{m+2}}$ respectively. What remains is to consider the cases of binary tetrahedral and binary octahedral groups.

Binary Polyhedral groups

As it is known any finite subgroup of $SO(3)$ is either a cyclic group, a dihedral group or one of the groups of a Platonic solid: tetrahedral group $\mathcal{T} \cong A_4$, cube/octahedral group $\mathcal{O} \cong S_4$, or icosahedral group $\mathcal{I} \cong A_5$. We consider the preimages of the latter groups under the covering homomorphism $S^3 \to SO(3)$.

Binary tetrahedral group

Binary tetrahedral group $2\mathcal{T}$ as the group of 24 units in the ring of Hurwitz integers $2\mathcal{T}$ is given by $\{\pm 1, \pm i, \pm j, \pm k, \frac{1}{2}(\pm 1 \pm i \pm j \pm k)\}$.

This group can be written as a semidirect product $2\mathcal{T} = Q_8 \rtimes \mathbb{Z}/3$, where Q_8 is the quaternion group consisting of the 8 Lipschitz units $\pm 1, \pm i, \pm j, \pm k$ and $\mathbb{Z}/3$ is the cyclic group generated by $-\frac{1}{2}(1+i+j+k)$. The cyclic group acts on the normal subgroup Q_8 by conjugation. So that the generator of $\mathbb{Z}/3$ cyclically rotates i, j, k.

Consider now Morava K-theory at 2. Then relations of Theorem 1.1 of [4] for $K(s)^*(BQ_8)$ imply that its subring of invariants under $\mathbb{Z}/3$ action is generated by c_2: the generator of $\mathbb{Z}/3$ cyclically rotates c, x and $c + x + v_s c^{2^{s-1}} x^{2^{s-1}}$. If ignoring the powers of v_s then the first and second elementary symmetric functions in these three symbols are equal to $c_2^{2^{s-1}}$ and $c_2^{2^s}$ respectively and the third is zero. It follows that $K(s)^*(B2\mathcal{T}) \cong [K(s)^*(BQ_8)]^{\mathbb{Z}/3}$.

Proposition 5. $K(s)^*(B2\mathcal{T}) \cong K(s)^*[c_2]/c_2^{(2^s+1)2^{s-1}}$, *where* $|c_2| = 4$.

Binary octahedral group $2\mathcal{O}$

This group is given as the union of the 24 Hurwitz units $\{\pm 1, \pm i, \pm j, \pm k, \frac{1}{2}(\pm 1 \pm i \pm j \pm k)\}$ with all 24 quaternions obtained from $\frac{1}{\sqrt{2}}(\pm 1 \pm i + 0j + 0k)$ by permutation of coordinates.

The generalized quaternion group Q_{16} forms a subgroup of $2\mathcal{O}$ and its conjugacy classes has 3 members. Therefore by the transfer argument $B2\mathcal{O}$ is a stable wedge summand of BQ_{16} after localized at 2, meaning $K(s)^*(B2\mathcal{O})$ is the subring in $K(s)^*(BQ_{16})$ at 2. We show that this is the subring generated by two symbols c and c_2 of Theorem 1.1 of [4]. Namely one has

Proposition 6. *$K(s)^*(B2\mathcal{O})$ is isomorphic to*

$$K(s)^*[c, c_2]/(c^{2^s}, c^2 + v_s c \sum_{i=1}^{s} c^{2^s - 2^i} c_2^{2^{i-1}}, c_2^{(2^s+1)2^{s-1}}),$$

where $|c| = 2$, $|c_2| = 4$.

Binary icosahedral group

$2\mathcal{I}$ is given as the union of the 24 Hutwitz units $\{\pm 1, \pm i, \pm j, \pm k, \frac{1}{2}(\pm 1 \pm i \pm j \pm k)\}$ with all 96 quaternions obtained from $\frac{1}{2}(0 \pm 1 \pm i \pm \varphi^{-1} j \pm \varphi k)$ by even permutation of coordinates. Here $\varphi = \frac{1}{2}(1 + \sqrt{5})$ is the golden ratio. This group is isomorphic to $SL_2(5)$-the group of all 2×2 matrices over $\mathbb{F}_5$ with unit determinant.

Among other subgroups the relevant subgroup is the binary tetrahedral group formed by Hurwitz units. Then coset $2\mathcal{I}/2\mathcal{O}$ has 5 members hence by the transfer argument again $B2\mathcal{I}$ splits off $B2\mathcal{O}$ after localized at 2. Thus we obtain

$$K(s)^* B(2\mathcal{I}) \cong K(s)^* B(2\mathcal{T}).$$

References

[1] J. F. Adams, *Infinite Loop Spaces*, Annals of Mathematics Studies, Princeton University Press, Princeton (1978).

[2] M. F. Atiyah, *Characters and cohomology of finite groups*, Publ. Math. of the I.H.E.S. **9** (1961), 23–64.

[3] M. Bakuradze and S. Priddy, *Transferred Chern classes in Morava K-theory*, Proc. Amer. Math. Soc., **132** (2004), 1855-1860.

[4] M. Bakuradze and V. V. Vershinin, *Morava K-theory rings for the dihedral, semidihedral and generalized quaternion groups in Chern classes*, Proc. Amer. Math. Soc., **134** (2006), 3707-3714.

[5] J. H. Conway and D. A. Smith, *On Quaternions and Octonions*, Natick, Massachusetts: AK Peters, Ltd. ISBN 1-56881-134-9, (2003).

[6] E. Jabara and P. Mayr, *Fröbenius complements of exponent dividing* $2^m \cdot 9$, Forum Math., **21** (2009), 217–220.

[7] D. Passman, *Permutation groups*, W. A. Benjamin, Inc., New York-Amsterdam, 1968.

[8] M. Tezuka and N. Yagita, *Cohomology of finite groups and Brown-Peterson cohomology*, Springer LNM **1370** (1989), 396-408.

This article may be accessed via WWW at `http://tcms.org.ge/Journals/JHRS/`

Malkhaz Bakuradze
`malkhaz.bakuradze@tsu.ge`

Tbilisi State University, Georgia

Journal of Homotopy and Related Structures, vol. 6(1), 2011, pp.71–101

HOMOLOGY FUNCTORS WITH CUBICAL BARS

VOLKER W. THÜREY

(*communicated by Ronald Brown*)

Abstract

This work arose from efforts to generalise the usual cubical boundary by using different 'weights' for opposite faces, but still to obtain a chain complex, and this method was found to generalise. We describe a variant of the classical singular cubical homology theory, in which the usual boundary $(n-1)$-cubes of each n-cube are replaced by combinations of internal $(n-1)$-cubes parallel to the boundary. This defines a generalised homology theory, but the usual singular homology can be recovered by taking the quotient by the degenerate singular cubes.

Contents

The author likes to thank Dr. Björn Rüffer, Prof. Dr. Eberhard Oeljeklaus, Dr. Nils Thürey, Jan Osmers, Walter Meyer, Prof. Dr. Rick Jardine and Prof. Dr. Hans-Eberhard Porst for interest and suggestions, discussions and technical help. Also we thank Prof. Dr. Ronald Brown for many hints and much patience, and Dr. Guentcho Skordev, who supported us with attentive listening and clever remarks, and not to forget the unknown referee for careful reading of the paper and suggesting many improvements.
Received May 05, 2009, revised January 18, 2011; published on March 08, 2011.
2000 Mathematics Subject Classification: 55N20
Key words and phrases: cubical homology, generalised homology

1. Introduction

There are different ways to define singular homology groups, for instance by using simplices, see e.g. [**4**], [**13**], [**10**], or cubes, see [**5, 9**]. Because the latter construction is only one of some possible ways to get this well known theory, it seems that today mathematicians only have historical interest in it, but less mathematical interest, because by using simplices instead of cubes one gets isomorphic homology groups, and the simplicial homology theory as it is introduced in [**3**] is well-understood and a common tool of topologists. Singular homology theory is a very useful and successful method not only for mathematicians but also in other fields of science. It is used for instance for digital image processing and nonlinear dynamics, where even the cubical variant is used, see [**7**]. Cubical methods are also essential in [**1**].

Here we show an easy way to generalise cubical singular homology. In the ordinary cubical singular homology theory the boundary operator is constructed by taking the topological boundary of an n-dimensional unit cube as a linear combination of $2 \cdot n$ cubes of dimension $(n-1)$, provided with alternating signs. We generalise this by 'drawing' in all n directions a linear combination of a fixed number $L+1$ of $(n-1)$-dimensional cubes parallel to the topological boundary, provided with a coefficient tuple $\vec{m} := (m_0, m_1, m_2, \dots, m_L)$. Note that for a fixed $L > 1$, 'our' boundary operator ${}_{\vec{m}}\partial_n$ is determined not only the by topological boundary but also by parts of the interior of the unit cube, in contrast to the classical cubical homology theory.

Let TOP^2 be the category of pairs of topological spaces and continuous maps as morphisms. That means that $(f : (X, A) \to (Y, B)) \in \mathsf{TOP}^2$ if and only if X and Y are topological spaces and $A \subset X, B \subset Y$ and A, B carry the subspace topology and f is continuous and $f(A) \subset B$. Let $\mathcal{R}$ be a commutative ring with unit $1_{\mathcal{R}}$. Let $\mathcal{R}$-MOD be the category of $\mathcal{R}$-modules.

As in the classical theory, our construction yields a chain complex with decreasing dimensions, i.e. we get a sequence of natural transformations $({}_{\vec{m}}\partial_n)_{n \geqslant 0}$ with the property ${}_{\vec{m}}\partial_n \circ {}_{\vec{m}}\partial_{n+1} = 0$. Hence we shall be able to define homology modules

$$
{}_{\vec{m}}\mathcal{H}_n(X, A) := \frac{\mathrm{kernel}({}_{\vec{m}}\partial_n)}{\mathrm{image}({}_{\vec{m}}\partial_{n+1})},
$$

and this will lead to a sequence of functors ${}_{\vec{m}}\mathcal{H}_n : \mathsf{TOP}^2 \longrightarrow \mathcal{R}\text{-}\mathsf{MOD}$, for all $n \geqslant 0$.

The exactness axiom follows immediately, and with an additional condition on the fixed coefficient tuple $\vec{m}$ the homotopy axiom holds. Unfortunately, so far the excision axiom could be verified only in the case of $L = 1$, but in that special case we get a class of extraordinary homology theories. For $L = 1$, our boundary operator can be regarded as a kind of 'weighted' topological boundary of a cube, with the weight (m_0, m_1).

In this way for every fixed $L \in \mathbb{N}$ and fixed tuple $\vec{m} \in \mathcal{R}^{L+1}$ a functor ${}_{\vec{m}}\mathcal{H}_n$: $\mathsf{TOP}^2 \longrightarrow \mathcal{R}\text{-}\mathsf{MOD}$ will be constructed for each $n \geqslant 0$. In the special case $\mathcal{R} := \mathbb{Z}$ we shall see that the homotopy axiom holds if and only if the greatest common divisor of $\{m_0, m_1, m_2, \dots, m_L\}$ is 1. If $L = 1$ and gcd $\{m_0, m_1\} = 1$ the excision axiom holds, so we get an extraordinary homology theory. Finally, the ordinary singular homology can be recovered by taking quotients by 'degenerate' singular cubes.

This new construction is primarily of theoretical interest, since all the homology modules which we can compute in our homology theory we can already compute in terms of the ordinary singular homology, by using the clever method from [**2**]. If we take the ring $\mathcal{R} := \mathbb{Z}$, we are able to compute the homology groups for a finite CW-complex, and we express them as a product of singular homology groups.

We assume that the reader is familiar both with the construction of the classical cubical singular homology, e.g. in [**9**], and also with homological algebra and ordinary singular homology theory, see e.g. [**10**, p.57 ff], or [**4**, p.97 ff].

2. General Definitions and Notations

We denote the natural numbers by $\mathbb{N}_0 := \{0, 1, 2, 3, \dots\}$, the positive integers by $\mathbb{N} := \{1, 2, 3, \dots\}$, the ring of integers by $\mathbb{Z}$ and the real numbers by $\mathbb{R}$.

The brackets $(\cdots)$ will be used for tuples and besides $[\cdots]$ to structure text and formulas, $[r, s]$ also for the closed interval, $[\mathsf{u}]$ for the equivalence class of a quotient module. The brackets $\langle\cdots\rangle$ will be needed for the boundary operator, $\|\cdots\|$ for the subdivision operator and $\{\cdots\}$ for sets.

As before let $\mathcal{R}$ be a commutative ring with unit $1_{\mathcal{R}}$. Let X be a topological space. All maps we shall use will be continuous.

For each $n \in \mathbb{N}$ let $\mathbf{I}^n$ be the *n-dimensional unit cube*, that means $\mathbf{I}^n := \{ (x_1, x_2, \dots, x_n) \in \mathbb{R}^n \mid x_i \in [0,1] \text{ for } 1 \leqslant i \leqslant n\}$, provided with the Euclidean topology, and let $\mathbf{I}^0 := \{0\}$. We write $\mathbf{I}^1 = \mathbf{I} = [0,1]$, the unit interval.

Definition 1. *Define the sets* $\mathcal{S}_n(X) := \{T : \mathbf{I}^n \to X \mid T \text{ is continuous}\}$, *and* $\mathcal{K}_n(X) :=$ *the free* $\mathcal{R}$*-module with the basis* $\mathcal{S}_n(X)$, *for all* $n \in \mathbb{N}_0$, *as well as* $\mathcal{K}_{-1}(X) := \{0\}$, *the trivial* $\mathcal{R}$*-module. Every* $\mathsf{u} \in \mathcal{K}_n(X)$ *is called a chain.*

Let us assume given for each topological space X and $n \in \mathbb{N}_0$ an $\mathcal{R}$-module morphism $\partial_n \colon \mathcal{K}_n(X) \longrightarrow \mathcal{K}_{n-1}(X)$. If we have the property $\partial_n \circ \partial_{n+1} = 0$ for all $n \geqslant 0$, we call the map ∂_n a *boundary operator*, and the sequence $(\partial_n)_{n\geqslant 0}$ of $\mathcal{R}$-module morphisms is called a *chain complex* $\mathcal{K}_*(X)$,

$$\mathcal{K}_*(X) := \quad \cdots\cdots \xrightarrow{\partial_{n+1}} \mathcal{K}_n(X) \xrightarrow{\partial_n} \mathcal{K}_{n-1}(X) \xrightarrow{\partial_{n-1}} \cdots \xrightarrow{\partial_1} \mathcal{K}_0(X) \xrightarrow{\partial_0} \{0\}.$$

An element $\mathsf{u} \in \text{kernel}({}_{\vec{m}}\partial_n)$ is called a *cycle*, an element $\mathsf{w} \in \text{image}({}_{\vec{m}}\partial_{n+1})$ is called a *boundary*. Because $\partial_n \circ \partial_{n+1} = 0$ the $\mathcal{R}$-module

$$\mathcal{H}_n(X) := \frac{kernel(\partial_n)}{image(\partial_{n+1})}$$

is well defined for all topological spaces X and all $n \in \mathbb{N}_0$; $\mathcal{H}_n(X)$ is called the nth*-homology* $\mathcal{R}$*-module of* X.

Two continuous functions $f, g \colon X \longrightarrow Y$ are *homotopic*, written $f \simeq g$, if and only if there is a continuous $H : X \times \mathbf{I} \longrightarrow Y$ such that for all $x \in X$ we have $H(x,0) = f(x)$ and $H(x,1) = g(x)$.

Definition 2. *Let* $\mathcal{H} := (\mathcal{H}_n)_{n\geqslant 0}$ *be a sequence of functors* $\mathcal{H}_n : \mathsf{TOP} \longrightarrow \mathcal{R}$*-*$\mathsf{MOD}$. *We say that the functors* $\mathcal{H}$ *satisfy the Homotopy Axiom if and only if for all* $f, g : X \longrightarrow Y$ *with* $f \simeq g$ *we have* $\mathcal{H}_n(f) = \mathcal{H}_n(g)$, *for all* $n \in \mathbb{N}_0$.

Let $\mathcal{M}$ be an arbitrary subset of the ring $\mathcal{R}$. We write $\mathrm{Span}_{\mathcal{R}}(\mathcal{M})$ for the ideal of $\mathcal{R}$ generated by $\mathcal{M}$. We have $\mathrm{Span}_{\mathcal{R}}(\mathcal{M}) = \mathcal{R}$ if and only if there are a number $k \in \mathbb{N}$ and sets $\{m_1, m_2, \ldots, m_k\} \subset \mathcal{M}$ and $\{r_1, r_2, \ldots, r_k\} \subset \mathcal{R}$ such that $\sum_{i=1}^{k} r_i \cdot m_i = 1_{\mathcal{R}}$.

Definition 3. *Let a, b be elements of the ring $\mathcal{R}$. We say that (a, b) fulfils the condition $\mathcal{NCD}$ if and only if for all $n \in \mathbb{N}$ there exists $x_n, y_n \in \mathcal{R}$ with*

$$x_n \cdot a^n + y_n \cdot b^n = 1_{\mathcal{R}}.$$

Of course, $\mathcal{NCD}$ is equivalent to $\mathrm{Span}_{\mathcal{R}}(a^n, b^n) = \mathcal{R}$ for all $n \in \mathbb{N}$. The letters $\mathcal{NCD}$ remind us of 'No Common Divisor'. In the ring $\mathbb{Z}$ we have that (a, b) has the property $\mathcal{NCD}$ if and only if the ideal generated by $\{a, b\}$ is $\mathbb{Z}$.

Definition 4. *Let $\Im$ be a set of indices, let $\mathcal{U} := \{U_i \mid i \in \Im\}$ be a family of subsets of X whose interiors cover X. Let*

$$\mathcal{S}_n(X, \mathcal{U}) := \{T \in \mathcal{S}_n(X) \mid \text{there is an } i \in \Im \text{ such that } T(\mathbf{I}^n) \subset U_i\}.$$

For all $n \in \mathbb{N}_0$ and for every topological space X we define $\mathcal{K}_n(X,\mathcal{U})$ to be the free $\mathcal{R}$-module with the basis $\mathcal{S}_n(X, \mathcal{U})$. The elements $\mathsf{u} \in \mathcal{K}_n(X,\mathcal{U})$ are called $\mathcal{U}$-small chains. For a subset $A \subset X$ with the canonical inclusion $i : A \hookrightarrow X$, the map i leads to a canonical inclusion

$$\widehat{i} : \mathcal{K}_n(A,\mathcal{U}) \hookrightarrow \mathcal{K}_n(X,\mathcal{U})$$

in $\mathcal{R}$-MOD. Define

$$\mathcal{K}_n(X, A,\mathcal{U}) := \frac{\mathcal{K}_n(X, \mathcal{U})}{\mathcal{K}_n(A, \mathcal{U})},$$

and this yields an inclusion

$$\mathcal{K}_n(X, A,\mathcal{U}) \overset{j}{\hookrightarrow} \mathcal{K}_n(X, A)$$

Definition 5. *For each $T \in \mathcal{S}_n(X)$, i.e. $T : \mathbf{I}^n \to X$, we construct the maps $\|T\|_{\alpha,\vec{e},\vec{v}} \in \mathcal{S}_n(X)$, which will be used for the excision axiom. First we need an auxiliary map $q_n : \mathbb{R}^n \to \mathbf{I}^n$, for all dimensions $n \in \mathbb{N}$. For $(y_1, y_2, \ldots, y_n) \in \mathbb{R}^n$ let $q_n(y_1, y_2, \ldots, y_n) := (z_1, z_2, \ldots, z_n) \in \mathbf{I}^n$, where*

$$z_i := \begin{cases} 0 & \text{if } y_i \leqslant 0, \\ y_i & \text{if } y_i \in [0,1], \\ 1 & \text{if } y_i \geqslant 1, \quad \text{for } i \in \{1, 2, \ldots, n\}. \end{cases}$$

For fixed $\alpha \in \mathbb{R}$ and fixed $\vec{v} := (v_1, v_2, \ldots, v_n), \vec{e} := (e_1, e_2, \ldots, e_n) \in \mathbb{R}^n$, define the map $H_{\alpha,\vec{e},\vec{v}} : \mathbf{I}^n \to \mathbb{R}^n$ by $H_{\alpha,\vec{e},\vec{v}}(x_1, x_2, \ldots, x_n) := (y_1, y_2, \ldots, y_n)$, where

$$y_i := \alpha \cdot (e_i + v_i \cdot x_i), \quad \text{for all } i \in \{1, 2, \ldots, n\}.$$

Finally for each $T : \mathbf{I}^n \to X$ we set $\|T\|_{\alpha,\vec{e},\vec{v}} := (T \circ q_n \circ H_{\alpha,\vec{e},\vec{v}}) : \mathbf{I}^n \to X$.

Definition 6. *Let $n \in \mathbb{N}$. Let $\mathcal{D}_n(X)$ be the subset of $\mathcal{S}_n(X)$ consisting of the degenerate cubes, i.e. those $T \in \mathcal{S}_n(X)$ such that there is a $j \in \{1, 2, \ldots, n\}$ and for $y, z \in [0, 1]$ we have*

$$T(x_1, \ldots, x_{j-1}, y, x_{j+1}, \ldots, x_n) = T(x_1, \ldots, x_{j-1}, z, x_{j+1}, \ldots, x_n),$$

i.e. T does not depend on the jth component. We shall write
$T(x_1, x_2, \ldots, x_{j-1}, *, x_{j+1}, \ldots, x_n) := T(x_1, x_2, \ldots, x_{j-1}, y, x_{j+1}, \ldots, x_n)$.

Definition 7. *We define the free $\mathcal{R}$-module $\mathcal{K}_{\mathcal{D},n}(X)$, which will be an ideal of $\mathcal{K}_n(X)$. A chain $\sum_{i=1}^{p} r_i \cdot T_i \in \mathcal{K}_n(X)$ is an element of $\mathcal{K}_{\mathcal{D},n}(X)$ if and only if for all $i = 1, 2, \ldots, p$ we have $T_i \in \mathcal{D}_n(X)$. That means that $\mathcal{K}_{\mathcal{D},n}(X)$ is the free $\mathcal{R}$-module generated by degenerate maps. We have $\mathcal{K}_{\mathcal{D},0}(X) = \{0\}$.*

Definition 8. *Define for a fixed $\alpha \in \mathcal{R}$ the submodule $\mathrm{Ideal}_{\alpha,n}(X)$ of $\mathcal{K}_n(X)$, generated by $\alpha\mathcal{R}$. That means a chain $\sum_{i=1}^{p} r_i \cdot T_i \in \mathcal{K}_n(X)$ belongs to $\mathrm{Ideal}_{\alpha,n}(X)$ if and only if for all $i = 1, 2, \ldots, p$ there is an element $y_i \in \mathcal{R}$ such that $r_i = y_i \cdot \alpha$.*

Definition 9. *For fixed $\alpha \in \mathcal{R}$ and for all $n \in \mathbb{N}_0$ let*

$$\Gamma_{\alpha,n}(X) := \mathrm{Ideal}_{\alpha,n}(X) + \mathcal{K}_{\mathcal{D},n}(X)$$

(generally this is not a direct sum). Then $\Gamma_{\alpha,n}(X)$ is an ideal of $\mathcal{K}_n(X)$, and a chain $\sum_{i=1}^{p} r_i \cdot T_i \in \mathcal{K}_n(X)$ belongs to $\Gamma_{\alpha,n}(X)$ if and only if for all $i = 1, 2, \ldots, p$ either T_i is degenerate, or r_i is a multiple of α.

Definition 10. *Correspondingly to the previous three definitions we define for pairs $(X, A) \in \mathsf{TOP}^2$ for a fixed $\alpha \in \mathcal{R}$ the quotients of $\mathcal{R}$-modules $\mathcal{K}_{\mathcal{D},n}(X, A)$, $\mathrm{Ideal}_{\alpha,n}(X, A)$ and $\Gamma_{\alpha,n}(X, A)$, for all $n \in \mathbb{N}_0$. That means that two chains u and w represent the same equivalence class if and only if the difference $\mathsf{u} - \mathsf{w}$ is a chain in A. Further, we define the quotients of $\mathcal{R}$-modules*

$$\mathcal{K}_n(X)_{\sim\Gamma,\alpha} := \frac{\mathcal{K}_n(X)}{\Gamma_{\alpha,n}(X)} \quad \text{and} \quad \mathcal{K}_n(X, A)_{\sim\Gamma,\alpha} := \frac{\mathcal{K}_n(X, A)}{\Gamma_{\alpha,n}(X, A)} .$$

3. The Boundary Operator

Fix a natural number $L \geqslant 1$ and an $(L+1)$-tuple $\vec{m}$ of ring elements, $\vec{m} := (m_0, m_1, m_2, \ldots, m_L) \in \mathcal{R}^{L+1}$. For each $n \in \mathbb{N}_0$ we shall define a 'boundary operator' ${}_{\vec{m}}\partial_n : \mathcal{K}_n(X) \longrightarrow \mathcal{K}_{n-1}(X)$. For integers $n \geqslant 1$ and $T \in \mathcal{S}_n(X)$ we shall define a map $\langle T \rangle_{n,i,j} \in \mathcal{S}_{n-1}(X)$, for all integers $0 \leqslant i \leqslant L$ and $1 \leqslant j \leqslant n$.

For $n = 1$ let $\langle T \rangle_{1,i,1} \in \mathcal{S}_0(X)$ with $\langle T \rangle_{1,i,1}(0) := T\left(\frac{i}{L}\right)$. For $n > 1$ and every $(n-1)$-tuple $(x_1, x_2, \ldots, x_{n-1}) \in \mathbf{I}^{n-1}$ we set

$$\langle T \rangle_{n,i,j}(x_1, x_2, \ldots, x_{j-1}, x_j, \ldots, x_{n-1}) := T\left(x_1, x_2, \ldots, x_{j-1}, \frac{i}{L}, x_j, \ldots, x_{n-1}\right).$$

Finally for every $n \in \mathbb{N}$ and all $T \in \mathcal{S}_n(X)$ let

$${}_{\vec{m}}\partial_n(T) := \sum_{j=1}^{n} (-1)^{j+1} \cdot \sum_{i=0}^{L} m_i \cdot \langle T \rangle_{n,i,j}, \tag{1}$$

and for $n = 0$ let ${}_{\vec{m}}\partial_0(T) := 0$, the only possible map.

See Figure 1, which illustrates the case $n := 2, L := 3, \vec{m} := (9, 1, 4, -3)$ and $T := id(\mathbf{I}^2)$. On the left hand side you see the two-dimensional unit cube $\mathbf{I}^2$, the right hand side shows ${}_{\vec{m}}\partial_2(T)$, i.e. the images of eight one-dimensional unit cubes $\langle T\rangle_{2,i,j}\,, i \in \{0, 1, 2, 3\}$ and $j \in \{1, 2\}$, multiplied by coefficients $9, 1, 4, -3$, elements of the ring $\mathcal{R} := \mathbb{Z}$.

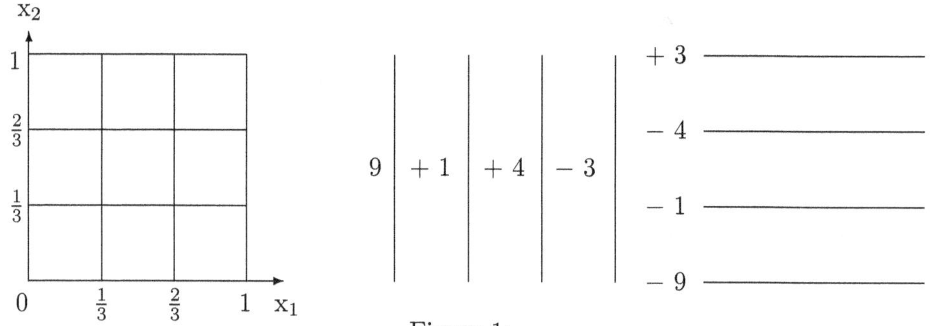

Figure 1:

For a 'chain' $\mathsf{u} = r_1 \cdot T_1 + r_2 \cdot T_2 \in \mathcal{K}_n(X)$ define ${}_{\vec{m}}\partial_n(\mathsf{u})$ by linearity,

$$ {}_{\vec{m}}\partial_n(r_1 \cdot T_1 + r_2 \cdot T_2) := r_1 \cdot {}_{\vec{m}}\partial_n(T_1) + r_2 \cdot {}_{\vec{m}}\partial_n(T_2). $$

Remark: For $L \geqslant 2$ the map ${}_{\vec{m}}\partial_n$ is defined not only on the topological boundary but also on parts of the interior of $\mathbf{I}^n$. We use the name 'boundary operator' only for historical reasons.

Theorem 1. *For arbitrary $L \in \mathbb{N}$ and $(L+1)$-tuples $\vec{m} = (m_0, m_1, m_2, \ldots, m_L) \in \mathcal{R}^{L+1}$ we have for all $n \in \mathbb{N}$: ${}_{\vec{m}}\partial_{n-1} \circ {}_{\vec{m}}\partial_n = 0$.*

Proof. For $n = 1$ the statement is trivial. For $n = 2$ the proof is similar to the cases with $n \geqslant 3$, which we shall show in detail. Thus, let $n \geqslant 3$. Because of the linearity of ${}_{\vec{m}}\partial_n$ it suffices to prove the theorem for the basis of $\mathcal{K}_n(X)$. Let $T : \mathbf{I}^n \to X$ be continuous, i.e. $T \in \mathcal{S}_n(X)$. We have

$$ {}_{\vec{m}}\partial_{n-1} \circ {}_{\vec{m}}\partial_n(T) = {}_{\vec{m}}\partial_{n-1}\left(\sum_{j=1}^{n}(-1)^{j+1} \cdot \sum_{i=0}^{L} m_i \cdot \langle T\rangle_{n,i,j}\right) \tag{2} $$

$$ = \sum_{j=1}^{n}(-1)^{j+1} \cdot \sum_{i=0}^{L} m_i \cdot {}_{\vec{m}}\partial_{n-1}\left(\langle T\rangle_{n,i,j}\right) \tag{3} $$

$$ = \sum_{j=1}^{n}(-1)^{j+1} \cdot \sum_{i=0}^{L} m_i \cdot \sum_{p=1}^{n-1}(-1)^{p+1} \sum_{k=0}^{L} m_k \cdot \left\langle \langle T\rangle_{n,i,j} \right\rangle_{n-1,k,p}. $$

Thus

$$ {}_{\vec{m}}\partial_{n-1} \circ {}_{\vec{m}}\partial_n(T) = \sum_{j=1}^{n}\sum_{p=1}^{n-1}(-1)^{j+p+2} \cdot \sum_{i=0}^{L}\sum_{k=0}^{L} m_i \cdot m_k \cdot \left\langle \langle T\rangle_{n,i,j} \right\rangle_{n-1,k,p}. \tag{4} $$

Note that the following display is not suited for special cases like $j = 1, j = n, p = 1, p = n-1, j = p$ or $j = p+1$. Nevertheless the claim of Theorem 1 remains true in all these cases. Let $(x_1,\ x_2,\ x_3,\ \dots\ , x_{n-2})$ be an arbitrary point in $\mathbf{I}^{n-2}$. For fixed $i, k \in \{0, 1, 2, \dots, L\}$ we have for $p \in \{1, 2, \dots, n-1\}$ and $j \in \{1, 2, \dots, n\}$

$$\begin{aligned}
&\left\langle \langle T\rangle_{n,i,j} \right\rangle_{n-1,k,p} (x_1, \dots, x_{p-1}, x_p, \dots, x_{n-2}) \\
&= \langle T\rangle_{n,i,j} \left(x_1, \dots, x_{p-1}, \frac{k}{L}, x_p, \dots, x_{n-2} \right) \\
&= \begin{cases} T\left(x_1, \dots, x_{j-1}, \frac{i}{L}, x_j, \dots, x_{p-1}, \frac{k}{L}, x_p, \dots, x_{n-2}\right) & \text{if } j \leqslant p, \\ T\left(x_1, \dots, x_{p-1}, \frac{k}{L}, x_p, \dots, x_{j-2}, \frac{i}{L}, x_{j-1}, \dots, x_{n-2}\right) & \text{if } j > p. \end{cases}
\end{aligned}$$

The sign of ${}_{\vec{m}}\partial_{n-1} \circ {}_{\vec{m}}\partial_n(T)$ depends on j and p only. It is easy to see that for $j \leqslant p$ we get

$$(-1)^{j+p+2}\, m_i\, m_k \left\langle \langle T\rangle_{n,i,j} \right\rangle_{n-1,k,p} + (-1)^{(p+1)+j+2}\, m_k\, m_i \left\langle \langle T\rangle_{n,k,p+1} \right\rangle_{n-1,i,j} = 0.$$

The set $M := \{1, 2, 3, \dots, n\} \times \{1, 2, 3, \dots, n-1\}$ contains $n \cdot (n-1)$ elements. With $M_{small} := \{(j, p) \in M | j \leqslant p\}$ and $M_{big} := \{(j, p) \in M | j > p\}$ we have $M = M_{small} \cup M_{big}$, and $M_{small} \cap M_{big} = \emptyset$. The map

$$M_{small} \rightarrow M_{big},\ (j, p) \mapsto (p+1, j)$$

is bijective. Thus the $n \cdot (n-1) \cdot (L+1)^2$ maps in (4) cancel pairwise. Hence ${}_{\vec{m}}\partial_{n-1} \circ {}_{\vec{m}}\partial_n(T) = 0$. □

For all topological spaces X the homology groups ${}_{\vec{m}}\mathcal{H}_n(X) = \frac{kernel({}_{\vec{m}}\partial_n)}{image({}_{\vec{m}}\partial_{n+1})}$ of the chain complex ${}_{\vec{m}}\mathcal{K}_*(X) =$

$$\cdots\cdots \xrightarrow{{}_{\vec{m}}\partial_{n+1}} \mathcal{K}_n(X) \xrightarrow{{}_{\vec{m}}\partial_n} \mathcal{K}_{n-1}(X) \xrightarrow{{}_{\vec{m}}\partial_{n-1}} \cdots \xrightarrow{{}_{\vec{m}}\partial_1} \mathcal{K}_0(X) \xrightarrow{{}_{\vec{m}}\partial_0} \{0\}$$

are well defined since ${}_{\vec{m}}\partial_n \circ {}_{\vec{m}}\partial_{n+1} = 0$, for all fixed $\vec{m} \in \mathcal{R}^{L+1}, n \in \mathbb{N}_0$. For a cycle u, i.e. ${}_{\vec{m}}\partial_n(\mathsf{u}) = 0$, we denote the equivalence class containing u by $[\mathsf{u}]_\sim \in {}_{\vec{m}}\mathcal{H}_n(X)$.

We use the abbreviation ${}_{\vec{m}}\mathcal{H} := ({}_{\vec{m}}\mathcal{H}_n)_{n \geqslant 0}$. We say that the fixed tuple $\vec{m} \in \mathcal{R}^{L+1}$ is the *weight*, the element $\sigma := \sum_{i=0}^{L} m_i \in \mathcal{R}$ is the *index* of ${}_{\vec{m}}\mathcal{H}$. The number $L \in \mathbb{N}$ is called the *length* of the weight $\vec{m}$.

Example: For the one-point space $\{p\}$ and for $n \in \mathbb{N}_0$ there is only one $T : \mathbf{I}^n \to \{p\}$, thus we have $\mathcal{K}_n(p) \cong \mathcal{R}$. And for the chain complex ${}_{\vec{m}}\mathcal{K}_*(p)$,

$${}_{\vec{m}}\mathcal{K}_*(p) \;=\; \cdots\cdots \xrightarrow{{}_{\vec{m}}\partial_4} \mathcal{K}_3(p) \xrightarrow{{}_{\vec{m}}\partial_3} \mathcal{K}_2(p) \xrightarrow{{}_{\vec{m}}\partial_2} \mathcal{K}_1(p) \xrightarrow{{}_{\vec{m}}\partial_1} \mathcal{K}_0(p) \xrightarrow{{}_{\vec{m}}\partial_0} \{0\},$$

we get $\quad {}_{\vec{m}}\mathcal{K}_*(p) \;\cong\; \cdots\cdots \xrightarrow{{}_{\vec{m}}\partial_4} \mathcal{R} \xrightarrow{{}_{\vec{m}}\partial_3} \mathcal{R} \xrightarrow{{}_{\vec{m}}\partial_2} \mathcal{R} \xrightarrow{{}_{\vec{m}}\partial_1} \mathcal{R} \xrightarrow{{}_{\vec{m}}\partial_0} \{0\}.$

If we define the map $\times\sigma : \mathcal{R} \to \mathcal{R}, x \mapsto \sigma \cdot x$, we can describe the boundary operators by

$${}_{\vec{m}}\partial_n \cong \begin{cases} 0 & \text{if } n \text{ is even ,} \\ \times\sigma & \text{if } n \text{ is odd.} \end{cases}$$

Explanation: Note the definition of ${}_{\vec{m}}\partial_n(T) = \sum_{j=1}^{n}(-1)^{j+1} \cdot \sum_{i=0}^{L} m_i \cdot \langle T\rangle_{n,i,j}$. Because of the alternating signs σ copies of the unique map from $\mathbf{I}^{n-1}$ to $\{p\}$ cancel pairwise. That means for an arbitrary index σ that

$$_{\vec{m}}\mathcal{H}_n(p) \cong \begin{cases} \{x \in \mathcal{R} \mid \sigma \cdot x = 0\} & \text{if } n \text{ is odd} \\ \mathcal{R}/(\sigma \cdot \mathcal{R}) & \text{if } n \text{ is even,} \end{cases}$$

i. e. for $\sigma = 0$ we get $_{\vec{m}}\mathcal{H}_n(p) \cong \mathcal{R}$ for all $n \in \mathbb{N}_0$.

The above construction of $_{\vec{m}}\mathcal{H}_n(X)$ yields a functor $_{\vec{m}}\mathcal{H}_n : \mathsf{TOP} \longrightarrow \mathcal{R}\text{-}\mathsf{MOD}$, for all $n \in \mathbb{N}_0$: Let $f : X \to Y$ be continuous and $T \in \mathcal{S}_n(X)$, then we have $f \circ T \in \mathcal{S}_n(Y)$. Let $\mathcal{K}_n(f) : \mathcal{S}_n(X) \longrightarrow \mathcal{S}_n(Y)$, for all basis elements $T \in \mathcal{S}_n(X)$ let $\mathcal{K}_n(f)(T) := f \circ T$. We create a functor $\mathcal{K}_n : \mathsf{TOP} \longrightarrow \mathcal{R}\text{-}\mathsf{MOD}$, with

$$\mathcal{K}_n\left(X \xrightarrow{f} Y\right) := \mathcal{K}_n(X) \xrightarrow{\mathcal{K}_n(f)} \mathcal{K}_n(Y),$$

$\mathcal{K}_n(f)$ is well defined by linearity, and for an arbitrary $(f : X \to Y) \in \mathsf{TOP}$ the following diagram commutes in $\mathcal{R}\text{-}\mathsf{MOD}$ for all $n \in \mathbb{N}$:

$$\begin{array}{ccccccccccc} \cdots & \xrightarrow{_{\vec{m}}\partial_{n+2}} & \mathcal{K}_{n+1}(X) & \xrightarrow{_{\vec{m}}\partial_{n+1}} & \mathcal{K}_n(X) & \xrightarrow{_{\vec{m}}\partial_n} & \mathcal{K}_{n-1}(X) & \xrightarrow{_{\vec{m}}\partial_{n-1}} & \mathcal{K}_{n-2}(X) & \xrightarrow{_{\vec{m}}\partial_{n-2}} & \cdots \\ & & \downarrow{\scriptstyle \mathcal{K}_{n+1}(f)} & & \downarrow{\scriptstyle \mathcal{K}_n(f)} & & \downarrow{\scriptstyle \mathcal{K}_{n-1}(f)} & & \downarrow{\scriptstyle \mathcal{K}_{n-2}(f)} & & \\ \cdots & \xrightarrow[_{\vec{m}}\partial_{n+2}]{} & \mathcal{K}_{n+1}(Y) & \xrightarrow[_{\vec{m}}\partial_{n+1}]{} & \mathcal{K}_n(Y) & \xrightarrow[_{\vec{m}}\partial_n]{} & \mathcal{K}_{n-1}(Y) & \xrightarrow[_{\vec{m}}\partial_{n-1}]{} & \mathcal{K}_{n-2}(Y) & \xrightarrow[_{\vec{m}}\partial_{n-2}]{} & \cdots \end{array}$$

Thus we get $\mathcal{K}_{n-1}(f) \circ {_{\vec{m}}\partial_n} = {_{\vec{m}}\partial_n} \circ \mathcal{K}_n(f)$ for all $n \in \mathbb{N}_0$, hence $\mathcal{K}_n(f)$ maps cycles to cycles and boundaries to boundaries. For an arbitrary map $(f : X \longrightarrow Y) \in \mathsf{TOP}$, for a cycle u, hence $[\mathsf{u}]_\sim \in {_{\vec{m}}\mathcal{H}_n(X)}$, let

$$_{\vec{m}}\mathcal{H}_n(f)\,([\mathsf{u}]_\sim) := [\mathcal{K}_n(f)(\mathsf{u})]_\sim \in {_{\vec{m}}\mathcal{H}_n(Y)},$$

and we define

$$_{\vec{m}}\mathcal{H}_n\left(X \xrightarrow{f} Y\right) := {_{\vec{m}}\mathcal{H}_n(X)} \xrightarrow{_{\vec{m}}\mathcal{H}_n(f)} {_{\vec{m}}\mathcal{H}_n(Y)}.$$

In this way $_{\vec{m}}\mathcal{H}_n$ is a functor $\mathsf{TOP} \longrightarrow \mathcal{R}\text{-}\mathsf{MOD}$.

In a similar way $_{\vec{m}}\mathcal{H}_n$ will be extended to a functor $\mathsf{TOP}^2 \longrightarrow \mathcal{R}\text{-}\mathsf{MOD}$: (The following description is rather brief. For more details the reader should study [**10**, p.95 ff], or [**9**, p.22 ff], or other books about singular homology theory).

If there is $(f : (X, A) \to (Y, B)) \in \mathsf{TOP}^2$, we have subspaces $A \hookrightarrow X$ and $B \hookrightarrow Y$ (in TOP), and submodules $\mathcal{K}_n(A) \hookrightarrow \mathcal{K}_n(X)$ as well as $\mathcal{K}_n(B) \hookrightarrow \mathcal{K}_n(Y)$ (in $\mathcal{R}$-MOD). Hence the $\mathcal{R}$-modules $\mathcal{K}_n(X, A) := \frac{\mathcal{K}_n(X)}{\mathcal{K}_n(A)}$ and $\mathcal{K}_n(Y, B) := \frac{\mathcal{K}_n(Y)}{\mathcal{K}_n(B)}$ are well defined for $n \in \mathbb{N}_0$, $\mathcal{K}_{-1}(X, A) := \{0\}$. For a 'chain' $\mathsf{u} \in \mathcal{K}_n(X)$ let $[\mathsf{u}] \in \mathcal{K}_n(X, A)$ be the equivalence class of u modulo $\mathcal{K}_n(A)$. Thus $[\mathsf{u}] = [\mathsf{w}]$ if and only if $\mathsf{u} - \mathsf{w} \in \mathcal{K}_n(A)$, that means that $\mathsf{u} - \mathsf{w}$ is a chain in A.

The just constructed boundary operator $_{\vec{m}}\partial_n : \mathcal{K}_n(X) \longrightarrow \mathcal{K}_{n-1}(X)$ also yields a map $\mathcal{K}_n(X, A) \longrightarrow \mathcal{K}_{n-1}(X, A)$, which we call $_{\vec{m}}\partial_n$, too. It has the property $_{\vec{m}}\partial_n \circ {_{\vec{m}}\partial_{n+1}} = 0$, for $n \geqslant 0$, as before. We define a corresponding chain complex $_{\vec{m}}\mathcal{K}_*(X, A)$ for pairs (X, A), to be

$$\cdots\cdots \xrightarrow{_{\vec{m}}\partial_{n+1}} \mathcal{K}_n(X, A) \xrightarrow{_{\vec{m}}\partial_n} \mathcal{K}_{n-1}(X, A) \xrightarrow{_{\vec{m}}\partial_{n-1}} \cdots \xrightarrow{_{\vec{m}}\partial_1} \mathcal{K}_0(X, A) \xrightarrow{_{\vec{m}}\partial_0} \{0\}.$$

For $(f : (X, A) \to (Y, B)) \in \mathsf{TOP}^2$ and $T \in \mathcal{S}_n(A)$ we have $f \circ T \in \mathcal{S}_n(B)$ (because $f(A) \subset B$). Let

$$\mathcal{K}_n(f)([\mathsf{u}]) := [\mathcal{K}_n(f)(\mathsf{u})] \in \mathcal{K}_n(Y, B) \ \text{ for } [\mathsf{u}] \in \mathcal{K}_n(X, A),$$

and $\mathcal{K}_n(f)$ is well defined. Hence $\mathcal{K}_n$ yields a functor $\mathsf{TOP} \longrightarrow \mathcal{R}\text{-}\mathsf{MOD}$ as well as a functor $\mathsf{TOP}^2 \longrightarrow \mathcal{R}\text{-}\mathsf{MOD}$.

As above, for an arbitrary $(f : (X, A) \longrightarrow (Y, B)) \in \mathsf{TOP}^2$, the following diagram commutes in $\mathcal{R}\text{-}\mathsf{MOD}$ for $n \in \mathbb{N}_0$:

$$\begin{array}{ccccccccc} \cdots & \xrightarrow{{}_{\vec{m}}\partial_{n+2}} & \mathcal{K}_{n+1}(X,A) & \xrightarrow{{}_{\vec{m}}\partial_{n+1}} & \mathcal{K}_n(X,A) & \xrightarrow{{}_{\vec{m}}\partial_{n}} & \mathcal{K}_{n-1}(X,A) & \xrightarrow{{}_{\vec{m}}\partial_{n-1}} & \cdots \\ & & \downarrow{\scriptstyle \mathcal{K}_{n+1}(f)} & & \downarrow{\scriptstyle \mathcal{K}_{n}(f)} & & \downarrow{\scriptstyle \mathcal{K}_{n-1}(f)} & & \\ \cdots & \xrightarrow[{}_{\vec{m}}\partial_{n+2}]{} & \mathcal{K}_{n+1}(Y,B) & \xrightarrow[{}_{\vec{m}}\partial_{n+1}]{} & \mathcal{K}_n(Y,B) & \xrightarrow[{}_{\vec{m}}\partial_{n}]{} & \mathcal{K}_{n-1}(Y,B) & \xrightarrow[{}_{\vec{m}}\partial_{n-1}]{} & \cdots \end{array}$$

For a chain $\mathsf{u} \in \mathcal{K}_n(X)$, by abuse of notation we sometimes denote the equivalence class $[\mathsf{u}] \in \mathcal{K}_n(X, A)$ simply by u. If ${}_{\vec{m}}\partial_n(\mathsf{u}) \in \mathcal{K}_{n-1}(A)$ let from now on $[\mathsf{u}]_\sim \in {}_{\vec{m}}\mathcal{H}_n(X, A)$ be the equivalence class modulo $\text{image}({}_{\vec{m}}\partial_{n+1})$. We call such a chain u a *cycle*. For a cycle u and $f : (X, A) \to (Y, B)$, let

$${}_{\vec{m}}\mathcal{H}_n(f)\ ([\mathsf{u}\,]_\sim) := [\mathcal{K}_n(f)(\mathsf{u})]_\sim\,,$$

and we have

$${}_{\vec{m}}\partial_n \circ \mathcal{K}_n(f)(\mathsf{u}) = \mathcal{K}_{n-1}(f) \circ {}_{\vec{m}}\partial_n(\mathsf{u}) \in \mathcal{K}_{n-1}(B).$$

Hence $\mathcal{K}_n(f)$ maps cycles to cycles and boundaries to boundaries as above. Therefore $[\mathcal{K}_n(f)(\mathsf{u})]_\sim \in {}_{\vec{m}}\mathcal{H}_n(Y, B)$, and the functor ${}_{\vec{m}}\mathcal{H}_n : \mathsf{TOP}^2 \longrightarrow \mathcal{R}\text{-}\mathsf{MOD}$ is well defined for all $n \in \mathbb{N}_0$. As an abbreviation take

$${}_{\vec{m}}\mathcal{H}_n\left[(X,A) \xrightarrow{f} (Y,B)\right] =: {}_{\vec{m}}\mathcal{H}_n(X,A) \xrightarrow{f_*} {}_{\vec{m}}\mathcal{H}_n(Y,B).$$

4. The Homotopy Axiom

The reader should note that in the following we shall omit the weight '$\vec{m}$' in the boundary operator ${}_{\vec{m}}\partial_n$ for an easier display.

Lemma 1. *${}_{\vec{m}}\mathcal{H}$ satisfies the homotopy axiom if and only if for all topological spaces X and for $e_0, e_1 : X \to X \times \mathbf{I}$, $e_0(x) := (x, 0)$ and $e_1(x) := (x, 1)$, the equation ${}_{\vec{m}}\mathcal{H}_n(e_0) = {}_{\vec{m}}\mathcal{H}_n(e_1)$ holds for every $n \in \mathbb{N}_0$.*

Proof. '$\Longrightarrow$': Since $e_0 = Id_{X\times\mathbf{I}} \circ e_0$ and $e_1 = Id_{X\times\mathbf{I}} \circ e_1$, we have $e_0 \simeq e_1$.
'$\Longleftarrow$': If we assume $f \simeq g$ there is a continuous H with $f = H \circ e_0$ and $g = H \circ e_1$, and ${}_{\vec{m}}\mathcal{H}_n$ is a functor, hence

$${}_{\vec{m}}\mathcal{H}_n(f) = {}_{\vec{m}}\mathcal{H}_n(H) \circ {}_{\vec{m}}\mathcal{H}_n(e_0) = {}_{\vec{m}}\mathcal{H}_n(H) \circ {}_{\vec{m}}\mathcal{H}_n(e_1) = {}_{\vec{m}}\mathcal{H}_n(g).$$

□

Note that the maps $e_0, e_1 : X \to X \times \mathbf{I}$ induce canonically two maps

$$e_0, e_1 : \mathcal{S}_n(X) \to \mathcal{S}_n(X \times \mathbf{I}), \quad \text{by } e_i(T) := e_i \circ T, \text{ for } i \in \{0, 1\},$$

and by linearity two maps $e_0, e_1 : \mathcal{K}_n(X) \to \mathcal{K}_n(X \times \mathbf{I})$.

Theorem 2 (Homotopy Axiom). *Let $L \in \mathbb{N}$ and let $\vec{m} = (m_0, m_1, \ldots, m_L) \in \mathcal{R}^{L+1}$ be the weight of ${}_{\vec{m}}\mathcal{H}$. Then ${}_{\vec{m}}\mathcal{H}$ satisfies the homotopy axiom if and only if $\text{Span}_{\mathcal{R}}(\vec{m}) = \mathcal{R}$.*

Proof. '$\Longleftarrow$': We assume that $\mathrm{Span}_{\mathcal{R}}(\vec{m}) = \mathcal{R}$. Because of this assumption a set $\{r_0, r_1, \ldots, r_L\} \subset \mathcal{R}$ exists with $\sum_{k=0}^{L} r_k \cdot m_k = 1_{\mathcal{R}}$.

First we construct $L+1$ continuous auxiliary functions. For fixed $L \in \mathbb{N}$ and for all $k \in \{0, 1, 2, \ldots, L\}$ we define a map $\chi_k : [0,1] \to [0,1]$. The functions χ_k are mostly 0 and they have a 'jag' of height 1 at $\frac{k}{L}$. More precisely, for $k = 0$ and $k = L$ the formulas are

$$\chi_0(x) := \begin{cases} 1 - L \cdot x & \text{for} \quad x \in \left[0, \frac{1}{L}\right], \\ 0 & \text{for} \quad x \in \left[\frac{1}{L}, 1\right], \end{cases}$$

and
$$\chi_L(x) := \begin{cases} 0 & \text{for} \quad x \in \left[0, \frac{L-1}{L}\right], \\ L \cdot x - L + 1 & \text{for} \quad x \in \left[\frac{L-1}{L}, 1\right]. \end{cases}$$

For $L > 1$ and $k \in \{1, 2, \ldots, L-1\}$ we define χ_k to be the polygon in $\mathbb{R}^2$ through the five points (0,0), $\left(\frac{k-1}{L}, 0\right)$, $\left(\frac{k}{L}, 1\right)$, $\left(\frac{k+1}{L}, 0\right)$ and $(1, 0)$, as given by the following formula and Figure 2 for the case $(k, L) = (2, 5)$.

$$\chi_k(x) := \begin{cases} 0 & \text{for} \quad x \in \left[0, \frac{k-1}{L}\right] \cup \left[\frac{k+1}{L}, 1\right], \\ L \cdot x - k + 1 & \text{for} \quad x \in \left[\frac{k-1}{L}, \frac{k}{L}\right], \\ 1 - L \cdot x + k & \text{for} \quad x \in \left[\frac{k}{L}, \frac{k+1}{L}\right]. \end{cases}$$

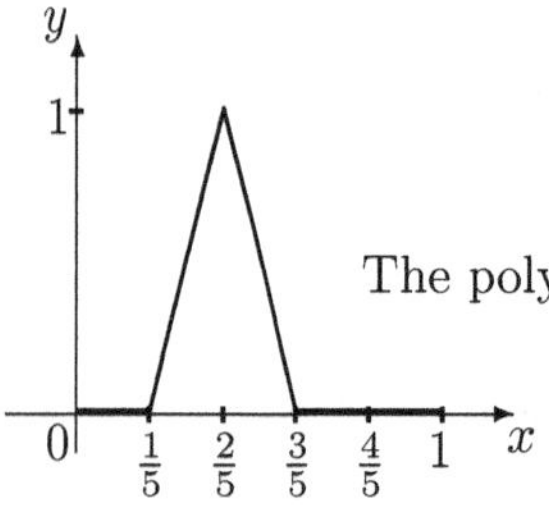

The polygon χ_2, $L = 5$

Figure 2:

Note that for $j, k \in \{0, 1, 2, \ldots, L\}$ we have $\chi_k(\frac{j}{L}) = \delta_{j,k}$ (that means $\chi_k(\frac{j}{L}) = 1$ if $k = j$ and $\chi_k(\frac{j}{L}) = 0$ if $k \neq j$).

Now we define 'chain homotopies', $\Theta_n : \mathcal{K}_n(X) \longrightarrow \mathcal{K}_{n+1}(X \times \mathbf{I})$, which means that the Θ_n's will satisfy the equation $\partial_{n+1} \circ \Theta_n = \pm(e_0 - e_1) + \Theta_{n-1} \circ \partial_n$. More precisely, for $n \in \mathbb{N}$ and $k \in \{0, 1, 2, \ldots, L\}$ define $\xi_n, \psi_{n,k} : \mathcal{S}_n(X) \longrightarrow \mathcal{S}_{n+1}(X \times \mathbf{I})$ as follows. For every $T : \mathbf{I}^n \longrightarrow X$, for all $(x_1, x_2, \ldots, x_n, x_{n+1}) \in \mathbf{I}^{n+1}$ let

$$\xi_n(T)(x_1, x_2, \ldots, x_n, x_{n+1}) := (T(x_1, x_2, \ldots, x_n), 0),$$
$$\psi_{n,k}(T)(x_1, x_2, \ldots, x_n, x_{n+1}) := (T(x_1, x_2, \ldots, x_n), \chi_k(x_{n+1})),$$

and for $n = 0$ let $\xi_0(T)(x) := (T(0), 0)$ and $\psi_{0,k}(T)(x) := (T(0), \chi_k(x))$, for $x \in \mathbf{I}$. Finally let $\Theta_{-1} := 0$, and for all $n \in \mathbb{N}_0$ we set

$$\Theta_n(T) := \sum_{k=0}^{L} r_k \cdot (\xi_n(T) - \psi_{n,k}(T)). \tag{5}$$

For $\mathsf{u} \in \mathcal{K}_n(X)$ let $\Theta_n(\mathsf{u})$ be defined by linearity. Hence we get for all integers $n \geqslant -1$ an $\mathcal{R}$-linear map $\Theta_n : \mathcal{K}_n(X) \longrightarrow \mathcal{K}_{n+1}(X \times \mathbf{I})$. Thus we get the following (noncommutative) diagram in $\mathcal{R}$-MOD.

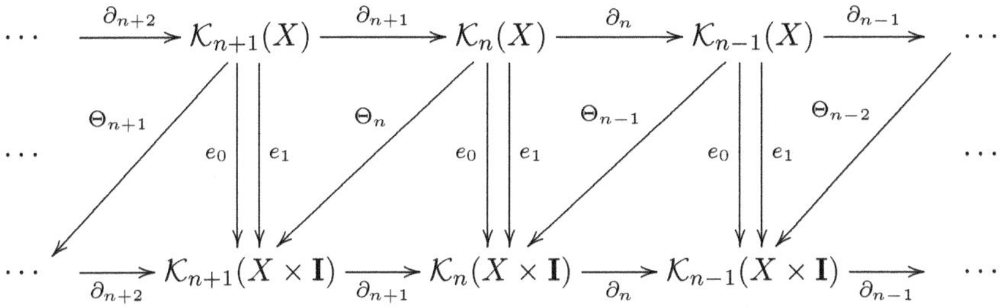

Lemma 2. *For all $n \in \mathbb{N}_0$ and all $T \in \mathcal{S}_n(X)$ we get*

$$\left[\partial_{n+1} \circ \Theta_n\right](T) = \left[(-1)^{n+2} \cdot (e_0 - e_1) + \Theta_{n-1} \circ \partial_n\right](T).$$

Proof. For $n = 0$ the proof is a simpler version of the following one and will be omitted. For $n \in \mathbb{N}$ we have:

$$\begin{aligned}
\left[\partial_{n+1} \circ \Theta_n\right](T) &= \partial_{n+1}\left[\sum_{k=0}^{L} r_k \cdot (\xi_n(T) - \psi_{n,k}(T))\right] \\
&= \sum_{k=0}^{L} r_k \cdot \sum_{j=1}^{n+1} (-1)^{j+1} \cdot \sum_{i=0}^{L} m_i \cdot \left[\langle \xi_n(T)\rangle_{n+1,i,j} - \langle \psi_{n,k}(T)\rangle_{n+1,i,j}\right] \\
&= Rest + \mathcal{D}, \quad \text{where}
\end{aligned}$$

$$Rest := \sum_{j=1}^{n} (-1)^{j+1} \cdot \sum_{i,k=0}^{L} r_k \cdot m_i \cdot \left[\langle \xi_n(T)\rangle_{n+1,i,j} - \langle \psi_{n,k}(T)\rangle_{n+1,i,j}\right], \tag{6}$$

$$\text{and} \quad \mathcal{D} := (-1)^{n+2} \cdot \sum_{i,k=0}^{L} r_k \cdot m_i \cdot \left[\langle \xi_n(T)\rangle_{n+1,i,n+1} - \langle \psi_{n,k}(T)\rangle_{n+1,i,n+1}\right]. \tag{7}$$

We have for $i \in \{0, 1, \ldots, L\}$ and all tuples $(x_1, x_2, \ldots, x_n) \in \mathbf{I}^n$:

$$\begin{aligned}
&\left[\langle \xi_n(T)\rangle_{n+1,i,n+1} - \langle \psi_{n,k}(T)\rangle_{n+1,i,n+1}\right](x_1, x_2, \ldots, x_n) \\
&= \xi_n(T)\left(x_1, x_2, \ldots, x_n, \frac{i}{L}\right) - \psi_{n,k}(T)\left(x_1, x_2, \ldots, x_n, \frac{i}{L}\right) \\
&= (T(x_1, x_2, \ldots, x_n), 0) - \left(T(x_1, x_2, \ldots, x_n), \chi_k(\frac{i}{L})\right).
\end{aligned}$$

Since $\chi_k\left(\frac{i}{L}\right) = \delta_{i,k}$ and $\sum_{k=0}^{L} r_k \cdot m_k = 1_{\mathcal{R}}$ it follows that

$$\begin{aligned}
\mathcal{D} &= (-1)^{n+2} \cdot \sum_{k=0}^{L} r_k \cdot m_k \cdot \left[\langle \xi_n(T)\rangle_{n+1,k,n+1} - \langle \psi_{n,k}(T)\rangle_{n+1,k,n+1}\right] \\
&= (-1)^{n+2} \cdot \sum_{k=0}^{L} r_k \cdot m_k \cdot [e_0 \circ T - e_1 \circ T] \\
&= (-1)^{n+2} \cdot (e_0 \circ T - e_1 \circ T) \cdot \sum_{k=0}^{L} r_k \cdot m_k \\
&= (-1)^{n+2} \cdot (e_0 \circ T - e_1 \circ T) \;=\; (-1)^{n+2} \cdot (e_0(T) - e_1(T)).
\end{aligned}$$

It remains to show that $[\Theta_{n-1} \circ \partial_n](T) = Rest$. We have

$$\begin{aligned}
&[\Theta_{n-1} \circ \partial_n](T) = \Theta_{n-1}\left(\sum_{j=1}^{n}(-1)^{j+1} \cdot \sum_{i=0}^{L} m_i \cdot \langle T\rangle_{n,i,j}\right) \\
&= \sum_{j=1}^{n}(-1)^{j+1} \cdot \sum_{i=0}^{L} m_i \cdot \Theta_{n-1}\left(\langle T\rangle_{n,i,j}\right) \\
&= \sum_{j=1}^{n}(-1)^{j+1} \cdot \sum_{i=0}^{L} m_i \cdot \sum_{k=0}^{L} r_k \cdot \left(\xi_{n-1}(\langle T\rangle_{n,i,j}) - \psi_{n-1,k}(\langle T\rangle_{n,i,j})\right) \\
&= \sum_{j=1}^{n}(-1)^{j+1} \cdot \sum_{i,k=0}^{L} m_i \cdot r_k \cdot \left(\xi_{n-1}(\langle T\rangle_{n,i,j}) - \psi_{n-1,k}(\langle T\rangle_{n,i,j})\right).
\end{aligned}$$

We consider the maps ξ_{n-1} and $\psi_{n-1,k}$ more carefully. We have for all integers $n > 1$, for $T \in \mathcal{S}_n(X)$, for $j \in \{1, 2, \ldots, n\}$, and $i, k \in \{0, 1, 2, \ldots, L\}$ the following two equations for each n-tuple $(x_1, x_2, \ldots, x_{n-1}, x_n) \in \mathbf{I}^n$:

$$\begin{aligned}
\xi_{n-1}\left(\langle T\rangle_{n,i,j}\right)(x_1, x_2, \ldots, x_n) &= \left(\langle T\rangle_{n,i,j}(x_1, x_2, \ldots, x_{n-1}), 0\right) \\
&= \left(T(x_1, x_2, \ldots, x_{j-1}, \frac{i}{L}, x_j, \ldots, x_{n-1}), 0\right) \\
&= \xi_n(T)\left(x_1, x_2, \ldots, x_{j-1}, \frac{i}{L}, x_j, \ldots, x_n\right) \\
&= \langle \xi_n(T)\rangle_{n+1,i,j}(x_1, x_2, \ldots, x_n).
\end{aligned}$$

Shortly, we have $\xi_{n-1}\left(\langle T\rangle_{n,i,j}\right) = \langle \xi_n(T) \rangle_{n+1,i,j}$.

In the same way we find that

$$\begin{aligned}\psi_{n-1,k}\left(\langle T\rangle_{n,i,j}\right)(x_1,x_2,\ldots,x_n) &= \left(\langle T\rangle_{n,i,j}(x_1,x_2,\ldots,x_{n-1}),\chi_k(x_n)\right)\\ &= \left(T(x_1,x_2,\ldots,x_{j-1},\frac{i}{L},x_j,\ldots,x_{n-1}),\chi_k(x_n)\right)\\ &= \psi_{n,k}(T)\left(x_1,x_2,\ldots,x_{j-1},\frac{i}{L},x_j,\ldots,x_n\right)\\ &= \langle\,\psi_{n,k}(T)\,\rangle_{n+1,i,j}(x_1,x_2,\ldots,x_n),\end{aligned}$$

therefore $\psi_{n-1,k}\left(\langle T\rangle_{n,i,j}\right) = \langle\psi_{n,k}(T)\rangle_{n+1,i,j}$, and finally $[\Theta_{n-1}\circ\partial_n](T) = Rest$, as defined in equation (6) follows. This ends the proof of Lemma 2. □

We have just proved that $[\partial_{n+1}\circ\Theta_n](T) = [(-1)^{n+2}\cdot(e_0-e_1)+\Theta_{n-1}\circ\partial_n](T)$. Take a cycle $\mathsf{u}\in\mathcal{K}_n(X)$ (i.e. $\partial_n(\mathsf{u})=0$) instead of T. The fact that

$$[\partial_{n+1}\circ\Theta_n](\mathsf{u}) = [(-1)^{n+2}(e_0-e_1)](\mathsf{u}) \ \in\ \mathrm{image}(\partial_{n+1})$$

means that $(e_0-e_1)(\mathsf{u})$ is a boundary, hence we can deduce for the equivalence class $[\mathsf{u}]_\sim\in{}_{\vec{m}}\mathcal{H}_n(X)$ that ${}_{\vec{m}}\mathcal{H}_n(e_0-e_1)([\mathsf{u}]_\sim) = 0 = {}_{\vec{m}}\mathcal{H}_n(e_0)([\mathsf{u}]_\sim) - {}_{\vec{m}}\mathcal{H}_n(e_1)([\mathsf{u}]_\sim)$, and therefore ${}_{\vec{m}}\mathcal{H}_n(e_0) = {}_{\vec{m}}\mathcal{H}_n(e_1)$. By Lemma 1 the homotopy axiom is satisfied.

'$\Longrightarrow$': We assume that ${}_{\vec{m}}\mathcal{H}$ satisfies the homotopy axiom. We fix an $n\in\mathbb{N}$. Let $X := \mathbf{I}^2$, and let $T_1,T_2:\mathbf{I}^n\to\mathbf{I}^2$ with $T_1\neq T_2$, but $\partial_n(T_1)=\partial_n(T_2)$. Let

$$\mathsf{u} := T_1 - T_2\in\mathcal{K}_n(X).$$

Because $\partial_n\ (\mathsf{u}) = 0$, u is a cycle, hence $[\mathsf{u}]_\sim\in{}_{\vec{m}}\mathcal{H}_n(X)$. For each n, for $i\in\{0,1\}$ and $\vec{x}\in\mathbf{I}^n, T\in\mathcal{S}_n(X)$ let us again use the linear maps $e_{n,i}$,

$$e_{n,i}:\mathcal{K}_n(X)\longrightarrow\mathcal{K}_n(X\times\mathbf{I}),\ e_{n,i}(T)(\vec{x}) := (T(\vec{x}),i),\ \text{ i.e. } e_{n,i}(T) = e_{n,i}\circ T,$$

see the definitions of e_0 and e_1 at the beginning of this section. The boundary operator ∂_n is a natural transformation, hence for $i\in\{0,1\}$ the following diagram commutes, i.e. $\partial_n\circ e_{n,i} = e_{n-1,i}\circ\partial_n$ for all $n\in\mathbb{N}$:

$$\begin{array}{ccccccccc}\cdots & \xrightarrow{\partial_{n+2}} & \mathcal{K}_{n+1}(X) & \xrightarrow{\partial_{n+1}} & \mathcal{K}_n(X) & \xrightarrow{\partial_n} & \mathcal{K}_{n-1}(X) & \xrightarrow{\partial_{n-1}} & \cdots\\ & & \Big\downarrow{\scriptstyle e_{n+1,i}} & & \Big\downarrow{\scriptstyle e_{n,i}} & & \Big\downarrow{\scriptstyle e_{n-1,i}} & & \\ \cdots & \xrightarrow[\partial_{n+2}]{} & \mathcal{K}_{n+1}(X\times\mathbf{I}) & \xrightarrow[\partial_{n+1}]{} & \mathcal{K}_n(X\times\mathbf{I}) & \xrightarrow[\partial_n]{} & \mathcal{K}_{n-1}(X\times\mathbf{I}) & \xrightarrow[\partial_{n-1}]{} & \cdots\end{array}$$

The fact that $\partial_n\ (\mathsf{u}) = 0$ implies $[e_{n-1,i}\circ\partial_n]\ (\mathsf{u}) = 0 = [\partial_n\circ e_{n,i}]\ (\mathsf{u})$, and we get that $e_{n,i}(\mathsf{u})$ is a cycle in $X\times\mathbf{I}$, thus $[e_{n,i}(\mathsf{u})]_\sim\in{}_{\vec{m}}\mathcal{H}_n(X\times\mathbf{I})$.

Since we assumed that ${}_{\vec{m}}\mathcal{H}$ satisfies the homotopy axiom, by Lemma 1 the equivalence classes of $e_{n,0}(\mathsf{u})$ and $e_{n,1}(\mathsf{u})$ in ${}_{\vec{m}}\mathcal{H}_n(X\times\mathbf{I})$ are the same. This means we get the equality $[e_{n,0}(\mathsf{u})]_\sim = [e_{n,1}(\mathsf{u})]_\sim$, hence $[e_{n,0}(\mathsf{u}) - e_{n,1}(\mathsf{u})]_\sim = 0$. It follows that $e_{n,0}(\mathsf{u}) - e_{n,1}(\mathsf{u})$ is a boundary, i.e. there is a chain $\varphi\in\mathcal{K}_{n+1}(X\times\mathbf{I})$ and $e_{n,0}(\mathsf{u}) - e_{n,1}(\mathsf{u}) = \partial_{n+1}(\varphi)$. Let $\varphi = \sum_{t=1}^{p} r_t\cdot\varphi_t\in\mathcal{K}_{n+1}(X\times\mathbf{I})$ where $p\in\mathbb{N}$, $r_1,r_2,\ldots,r_p\in\mathcal{R}$, $\varphi_1,\varphi_2,\ldots,\varphi_p\in\mathcal{S}_{n+1}(X\times\mathbf{I})$. Now we define four maps

$T_{1,0}, T_{2,0}, T_{1,1}, T_{2,1} \in \mathcal{S}_n(X \times \mathbf{I})$. Let

$$T_{1,0} := e_{n,0} \circ T_1,\ T_{2,0} := e_{n,0} \circ T_2,\ T_{1,1} := e_{n,1} \circ T_1 \text{ and } T_{2,1} := e_{n,1} \circ T_2.$$

With $\mathsf{u} = T_1 - T_2$ we have $e_{n,0}(\mathsf{u}) - e_{n,1}(\mathsf{u}) = T_{1,0} - T_{2,0} - T_{1,1} + T_{2,1}$. Since $T_1 \neq T_2$ the four maps $T_{1,0}, T_{2,0}, T_{1,1}$ and $T_{2,1}$ are pairwise distinct. We get:

$$T_{1,0} - T_{2,0} - T_{1,1} + T_{2,1} = e_{n,0}(T_1 - T_2) - e_{n,1}(T_1 - T_2) = e_{n,0}(\mathsf{u}) - e_{n,1}(\mathsf{u})$$
$$= \partial_{n+1}(\varphi) = \sum_{t=1}^{p} r_t \cdot \partial_{n+1}(\varphi_t) = \sum_{t=1}^{p} r_t \cdot \sum_{j=1}^{n+1} (-1)^{j+1} \cdot \sum_{i=0}^{L} m_i \cdot \langle \varphi_t \rangle_{n+1,i,j} \,.$$

The summands $\langle \varphi_t \rangle_{n+1,i,j}$ on the right hand side are elements of $\mathcal{S}_n(X \times \mathbf{I})$ (with coefficients r_t, m_i), which generate the four summands $T_{1,0}, T_{2,0}, T_{1,1}, T_{2,1}$ on the left hand side. Let us take the set B of triples,

$$B := \{(t,j,i) \mid t \in \{1,2,\ldots,p\}, j \in \{1,2,\ldots,n{+}1\}, i \in \{0,1,\ldots,L\} \wedge \langle \varphi_t \rangle_{n+1,i,j} = T_{1,0}\}.$$

Then we have

$$1_{\mathcal{R}} \cdot T_{1,0} = \sum_{(t,j,i) \in B} r_t \cdot (-1)^{j+1} \cdot m_i \cdot \langle \varphi_t \rangle_{n+1,i,j} = \sum_{(t,j,i) \in B} r_t \cdot (-1)^{j+1} \cdot m_i \cdot T_{1,0} \,.$$

This means that $\mathrm{Span}_{\mathcal{R}}(\vec{m}) = \mathcal{R}$, and the proof of Theorem 2 is complete. □

5. The Exact Sequence of a Pair

As we mentioned before, for all $n \in \mathbb{N}_0$ the boundary operator yields a functor ${}_{\vec{m}}\mathcal{H}_n : \mathsf{TOP}^2 \longrightarrow \mathcal{R}\text{-}\mathsf{MOD}$, that means for any $(f : (X,A) \longrightarrow (Y,B)) \in \mathsf{TOP}^2$ we have a morphism of $\mathcal{R}$-modules ${}_{\vec{m}}\mathcal{H}_n(f) : {}_{\vec{m}}\mathcal{H}_n(X,A) \longrightarrow {}_{\vec{m}}\mathcal{H}_n(Y,B)$.

For any cycle $\mathsf{u} \in \mathcal{K}_n(X,A)$ (i.e. $\partial_n(\mathsf{u})$ is a chain in A, hence we have an equivalence class $[\mathsf{u}]_\sim \in {}_{\vec{m}}\mathcal{H}_n(X,A)$), we had abbreviated (at the end of section 3)

$${}_{\vec{m}}\mathcal{H}_n(f)([\mathsf{u}]_\sim) = f_*([\mathsf{u}]_\sim) = [\mathcal{K}_n(f)(\mathsf{u})]_\sim \in {}_{\vec{m}}\mathcal{H}_n(Y,B).$$

For a subspace $A \subset X$ we get a short exact sequence of $\mathcal{R}$-modules

$$\{0\} \longrightarrow \mathcal{K}_n(A) \longrightarrow \mathcal{K}_n(X) \longrightarrow \mathcal{K}_n(X,A) \longrightarrow \{0\}.$$

Together with the boundary operators $(\partial_n)_{n \geqslant 0}$ we get a short exact sequence of chain complexes

$$\{0\} \longrightarrow {}_{\vec{m}}\mathcal{K}_*(A) \longrightarrow {}_{\vec{m}}\mathcal{K}_*(X) \longrightarrow {}_{\vec{m}}\mathcal{K}_*(X,A) \longrightarrow \{0\}.$$

Let $i : A \hookrightarrow X$ and $j : (X,\emptyset) \hookrightarrow (X,A)$ be the canonical topological inclusions. Now we are able to construct for all $n \in \mathbb{N}_0$ a morphism k_* of $\mathcal{R}$-modules,

$$k_* : {}_{\vec{m}}\mathcal{H}_n(X,A) \longrightarrow {}_{\vec{m}}\mathcal{H}_{n-1}(A),$$

the *connecting homomorphism*. Finally this yields a long exact sequence of $\mathcal{R}$-module morphisms:

$$\cdots \xrightarrow{j_*} {}_{\vec{m}}\mathcal{H}_{n+1}(X,A) \xrightarrow{k_*} {}_{\vec{m}}\mathcal{H}_n(A) \xrightarrow{i_*} {}_{\vec{m}}\mathcal{H}_n(X) \xrightarrow{j_*} {}_{\vec{m}}\mathcal{H}_n(X,A) \xrightarrow{k_*} \cdots$$

For details see any book about singular homology theory, for instance [**10**, p.93 ff], or [**13**, p.18 ff], but there is no necessity for us to repeat all these well known facts.

6. The Excision Axiom

For the next section it is very useful to compare the corresponding section in [**9**, p.26 ff]. The reader should note that we are able to prove the excision axiom only in the case of $L = 1$. Furthermore, the reader may find perhaps an easier way to prove the excision axiom, e.g. in [**11**]. For a topological space X and a subset A, $Int(A)$ means the interior of A and $Cl(A)$ the closure of A. To prove the following theorem we need an extra assumption called $\mathcal{NCD}$, see Definition 3.

Theorem 3 (Excision Axiom). *Let X be a topological space and let B and A be subsets of X such that $Cl(B) \subseteq Int(A)$. Let*

$$i : (X \backslash B, A \backslash B) \hookrightarrow (X, A)$$

be the canonical inclusion in TOP^2. Let us fix the length $L := 1$, let the weight $\vec{m}$ be $\vec{m} := (m_0, m_1) := (a, b) \in \mathcal{R}^2$, let (a, b) satisfy the condition $\mathcal{NCD}$. Then for each $n \in \mathbb{N}_0$ the morphism

$$i_* : {}_{\vec{m}}\mathcal{H}_n(X \backslash B, A \backslash B) \rightarrow {}_{\vec{m}}\mathcal{H}_n(X, A)$$

induced by the inclusion i is an isomorphism.

This theorem follows directly from the following Proposition 1. Let $\mathcal{U} := \{U_i \mid i \in \Im\}$ be an indexed family of subsets of X whose interiors cover X. Then $\mathcal{U}$ is called a *generalised open covering of X*. (The sets U_i need not be open). See Definition 4. Note that the boundary operator ∂_n commutes with the inclusion $\widehat{i}$, that means $\partial_n \circ \widehat{i}(T) = \widehat{i} \circ \partial_n(T)$ for all $T \in \mathcal{S}_n(A, \mathcal{U})$. Therefore ∂_n induces a linear map $\mathcal{K}_n(X, A, \mathcal{U}) \longrightarrow \mathcal{K}_{n-1}(X, A, \mathcal{U})$, which we call ∂_n, too. Because $\partial_n \circ \partial_{n+1} = 0$ it leads to *$\mathcal{U}$-small homology $\mathcal{R}$-modules*

$$ {}_{\vec{m}}\mathcal{H}_n(X, A, \mathcal{U}) := \frac{kernel(\partial_n)}{image(\partial_{n+1})} .$$

For more details see [**9**, p.29,30]. Now we are able to formulate and to prove the following proposition.

Proposition 1. *Let X be a topological space and A a subspace with the inclusion $A \overset{i}{\hookrightarrow} X$, and let $\mathcal{U}$ be a generalised open covering of X with the canonical inclusion $\mathcal{K}_n(X, A, \mathcal{U}) \overset{j}{\hookrightarrow} \mathcal{K}_n(X, A)$. Let us assume a weight $\vec{m} := (a, b) \in \mathcal{R}^2$ and let (a, b) satisfy the condition $\mathcal{NCD}$. Then for all $n \in \mathbb{N}_0$ the morphism*

$$j_* : {}_{\vec{m}}\mathcal{H}_n(X, A, \mathcal{U}) \xrightarrow{\cong} {}_{\vec{m}}\mathcal{H}_n(X, A)$$

induced by j is an isomorphism.

The proof is rather lengthy and will need the entire section. With this proposition the excision axiom easily follows, see [**9**, p.30,31]. It remains to prove the proposition.

Proof. First we shall present the proof with $A = \emptyset$, the empty set. Afterwards the general case $A \neq \emptyset$ is an easy application of the Five-Lemma. Hence let $A := \emptyset$.

We have to define for all integers $n \geqslant -1$ 'subdivision maps'

$$\mathcal{SD}_n : \mathcal{K}_n(X) \longrightarrow \mathcal{K}_n(X).$$

We need some preparations. We define the $\mathcal{SD}_n$'s on the basis $\mathcal{S}_n(X)$ and we extend the definition on $\mathcal{K}_n(X)$ by linearity. For $n = -1$, $\mathcal{SD}_{-1}$ is the 0-map; for $n = 0$ let $\mathcal{SD}_0(T) = -T$ for all $T : \mathbf{I}^0 \to X$. For $n > 0$ a map $T \in \mathcal{S}_n(X)$ will be 'subdivided' into smaller ones.

Now we use the sets $\mathcal{E} := \{0, 2\}, \mathcal{V} := \{-1, 1\}$. Further the reader should recall a map $\|T\|_{\alpha,\vec{e},\vec{v}} \in \mathcal{S}_n(X)$ of Definition 5. In the following we set $\alpha := \frac{1}{3}$. We take $e_i \in \mathcal{E}$ and $v_i \in \mathcal{V}$ for $i = 1, 2, \ldots, n$, and q_n will be the identity on $\mathbf{I}^n$. Define for all $n \in \mathbb{N}$ and for any continuous $T : \mathbf{I}^n \to X$ (i.e. $T \in \mathcal{S}_n(X)$) :

$$\mathcal{SD}_n(T) := \sum_{\vec{e}\in\mathcal{E}^n} \sum_{\vec{v}\in\mathcal{V}_{\vec{e},n}} (-\prod_{i=1}^{n} v_i) \cdot \|T\|_{\frac{1}{3},\vec{e},\vec{v}} \,, \tag{8}$$

where $\mathcal{V}_{\vec{e},n}$ is the set of $(v_1, v_2, \ldots, v_n) \in \mathcal{V}^n$ such that for all $i = 1, 2, \ldots, n$, $v_i = 1$ if $e_i = 0$ and $v_i \in \{-1, 1\}$ if $e_i = 2$. We get a map $\mathcal{SD}_n : \mathcal{K}_n(X) \longrightarrow \mathcal{K}_n(X)$ by linearity.

Examples: Let $n := 1$. For $T \in \mathcal{S}_1(X)$, i.e. $T : \mathbf{I} \to X$ we have

$$\mathcal{SD}_1(T) \;=\; -\|T\|_{\frac{1}{3},0,1} + \|T\|_{\frac{1}{3},2,-1} - \|T\|_{\frac{1}{3},2,1} \quad \text{(see Figure 3)},$$

and for $n := 2$, for $T \in \mathcal{S}_2(X)$ we get the linear combination (see the figure, too)

$$\begin{aligned}\mathcal{SD}_2(T) = &-\|T\|_{\frac{1}{3},\binom{0}{0},\binom{1}{1}} - \|T\|_{\frac{1}{3},\binom{2}{0},\binom{1}{1}} + \|T\|_{\frac{1}{3},\binom{2}{0},\binom{-1}{1}} - \|T\|_{\frac{1}{3},\binom{0}{2},\binom{1}{1}} \\ &+ \|T\|_{\frac{1}{3},\binom{0}{2},\binom{1}{-1}} + \|T\|_{\frac{1}{3},\binom{2}{2},\binom{-1}{1}} - \|T\|_{\frac{1}{3},\binom{2}{2},\binom{1}{1}} + \|T\|_{\frac{1}{3},\binom{2}{2},\binom{1}{-1}} - \|T\|_{\frac{1}{3},\binom{2}{2},\binom{-1}{-1}} \,.\end{aligned}$$

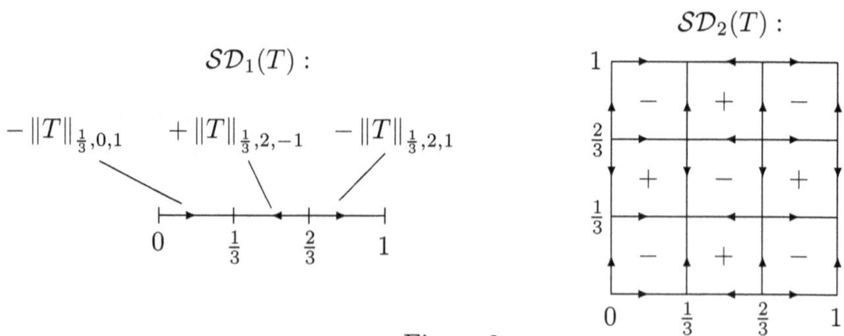

Figure 3:

Generally for $n \in \mathbb{N}_0$ and $T \in \mathcal{S}_n(X)$, $\mathcal{SD}_n(T)$ is a linear combination of 3^n maps in $\mathcal{S}_n(X)$.

Lemma 3. *For all $n \in \mathbb{N}_0$ the map $\mathcal{SD}_n$ commutes with the boundary operator ∂_n, i.e. the following diagram commutes:*

$$\begin{array}{ccccccccccc}
\cdots & \xrightarrow{\partial_{n+2}} & \mathcal{K}_{n+1}(X) & \xrightarrow{\partial_{n+1}} & \mathcal{K}_n(X) & \xrightarrow{\partial_n} & \mathcal{K}_{n-1}(X) & \xrightarrow{\partial_{n-1}} & \mathcal{K}_{n-2}(X) & \xrightarrow{\partial_{n-2}} & \cdots \\
 & & \downarrow{\scriptstyle \mathcal{SD}_{n+1}} & & \downarrow{\scriptstyle \mathcal{SD}_n} & & \downarrow{\scriptstyle \mathcal{SD}_{n-1}} & & \downarrow{\scriptstyle \mathcal{SD}_{n-2}} & & \\
\cdots & \xrightarrow[\partial_{n+2}]{} & \mathcal{K}_{n+1}(X) & \xrightarrow[\partial_{n+1}]{} & \mathcal{K}_n(X) & \xrightarrow[\partial_n]{} & \mathcal{K}_{n-1}(X) & \xrightarrow[\partial_{n-1}]{} & \mathcal{K}_{n-2}(X) & \xrightarrow[\partial_{n-2}]{} & \cdots
\end{array}$$

Proof. We have to prove $\partial_n \circ \mathcal{SD}_n(T) = \mathcal{SD}_{n-1} \circ \partial_n(T)$ for each $n \in \mathbb{N}_0$ and $T \in \mathcal{S}_n(X)$. This is trivial for $n = 0$ and easy for $n = 1$, so let $n \geqslant 2$. Let $T \in \mathcal{S}_n(X)$. Note that in the following we shall use '$\langle T \rangle_{i,\,j}$' instead of the expression '$\langle T \rangle_{n,\,i,\,j}$', to make it better readable. We have:

$$\partial_n \circ \mathcal{SD}_n(T) = \partial_n \left[\sum_{\vec{e} \in \mathcal{E}^n} \sum_{\vec{v} \in \mathcal{V}_{\vec{e},n}} (-\prod_{i=1}^{n} v_i) \cdot \|T\|_{\frac{1}{3},\vec{e},\vec{v}} \right]$$

$$= \sum_{j=1}^{n} (-1)^{j+1} \sum_{\vec{e} \in \mathcal{E}^n} \sum_{\vec{v} \in \mathcal{V}_{\vec{e},n}} (-\prod_{i=1}^{n} v_i) \cdot \left[a \cdot \left\langle \|T\|_{\frac{1}{3},\vec{e},\vec{v}} \right\rangle_{0,j} + b \cdot \left\langle \|T\|_{\frac{1}{3},\vec{e},\vec{v}} \right\rangle_{1,j} \right].$$

As well as

$$\mathcal{SD}_{n-1} \circ \partial_n(T) = \mathcal{SD}_{n-1} \left[\sum_{j=1}^{n} (-1)^{j+1} \left(a \cdot \langle T \rangle_{0,j} + b \cdot \langle T \rangle_{1,j} \right) \right]$$

$$= \sum_{j=1}^{n} (-1)^{j+1} \sum_{\vec{e} \in \mathcal{E}^{n-1}} \sum_{\vec{v} \in \mathcal{V}_{\vec{e},n-1}} (-\prod_{i=1}^{n-1} v_i) \cdot \left[a \cdot \left\| \langle T \rangle_{0,j} \right\|_{\frac{1}{3},\vec{e},\vec{v}} + b \cdot \left\| \langle T \rangle_{1,j} \right\|_{\frac{1}{3},\vec{e},\vec{v}} \right].$$

The equality is not obvious; so we have to calculate. It seems that the first sum is 'bigger'. But many elements cancel pairwise, and the rest is equal to the second sum.

We fix an arbitrary $j \in \{1, 2, \dots, n\}$. We take

$$\vec{e_2} = (e_1,\ e_2, \dots, e_{j-1},\ 2,\ e_{j+1}, \dots, e_n) \in \mathcal{E}^n$$

and $\vec{\vartheta_1}, \vec{\vartheta_{-1}} \in \mathcal{V}_{\vec{e_2},n}$ given by

$$\vec{\vartheta_1} = (v_1, v_2, \dots, v_{j-1},\ 1,\ v_{j+1}, \dots, v_n),\ \vec{\vartheta_{-1}} = (v_1, v_2, \dots, v_{j-1},\ -1,\ v_{j+1}, \dots, v_n).$$

Then

$$\left\langle \|T\|_{\frac{1}{3},\vec{e_2},\vec{\vartheta_1}} \right\rangle_{0,j}, \left\langle \|T\|_{\frac{1}{3},\vec{e_2},\vec{\vartheta_{-1}}} \right\rangle_{0,j} \in \mathcal{S}_{n-1}(X).$$

For a point $(x_1, x_2, \dots, x_{j-1}, x_j, \dots, x_{n-1}) \in \mathbf{I}^{n-1}$ we get

$$\left\langle \|T\|_{\frac{1}{3},\vec{e_2},\vec{\vartheta_1}} \right\rangle_{0,j} (x_1, x_2, \dots, x_{j-1}, x_j, \dots, x_{n-1})$$

$$= \|T\|_{\frac{1}{3},\vec{e_2},\vec{\vartheta_1}} (x_1, x_2, \dots, x_{j-1},\ 0,\ x_j, \dots, x_{n-1})$$

$$= T\left(\frac{1}{3} \cdot [e_1 + v_1 \cdot x_1, \dots, e_{j-1} + v_{j-1} \cdot x_{j-1},\ 2,\ e_{j+1} + v_{j+1} \cdot x_j, \dots, e_n + v_n \cdot x_{n-1}] \right),$$

and

$$\begin{aligned}\left\langle \|T\|_{\frac{1}{3},\vec{e_2},\vec{\vartheta_{-1}}} \right\rangle_{0,j} &= \|T\|_{\frac{1}{3},\vec{e_2},\vec{\vartheta_{-1}}} (x_1,\,\ldots,\,x_{j-1}\,,\,0\,,\,x_j\,,\,\ldots\,,\,x_{n-1}) \\ &= T\left(\frac{1}{3}\cdot[e_1+v_1\cdot x_1,\,\ldots\ldots\,,2\,,\;\ldots\ldots\,,e_n+v_n\cdot x_{n-1}\,]\right)\end{aligned}$$

whence

$$\left\langle \|T\|_{\frac{1}{3},\vec{e_2},\vec{\vartheta_1}} \right\rangle_{0,j} = \left\langle \|T\|_{\frac{1}{3},\vec{e_2},\vec{\vartheta_{-1}}} \right\rangle_{0,j}.$$

Now note that $(\prod_{v_i\in\vec{\vartheta_1}} v_i)\cdot(\prod_{v_i\in\vec{\vartheta_{-1}}} v_i) = -1$, from which it follows that

$$a\cdot(-\prod_{v_i\in\vec{\vartheta_1}} v_i)\cdot \left\langle \|T\|_{\frac{1}{3},\vec{e_2},\vec{\vartheta_1}} \right\rangle_{0,j} + a\cdot(-\prod_{v_i\in\vec{\vartheta_{-1}}} v_i)\cdot \left\langle \|T\|_{\frac{1}{3},\vec{e_2},\vec{\vartheta_{-1}}} \right\rangle_{0,j} = 0.$$

In the same way with

$$\begin{aligned}\vec{e_0} &:= (e_1,e_2,\ldots,e_{j-1},\,0,\,e_{j+1},\ldots,e_n),\\ \vec{e_2} &:= (e_1,e_2,\ldots,e_{j-1},\,2,\,e_{j+1},\ldots,e_n)\in\mathcal{E}^n\\ \vec{\vartheta_1} &:= (v_1,\ldots,v_{j-1},\,1,\,v_{j+1},\ldots,v_n),\\ \vec{\vartheta_{-1}} &:= (v_1,\ldots,v_{j-1},\,-1,\,v_{j+1},\ldots,v_n)\in\mathcal{V}_{\vec{e_2},n}\end{aligned}$$

we get

$$\begin{aligned}&\left\langle \|T\|_{\frac{1}{3},\vec{e_0},\vec{\vartheta_1}} \right\rangle_{1,j} (x_1,x_2,\ldots,x_{n-1})\\ &= \|T\|_{\frac{1}{3},\vec{e_0},\vec{\vartheta_1}} (x_1,x_2,\ldots,x_{j-1},\,1,\,x_j,\ldots,x_{n-1})\\ &= T\left(\frac{1}{3}\cdot[e_1+v_1\cdot x_1,\ldots,e_{j-1}+v_{j-1}\cdot x_{j-1},\,1,\,e_{j+1}+v_{j+1}\cdot x_j,\ldots,e_n+v_n\cdot x_{n-1}]\right)\\ &= \|T\|_{\frac{1}{3},\vec{e_2},\vec{\vartheta_{-1}}} (x_1,x_2,\ldots,x_{j-1},\,1,\,x_j,\ldots,x_{n-1})\\ &= \left\langle \|T\|_{\frac{1}{3},\vec{e_2},\vec{\vartheta_{-1}}} \right\rangle_{1,j} (x_1,x_2,\ldots,x_{n-1}).\end{aligned}$$

Hence

$$b\cdot(-\prod_{v_i\in\vec{\vartheta_1}} v_i)\cdot \left\langle \|T\|_{\frac{1}{3},\vec{e_0},\vec{\vartheta_1}} \right\rangle_{1,j} + b\cdot(-\prod_{v_i\in\vec{\vartheta_{-1}}} v_i)\cdot \left\langle \|T\|_{\frac{1}{3},\vec{e_2},\vec{\vartheta_{-1}}} \right\rangle_{1,j} = 0.$$

Now take again

$\vec{e_0} = (e_1,\ldots,e_{j-1},0,e_{j+1},\ldots,e_n)\in\mathcal{E}^n,\quad \vec{\vartheta_1} = (v_1,\ldots,v_{j-1},1,v_{j+1},\ldots,v_n)\in\mathcal{V}_{\vec{e_0},n}$

as above, and define

$$\widetilde{e} = (e_1,e_2,\ldots,e_{j-1},e_{j+1},\ldots,e_n)\in\mathcal{E}^{n-1},$$

$$\widetilde{\vartheta} = (v_1,v_2,\ldots,v_{j-1},v_{j+1},\ldots,v_n)\in\mathcal{V}_{\widetilde{e},n-1}.$$

Then

$$\left\langle \|T\|_{\frac{1}{3},\vec{e_0},\vec{\vartheta_1}} \right\rangle_{0,j}, \left\| \langle T\rangle_{0,j} \right\|_{\frac{1}{3},\widetilde{e},\widetilde{\vartheta}} \in \mathcal{S}_{n-1}(X),$$

and for all points $(x_1, x_2, \dots, x_{j-1}, x_j, \dots, x_{n-1}) \in \mathbf{I}^{n-1}$ we calculate

$$\begin{aligned}
&\left\langle \|T\|_{\frac{1}{3},\vec{e_0},\vec{\vartheta_1}} \right\rangle_{0,j} (x_1, x_2, \dots, x_{n-1}) \\
&= \|T\|_{\frac{1}{3},\vec{e_0},\vec{\vartheta_1}} (x_1, x_2, \dots, x_{j-1},\ 0,\ x_j, \dots, x_{n-1}) \\
&= T\left(\frac{1}{3}\cdot[e_1 + v_1\cdot x_1, \dots, e_{j-1} + v_{j-1}\cdot x_{j-1},\ 0,\ e_{j+1} + v_{j+1}\cdot x_j, \dots, e_n + v_n\cdot x_{n-1}]\right) \\
&= \langle T\rangle_{0,j}\left(\frac{1}{3}\cdot[e_1 + v_1\cdot x_1, \dots, e_{j-1} + v_{j-1}\cdot x_{j-1},\ e_{j+1} + v_{j+1}\cdot x_j, \dots, e_n + v_n\cdot x_{n-1}]\right) \\
&= \left\| \langle T\rangle_{0,j} \right\|_{\frac{1}{3},\widetilde{e},\widetilde{\vartheta}} (x_1, x_2, \dots, x_{j-1}, x_j, \dots, x_{n-1}).
\end{aligned}$$

Hence

$$a\cdot(-\prod_{v_i\in\vec{\vartheta_1}} v_i)\cdot \left\langle \|T\|_{\frac{1}{3},\vec{e_0},\vec{\vartheta_1}} \right\rangle_{0,j} = a\cdot(-\prod_{v_i\in\widetilde{\vartheta}} v_i)\cdot \left\| \langle T\rangle_{0,j} \right\|_{\frac{1}{3},\widetilde{e},\widetilde{\vartheta}}.$$

If we take as above,

$$\vec{e_2} = (e_1, e_2, \dots, e_{j-1},\ 2,\ e_{j+1}, \dots, e_n),\ \widetilde{e} = (e_1, e_2, \dots, e_{j-1}, e_{j+1}, \dots, e_n)$$

and

$$\vec{\vartheta_1} = (v_1, v_2, \dots, v_{j-1},\ 1,\ v_{j+1}, \dots, v_n) \in \mathcal{V}_{\vec{e_2},n},$$
$$\widetilde{\vartheta} = (v_1, v_2, \dots, v_{j-1}, v_{j+1}, \dots, v_n) \in \mathcal{V}_{\widetilde{e},n-1}$$

we have

$$\left\langle \|T\|_{\frac{1}{3},\vec{e_2},\vec{\vartheta_1}} \right\rangle_{1,j}, \left\| \langle T\rangle_{1,j} \right\|_{\frac{1}{3},\widetilde{e},\widetilde{\vartheta}} \in \mathcal{S}_{n-1}(X).$$

We compute

$$\begin{aligned}
&\left\langle \|T\|_{\frac{1}{3},\vec{e_2},\vec{\vartheta_1}} \right\rangle_{1,j} (x_1, x_2, \dots, x_{n-1}) \\
&= \|T\|_{\frac{1}{3},\vec{e_2},\vec{\vartheta_1}} (x_1, x_2, \dots, x_{j-1},\ 1,\ x_j, \dots, x_{n-1}) \\
&= T\left(\frac{1}{3}\cdot[e_1 + v_1\cdot x_1, \dots, e_{j-1} + v_{j-1}\cdot x_{j-1},\ 3,\ e_{j+1} + v_{j+1}\cdot x_j, \dots, e_n + v_n\cdot x_{n-1}]\right) \\
&= \langle T\rangle_{1,j}\left(\frac{1}{3}\cdot[e_1 + v_1\cdot x_1, \dots, e_{j-1} + v_{j-1}\cdot x_{j-1}, e_{j+1} + v_{j+1}\cdot x_j, \dots, e_n + v_n\cdot x_{n-1}]\right) \\
&= \left\| \langle T\rangle_{1,j} \right\|_{\frac{1}{3},\widetilde{e},\widetilde{\vartheta}} (x_1, x_2, \dots, x_{j-1}, x_j, \dots, x_{n-1}).
\end{aligned}$$

Hence we get

$$b\cdot(-\prod_{v_i\in\vec{\vartheta_1}} v_i)\cdot \left\langle \|T\|_{\frac{1}{3},\vec{e_2},\vec{\vartheta_1}} \right\rangle_{1,j} = b\cdot(-\prod_{v_i\in\widetilde{\vartheta}} v_i)\cdot \left\| \langle T\rangle_{1,j} \right\|_{\frac{1}{3},\widetilde{e},\widetilde{\vartheta}}.$$

This is all we need to show that $\partial_n \circ \mathcal{SD}_n(T) = \mathcal{SD}_{n-1} \circ \partial_n(T)$, and Lemma 3 has been proved. □

Because of Lemma 3, the map $\mathcal{SD}_n : \mathcal{K}_n(X) \longrightarrow \mathcal{K}_n(X)$ induces an $\mathcal{R}$-module endomorphism of the space ${}_{\vec{m}}\mathcal{H}_n(X)$, which we also call $\mathcal{SD}_n$.

Now we shall show that for the weight $\vec{m} = (a, b)$ we have

$$[a \cdot \mathcal{SD}_n(\mathsf{u})]_\sim = [b \cdot \mathsf{u}]_\sim$$

on the level of homology classes, for all $\mathsf{u} \in \mathcal{K}_n(X)$ with $\mathsf{u} \in$ kernel(∂_n). We are able to do this by the help of a chain homotopy in the same way we used it for the proof of the homotopy axiom. This means for $n \geqslant -1$ the construction of a linear map $\Theta_n : \mathcal{K}_n(X) \longrightarrow \mathcal{K}_{n+1}(X)$, which yields a (noncommutative) diagram

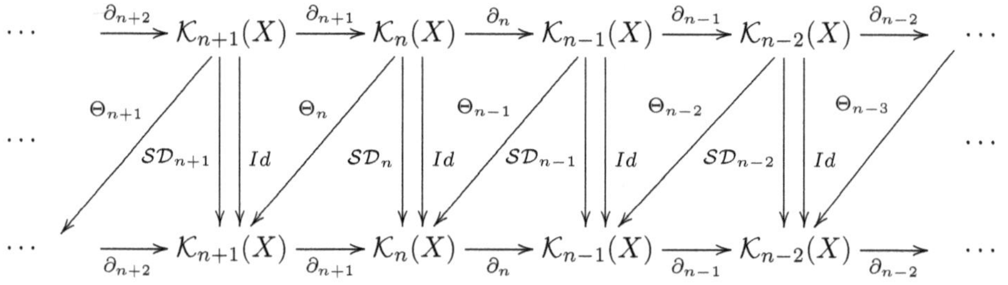

Let Id be the identity map on $\mathcal{K}_n(X)$, for all n. Our aim is, for $\mathsf{u} \in \mathcal{K}_n(X)$, to get the equation

$$(\partial_{n+1} \circ \Theta_n)(\mathsf{u}) = \pm(b \cdot Id - a \cdot \mathcal{SD}_n)(\mathsf{u}) + (\Theta_{n-1} \circ \partial_n)(\mathsf{u}), \text{ for } n \in \mathbb{N}_0. \quad (9)$$

Of course $\Theta_{-1} := 0$. For $n := 0$ for every $T : \{0\} \to X$ and for $x \in \mathbf{I} = [0, 1]$ define $\Theta_0(T)(x) := T(0)$. Then $(\partial_1 \circ \Theta_0)(T) = \ a \cdot T + b \cdot T = +(b \cdot T - a \cdot \mathcal{SD}_0(T))$, as required. Let $n \geqslant 1$. We need three auxiliary functions $\eta_0, \eta_1, \eta_2 : \ \mathbf{I}^2 \to \mathbf{I}$: for all $x, y \in [0, 1]$ let

$$\eta_0(x, y) := \frac{x}{3 - 2 \cdot y},$$

$$\eta_1(x, y) := \begin{cases} \frac{2-x}{3-2\cdot y} & \text{for } y \leqslant \frac{1}{2} + \frac{1}{2} \cdot x, \\ 1 & \text{otherwise }, \end{cases}$$

$$\eta_2(x, y) := \begin{cases} \frac{2+x}{3-2\cdot y} & \text{for } y \leqslant \frac{1}{2} - \frac{1}{2} \cdot x, \\ 1 & \text{otherwise }. \end{cases}$$

The maps η_0, η_1, η_2 are continuous. We use the set $\Upsilon := \{0, 1, 2\}$. For all tuples $\vec{z} = (z_1, z_2, \ldots, z_n) \in \Upsilon^n$ and all $T \in \mathcal{S}_n(X)$ define $\ G_{\vec{z}}(T) : \mathbf{I}^{n+1} \to X$ by the equation

$$G_{\vec{z}}(T)(x_1, \ldots, x_n, x_{n+1}) := \ T(\eta_{z_1}(x_1, x_{n+1}), \eta_{z_2}(x_2, x_{n+1}), \ldots\ldots, \eta_{z_n}(x_n, x_{n+1}))\,,$$

for all $(n+1)$-tuples $(x_1, \ldots, x_n, x_{n+1}) \in \mathbf{I}^{n+1}$. $G_{\vec{z}}(T)$ is an element of $\mathcal{S}_{n+1}(X)$. Let for all $\vec{z} := (z_1, z_2, \ldots, z_n) \in \Upsilon^n$ the number $v_{\vec{z}} \in \{-1, +1\}$ by $v_{\vec{z}} := (-1)^{\sum_{i=1}^{n} z_i}$, hence $v_{\vec{z}} = (-1)^\alpha$ where α is the number of 1's in $\vec{z}$, and finally define:

$$\Theta_n(T) := \sum_{\vec{z}\in\Upsilon^n} v_{\vec{z}} \cdot G_{\vec{z}}(T). \tag{10}$$

We describe an example for $n = 1$. For $T \in \mathcal{S}_1(X)$ we have for all pairs $(x, y) \in \mathbf{I}^2$:

$$\begin{aligned}\Theta_1(T)(x,y) &= +G_0(T)(x,y) - G_1(T)(x,y) + G_2(T)(x,y)\\ &= +T(\eta_0(x,y)) - T(\eta_1(x,y)) + T(\eta_2(x,y)).\end{aligned}$$

We hope that the following Figure 4 will help to get a better understanding. By definition, $\Theta_1(T)$ generates three maps $T(\eta_0), T(\eta_1), T(\eta_2) \in \mathcal{S}_2(X)$, whose images are indicated by squares. Later we shall prove that the top of the left square equals the image of T, while the bottoms of all squares are equal to $-\mathcal{SD}_1(T)$. Further, the upper fourth (diagonal) section of the middle square is constant equal $-T(1)$, and the upper three fourths section of the right square is constant equal $T(1)$.

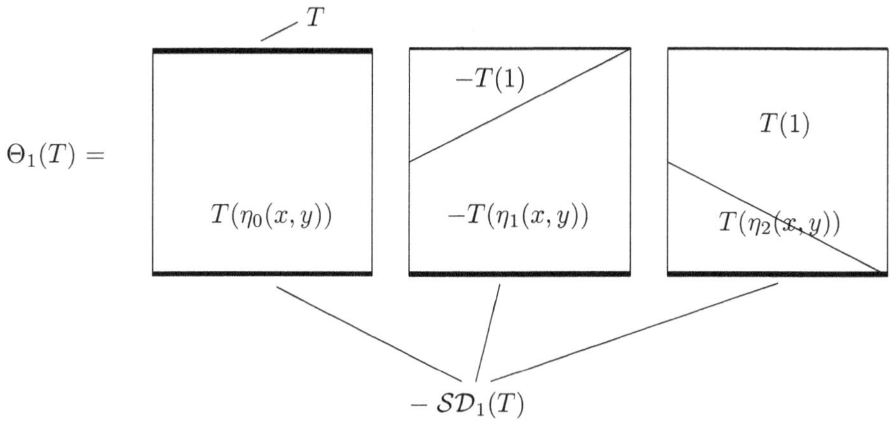

Figure 4:

We want to show that equation (9) holds for all basis elements $T \in \mathcal{S}_n(X)$. We have that

$$\begin{aligned}(\partial_{n+1} \circ \Theta_n)(T) &= \partial_{n+1}\left(\sum_{\vec{z}\in\Upsilon^n} v_{\vec{z}} \cdot G_{\vec{z}}(T)\right)\\ &= \sum_{j=1}^{n+1}(-1)^{j+1} \sum_{\vec{z}\in\Upsilon^n} v_{\vec{z}} \cdot \Big(a \cdot \langle\, G_{\vec{z}}(T)\,\rangle_{0,j} + b \cdot \langle\, G_{\vec{z}}(T)\,\rangle_{1,j}\Big).\end{aligned}$$

In the beginning let us consider the special case $j := n+1$. For $\vec{z} := (0, 0, 0, \ldots, 0) \in \Upsilon^n$ we compute $\langle G_{\vec{z}}(T)\rangle_{1,n+1}$ for $(x_1, x_2, \ldots, x_n) \in \mathbf{I}^n$:

$$\begin{aligned}\langle G_{\vec{z}}(T)\rangle_{1,n+1}(x_1, x_2, \ldots, x_n) &= G_{\vec{z}}(T)(x_1, x_2, \ldots, x_n, 1)\\ = T(\eta_0(x_1, 1), \eta_0(x_2, 1), \ldots, \eta_0(x_n, 1)) &= T(x_1, x_2, \ldots, x_n).\end{aligned}$$

Hence $b \cdot \langle G_{\vec{z}}(T)\rangle_{1,n+1} = b \cdot T$. (We are just computing the top of the squares, see again the previous figure!) We shall see that for the other $\vec{z} \in \Upsilon^n$ the corresponding elements of $\langle G_{\vec{z}}(T)\rangle_{1,n+1}$ cancel pairwise. For a fixed $k \in \{1, 2, \ldots, n\}$ let
$\vec{\lambda} := (z_1, z_2, \ldots, z_{k-1}, 1, z_{k+1}, \ldots, z_n), \vec{\zeta} := (z_1, z_2, \ldots, z_{k-1}, 2, z_{k+1}, \ldots, z_n) \in \Upsilon^n$.
For an arbitrary element $(x_1, \ldots, x_k, \ldots, x_n) \in \mathbf{I}^n$ we get

$$\begin{aligned}
&\langle G_{\vec{\lambda}}(T)\rangle_{1,n+1}(x_1, \ldots, x_k, \ldots, x_n) = G_{\vec{\lambda}}(T)(x_1, \ldots, x_k, \ldots, x_n, 1)\\
&= T(\ldots\ldots, \eta_1(x_k, 1), \ldots\ldots) \; = T(\ldots\ldots, 1, \ldots\ldots)\\
&= T(\ldots\ldots, \eta_2(x_k, 1), \ldots\ldots) \; = G_{\vec{\zeta}}(T)(x_1, \ldots, x_k, \ldots, x_n, 1)\\
&= \left\langle G_{\vec{\zeta}}(T)\right\rangle_{1,n+1}(x_1, \ldots, x_k, \ldots, x_n).
\end{aligned}$$

Because $v_{\vec{\lambda}} \cdot v_{\vec{\zeta}} = -1$ it follows that

$$b \cdot v_{\vec{\lambda}} \cdot \langle G_{\vec{\lambda}}(T)\rangle_{1,n+1} + b \cdot v_{\vec{\zeta}} \cdot \left\langle G_{\vec{\zeta}}(T)\right\rangle_{1,n+1} = 0.$$

(Now we compute the bottom of the squares, see again the previous figure.) We still have $j = n+1$. We get for all $(x_1, \ldots, x_n) \in \mathbf{I}^n$ and all $\vec{z} = (z_1, \ldots, z_n) \in \Upsilon^n$:

$$\begin{aligned}
&\langle G_{\vec{z}}(T)\rangle_{0,n+1}(x_1, x_2, \ldots, x_n) = G_{\vec{z}}(T)\,(x_1, x_2, \ldots, x_n, 0)\\
&= T(\eta_{z_1}(x_1, 0), \eta_{z_2}(x_2, 0), \ldots, \eta_{z_n}(x_n, 0)) = T(t_1, t_2, \ldots, t_n),
\end{aligned}$$

with $$t_i := \begin{cases} \frac{1}{3} \cdot x_i & \text{if } z_i = 0, \\ \frac{1}{3} \cdot (2 - x_i) & \text{if } z_i = 1, \\ \frac{1}{3} \cdot (2 + x_i) & \text{if } z_i = 2, \quad \text{for all } i = 1, 2, \ldots, n\,. \end{cases}$$

We define $\vec{e} := (e_1, e_2, \ldots, e_n)$, $\vec{v} := (v_1, v_2, \ldots, v_n)$, by setting for all $i \in \{1, 2, \ldots n\}$:

$$e_i := \begin{cases} 0 & \text{if } z_i = 0 \\ 2 & \text{if } z_i \in \{1, 2\}\,, \end{cases} \qquad v_i := \begin{cases} 1 & \text{if } z_i \in \{0, 2\} \text{ (hence } e_i \in \{0, 2\}) \\ -1 & \text{if } z_i = 1, \text{(hence } e_i = 2)\,. \end{cases}$$

We have $\vec{e} \in \mathcal{E}^n$ and $\vec{v} \in \mathcal{V}_{\vec{e},n}$. With a few calculations it is easy to see that

$$v_{\vec{z}} \cdot \langle G_{\vec{z}}(T)\rangle_{0,n+1} \;=\; \left(\prod_{i=1}^{n} v_i\right) \cdot \|T\|_{\frac{1}{3},\vec{e},\vec{v}}\,.$$

We compare this with the definition of $\mathcal{SD}_n(T)$ in (8). We get

$$\sum_{\vec{z} \in \Upsilon^n} a \cdot v_{\vec{z}} \cdot \langle G_{\vec{z}}(T)\rangle_{0,n+1} \;=\; -a \cdot \mathcal{SD}_n(T).$$

All in all for the fixed $j = n+1$ follows

$$\sum_{\vec{z} \in \Upsilon^n} v_{\vec{z}} \cdot \Big[a \cdot \langle G_{\vec{z}}(T)\rangle_{0,n+1} + b \cdot \langle G_{\vec{z}}(T)\rangle_{1,n+1}\Big] = b \cdot T - a \cdot \mathcal{SD}_n(T). \tag{11}$$

Now let j be an element of $\{1, 2, 3, \ldots, n\}$. Let
$\vec{\zeta_0} := (z_1, z_2, \ldots, z_{j-1}, 0, z_{j+1}, \ldots, z_n)$, $\vec{\zeta_1} := (z_1, z_2, \ldots, z_{j-1}, 1, z_{j+1}, \ldots, z_n) \in \Upsilon^n$,

and for all points $(x_1, x_2, \ldots, x_n) \in \mathbf{I}^n$ we get

$$\begin{aligned}
&\left\langle G_{\vec{\zeta_0}}(T)\right\rangle_{1,j}(x_1, x_2, \ldots, x_n) = G_{\vec{\zeta_0}}(T)(x_1, x_2, \ldots, x_{j-1}, 1, x_j, x_{j+1}, \ldots, x_n)\\
&= T\left(\eta_{z_1}(x_1, x_n), \ldots, \eta_{z_{j-1}}(x_{j-1}, x_n),\ \eta_0(1, x_n), \eta_{z_{j+1}}(x_j, x_n), \ldots, \eta_{z_n}(x_{n-1}, x_n)\right)\\
&= T\left(\eta_{z_1}(x_1, x_n),\ \ldots,\ \frac{1}{3-2\cdot x_n},\ \ldots\right) = T\left(\eta_{z_1}(x_1, x_n),\ \ldots,\ \eta_1(1, x_n), \ldots\right)\\
&= G_{\vec{\zeta_1}}(T)(x_1, x_2, \ldots, x_{j-1}, 1, x_j, x_{j+1}, \ldots, x_n) = \left\langle G_{\vec{\zeta_1}}(T)\right\rangle_{1,j}(x_1, x_2, \ldots, x_n).
\end{aligned}$$

Thus,

$$b \cdot v_{\vec{\zeta_0}} \cdot \left\langle G_{\vec{\zeta_0}}(T)\right\rangle_{1,j} + b \cdot v_{\vec{\zeta_1}} \cdot \left\langle G_{\vec{\zeta_1}}(T)\right\rangle_{1,j} = 0.$$

(See the previous figure: the right hand side of $T(\eta_0)$ cancels the right hand side of $-T(\eta_1)$.) With the same fixed number j we calculate with the tuples

$$\vec{\zeta_1} = (z_1, z_2, \ldots, z_{j-1}, 1, z_{j+1}, \ldots, z_n), \vec{\zeta_2} := (z_1, z_2, \ldots, z_{j-1}, 2, z_{j+1}, \ldots, z_n) \in \Upsilon^n$$

to obtain

$$\begin{aligned}
&\left\langle G_{\vec{\zeta_1}}(T)\right\rangle_{0,j}(x_1, x_2, \ldots, x_n)\\
&= G_{\vec{\zeta_1}}(T)(x_1, x_2, \ldots, x_{j-1}, 0, x_j, x_{j+1}, \ldots, x_n)\\
&= T(\eta_{z_1}(x_1, x_n), \ldots\ldots, \eta_1(0, x_n), \ldots\ldots, \eta_{z_n}(x_{n-1}, x_n))\\
&= T(\ldots\ldots, t_j, \ldots\ldots) \qquad \text{with } t_j := \begin{cases} \frac{2}{3-2\cdot x_n} & \text{if } x_n \in \left[0, \frac{1}{2}\right], \\ 1 & \text{if } x_n \in \left[\frac{1}{2}, 1\right], \end{cases}\\
&= T(\ldots\ldots, \eta_2(0, x_n), \ldots\ldots)\\
&= G_{\vec{\zeta_2}}(T)(x_1, x_2, \ldots, x_{j-1}, 0, x_j, x_{j+1}, \ldots, x_n)\\
&= \left\langle G_{\vec{\zeta_2}}(T)\right\rangle_{0,j}(x_1, x_2, \ldots, x_n).
\end{aligned}$$

Hence,

$$a \cdot v_{\vec{\zeta_1}} \cdot \left\langle G_{\vec{\zeta_1}}(T)\right\rangle_{0,j} + a \cdot v_{\vec{\zeta_2}} \cdot \left\langle G_{\vec{\zeta_2}}(T)\right\rangle_{0,j} = 0.$$

(See the previous figure again: The left side of $-T(\eta_1)$ cancels the left side of $T(\eta_2)$.) Now we take again $\vec{\zeta_0} = (z_1, z_2, \ldots, z_{j-1}, 0, z_{j+1}, \ldots, z_n) \in \Upsilon^n$, define

$$\vec{\mu} = (z_1, z_2, \ldots, z_{j-1}, z_{j+1}, \ldots, z_n) \in \Upsilon^{n-1},$$

and we get for $(x_1, \ldots, x_{j-1}, x_j, \ldots, x_n) \in \mathbf{I}^n$:

$$\begin{aligned}
&\left\langle G_{\vec{\zeta}_0}(T)\right\rangle_{0,j}(x_1, \ldots, x_{j-1}, x_j, \ldots, x_n)\\
&= G_{\vec{\zeta}_0}(T)(x_1, \ldots, x_{j-1}, 0, x_j, \ldots, x_n)\\
&= T(\eta_{z_1}(x_1, x_n), \ldots, \eta_{z_{j-1}}(x_{j-1}, x_n), \eta_0(0, x_n), \eta_{z_{j+1}}(x_j, x_n), \ldots, \eta_{z_n}(x_{n-1}, x_n))\\
&= T(\ldots\ldots, \eta_{z_{j-1}}(x_{j-1}, x_n), 0,\ \eta_{z_{j+1}}(x_j, x_n), \ldots\ldots)\\
&= \langle T\rangle_{0,j}(\eta_{z_1}(x_1, x_n), \ldots\ldots,\ \eta_{z_{j-1}}(x_{j-1}, x_n), \eta_{z_{j+1}}(x_j, x_n), \ldots\ldots)\\
&= G_{\vec{\mu}}\left(\langle T\rangle_{0,j}\right)(x_1, \ldots, x_{j-1}, x_j, \ldots, x_n).
\end{aligned}$$

Hence

$$a \cdot \left\langle G_{\vec{\zeta}_0}(T)\right\rangle_{0,j} = a \cdot G_{\vec{\mu}}\left(\langle T\rangle_{0,j}\right).$$

And, last but not least, we can show in the same way that

$$b \cdot \left\langle G_{\vec{\zeta}_2}(T)\right\rangle_{1,j} = b \cdot G_{\vec{\mu}}\left(\langle T\rangle_{1,j}\right)$$

for $\vec{\zeta}_2 = (z_1, z_2, \ldots, z_{j-1}, 2, z_{j+1}, \ldots, z_n) \in \Upsilon^n$.

Now we have collected all the needed facts to confirm for every $n \in \mathbb{N}$ the equation

$$(\partial_{n+1} \circ \Theta_n)(T) = (-1)^{n+2} \cdot (b \cdot Id - a \cdot \mathcal{SD}_n)(T) + (\Theta_{n-1} \circ \partial_n)(T).$$

Therefore, if we use a chain u instead of T, the equation (9) is proved, because all used maps are linear. It is a trivial consequence that for a cycle u (i.e. $\partial_n(\mathsf{u}) = 0$) the equation $[a \cdot \mathcal{SD}_n(\mathsf{u})]_\sim = [b \cdot \mathsf{u}]_\sim$ follows on the level of homology classes, because $b \cdot \mathsf{u} - a \cdot \mathcal{SD}_n(\mathsf{u})$ is in the image of ∂_{n+1}.

The next step is to show that for a cycle u the equation $[b \cdot \mathcal{SD}_n(\mathsf{u})]_\sim = [a \cdot \mathsf{u}]_\sim$ also holds on the level of homology classes. Looking at the previous proof this seems obvious, and we shall not explain it in all details. The proof is nearly the same, we only have to modify it by 'turning it upside down'. Instead of using the three auxiliary functions η_0, η_1, η_2, we need three others $\widetilde{\eta_0}, \widetilde{\eta_1}, \widetilde{\eta_2} : \mathbf{I}^2 \to \mathbf{I}$.

For $x, y \in [0,1]$ define:

$$\widetilde{\eta_0}(x, y) := \frac{x}{1 + 2 \cdot y},$$

$$\widetilde{\eta_1}(x, y) := \begin{cases} \frac{2-x}{1+2\cdot y} & \text{for } y \geqslant \frac{1}{2} - \frac{1}{2} \cdot x \quad, \\ 1 & \text{else} \quad, \end{cases}$$

$$\widetilde{\eta_2}(x, y) := \begin{cases} \frac{2+x}{1+2\cdot y} & \text{for } y \geqslant \frac{1}{2} + \frac{1}{2} \cdot x \quad, \\ 1 & \text{else} \quad. \end{cases}$$

Then $\widetilde{\eta_0}, \widetilde{\eta_1}, \widetilde{\eta_2}$ are continuous. For a fixed tuple $\vec{z} = (z_1, z_2, \ldots, z_n) \in \Upsilon^n$ and for $T \in \mathcal{S}_n(X)$ let us define the map $\widetilde{G_{\vec{z}}}(T) : \mathbf{I}^{n+1} \to X$ by setting for all $(x_1, \ldots, x_n, x_{n+1}) \in \mathbf{I}^{n+1}$:

$$\widetilde{G_{\vec{z}}}(T)(x_1, \ldots, x_n, x_{n+1}) := T\left(\widetilde{\eta_{z_1}}(x_1, x_{n+1}), \widetilde{\eta_{z_2}}(x_2, x_{n+1}), \ldots, \widetilde{\eta_{z_n}}(x_n, x_{n+1})\right).$$

Thus, $\widetilde{G_{\vec{z}}}(T) \in \mathcal{S}_{n+1}(X)$. Let for $\vec{z} = (z_1, z_2, \ldots, z_n) \in \Upsilon^n$ the sign $v_{\vec{z}} := (-1)^{\sum_{i=1}^n z_i}$

as before, and finally define

$$\widetilde{\Theta_n}(T) := \sum_{\vec{z}\in\Upsilon^n} v_{\vec{z}} \cdot \widetilde{G_{\vec{z}}}(T).$$

By a similar calculation as for the proof of equation (9) we show:

$$(\partial_{n+1} \circ \widetilde{\Theta_n})(T) = (-1)^{n+2} \cdot (a \cdot Id - b \cdot \mathcal{SD}_n)(T) + (\widetilde{\Theta_{n-1}} \circ \partial_n)(T), \tag{12}$$

and by using a cycle u instead of the map T, it leads directly to the desired formula

$$[b \cdot \mathcal{SD}_n(\mathsf{u})]_\sim = [a \cdot \mathsf{u}]_\sim.$$

Now let for all $k \in \mathbb{N}, n \in \mathbb{N}_0$ and all chains $\mathsf{u} \in \mathcal{K}_n(X)$

$$\mathcal{SD}_n^{(k)}(\mathsf{u}) := (\mathcal{SD}_n \circ \mathcal{SD}_n \circ \ldots \circ \mathcal{SD}_n)(\mathsf{u}), \text{ (with } k \text{ factors } \mathcal{SD}_n).$$

Lemma 4. *For all $k \in \mathbb{N}, n \in \mathbb{N}_0$, and all $\mathsf{u} \in kernel(\partial_n)$ we have*

$$[a^k \cdot \mathcal{SD}_n^{(k)}(\mathsf{u})]_\sim = [b^k \cdot \mathsf{u}]_\sim \quad \textit{and} \quad [b^k \cdot \mathcal{SD}_n^{(k)}(\mathsf{u})]_\sim = [a^k \cdot \mathsf{u}]_\sim .$$

Proof. We prove the first equation by induction on k. Note that, if $\partial_n(\mathsf{u}) = 0$, also $\partial_n(\mathcal{SD}_n^{(k)}(\mathsf{u})) = 0$ (because $\mathcal{SD}_n$ commutes with the boundary operator), and note that $\mathcal{SD}_n$ is a linear map. Assume for some $k \in \mathbb{N}$ for an $\mathsf{u} \in$ kernel(∂_n) : $[a^k \cdot \mathcal{SD}_n^{(k)}(\mathsf{u})]_\sim = [b^k \cdot \mathsf{u}]_\sim$. Further let $\mathsf{w} := a^k \cdot \mathcal{SD}_n^{(k)}(\mathsf{u})$, hence w is a cycle, too. Thus we get

$$\begin{aligned}[b^{k+1} \cdot \mathsf{u}]_\sim &= [b \cdot b^k \cdot \mathsf{u}]_\sim = [b \cdot a^k \cdot \mathcal{SD}_n^{(k)}(\mathsf{u})]_\sim = [b \cdot \mathsf{w}]_\sim = [a \cdot \mathcal{SD}_n(\mathsf{w})]_\sim \\ &= [a \cdot a^k \cdot \mathcal{SD}_n(\mathcal{SD}_n^{(k)}(\mathsf{u}))]_\sim = [a^{k+1} \cdot \mathcal{SD}_n^{(k+1)}(\mathsf{u})]_\sim.\end{aligned}$$

This proves the first equation of the lemma for all $k \in \mathbb{N}$. □

Lemma 5. *For all weights $\vec{m} = (a, b) \in \mathcal{R}^2$, (a, b) has the property $\mathcal{NCD}$, and for all $k \in \mathbb{N}$ there is an element $r_k \in \mathcal{R}$ such that the equation $r_k \cdot [\mathcal{SD}_n^{(k)}(\mathsf{u})]_\sim = [\mathsf{u}]_\sim$ holds for all $\mathsf{u} \in$ kernel(∂_n) and $n \in \mathbb{N}_0$ on the level of homology classes.*

Proof. The property $\mathcal{NCD}$ means that for all $k \in \mathbb{N}$ there are $x_k, y_k \in \mathcal{R}$ such that $x_k \cdot a^k + y_k \cdot b^k = 1_\mathcal{R}$. Now set $r_k := x_k \cdot b^k + y_k \cdot a^k$. Take the previous Lemma 4 and write

$$\begin{aligned}[\mathsf{u}]_\sim = [1_\mathcal{R} \cdot \mathsf{u}]_\sim &= [(x_k \cdot a^k + y_k \cdot b^k) \cdot \mathsf{u}]_\sim = x_k \cdot [a^k \cdot \mathsf{u}]_\sim + y_k \cdot [b^k \cdot \mathsf{u}]_\sim \\ &= x_k \cdot [b^k \cdot \mathcal{SD}_n^{(k)}(\mathsf{u})]_\sim + y_k \cdot [a^k \cdot \mathcal{SD}_n^{(k)}(\mathsf{u})]_\sim \\ &= (x_k \cdot b^k + y_k \cdot a^k) \cdot [\mathcal{SD}_n^{(k)}(\mathsf{u})]_\sim = r_k \cdot [\mathcal{SD}_n^{(k)}(\mathsf{u})]_\sim,\end{aligned}$$

and the lemma is proved. □

Now let's return to the statement of Proposition 1. We still are proving that for all $n \in \mathbb{N}$ the canonical inclusions $\mathcal{K}_n(X,\mathcal{U}) \overset{j}{\hookrightarrow} \mathcal{K}_n(X)$ induce isomorphisms

$$j_* : {}_{\vec{m}}\mathcal{H}_n(X, \mathcal{U}) \xrightarrow{\cong} {}_{\vec{m}}\mathcal{H}_n(X).$$

Note that, for a continuous $T : \mathbf{I}^n \to X$, the image $T(\mathbf{I}^n)$ is compact in X, and for the given open covering $\{Int(U_i) \mid i \in \Im\}$ of X a finite subset is sufficient to cover $T(\mathbf{I}^n)$. Further note that the diameters of the 3^n elements of the chain

$\mathcal{SD}_n(T)(\mathbf{I}^n)$ decrease to a third, compared with the diameter of $T(\mathbf{I}^n)$. Hence, by iterating $\mathcal{SD}_n$, there is a number $k_T \in \mathbb{N}$ such that $\mathcal{SD}_n^{(k_T)}(T) \in \mathcal{K}_n(X,\mathcal{U})$. And therefore for a chain $\mathsf{u} \in \mathcal{K}_n(X)$ (which is a finite linear combination of some $T's$), a number $k_{\mathsf{u}} \in \mathbb{N}$ exists such that $\mathcal{SD}_n^{(k_{\mathsf{u}})}(\mathsf{u}) \in \mathcal{K}_n(X,\mathcal{U})$.

Now let us look on the inclusion $j : \mathcal{K}_n(X,\mathcal{U}) \hookrightarrow \mathcal{K}_n(X)$, and the induced $\mathcal{R}$-module morphism $j_* : {}_{\vec{m}}\mathcal{H}_n(X, \mathcal{U}) \longrightarrow {}_{\vec{m}}\mathcal{H}_n(X)$. We have to show that j_* is an epimorphism and a monomorphism. (compare [**9**, p.36]).

j_* is an epimorphism:
Let $\mathbf{z} \in {}_{\vec{m}}\mathcal{H}_n(X)$. Then it follows that there is a chain $\mathsf{u} \in$ kernel$(\partial_n) \subset \mathcal{K}_n(X)$, with $[\mathsf{u}]_\sim = \mathbf{z}$. Hence we can deduce that there is a $k \in \mathbb{N}$ with $\mathcal{SD}_n^{(k)}(\mathsf{u}) \in \mathcal{K}_n(X,\mathcal{U})$. Take the factor $r_k \in \mathcal{R}$ (see Lemma 5) and write

$$[r_k \cdot \mathcal{SD}_n^{(k)}(\mathsf{u})]_\sim = [\mathsf{u}]_\sim \in {}_{\vec{m}}\mathcal{H}_n(X) \quad \text{with } r_k \cdot \mathcal{SD}_n^{(k)}(\mathsf{u}) \in \mathcal{K}_n(X,\mathcal{U}).$$

Hence $j_*([r_k \cdot \mathcal{SD}_n^{(k)}(\mathsf{u})]_\sim) = [j(r_k \cdot \mathcal{SD}_n^{(k)}(\mathsf{u}))]_\sim = [r_k \cdot \mathcal{SD}_n^{(k)}(\mathsf{u})]_\sim = [\mathsf{u}]_\sim = \mathbf{z}$.

j_* is a monomorphism:
Let $\mathbf{x} \in {}_{\vec{m}}\mathcal{H}_n(X,\mathcal{U})$ with $j_*(\mathbf{x}) = 0$. We must show that $\mathbf{x} = 0$.

For every $\mathbf{x} \in {}_{\vec{m}}\mathcal{H}_n(X, \mathcal{U})$ exists a cycle $\mathsf{v} \in \mathcal{K}_n(X,\mathcal{U})$ with $[\mathsf{v}]_\sim = \mathbf{x}$. We must show that v is a boundary, i.e. we have to show that there is a $\mathsf{w} \in \mathcal{K}_{n+1}(X,\mathcal{U})$ with $\partial_{n+1}(\mathsf{w}) = \mathsf{v}$. We have

$$[j(\mathsf{v})]_\sim = j_*([\mathsf{v}]_\sim) = j_*(\mathbf{x}) = 0 \in {}_{\vec{m}}\mathcal{H}_n(X).$$

The assumption $j_*(\mathbf{x}) = 0 \in {}_{\vec{m}}\mathcal{H}_n(X)$ means that $j_*(\mathbf{x})$ is the equivalence class of a cycle which is a boundary, i. e. that there is a chain $\widehat{\mathsf{w}} \in \mathcal{K}_{n+1}(X)$ with $\partial_{n+1}(\widehat{\mathsf{w}}) = \mathsf{v}$.

Choose a sufficient large number $k \in \mathbb{N}$ such that $\mathcal{SD}_{n+1}^{(k)}(\widehat{\mathsf{w}}) \in \mathcal{K}_{n+1}(X,\mathcal{U})$, and take the element $r_k \in \mathcal{R}$ from Lemma 5 and we have $[r_k \cdot \mathcal{SD}_n^{(k)}(\mathsf{v})]_\sim = [\mathsf{v}]_\sim$, (and note that $\mathcal{SD}_n^{(k)}(\mathsf{v}) \in \mathcal{K}_n(X,\mathcal{U})$ by triviality). Hence

$$\partial_{n+1}(r_k \cdot \mathcal{SD}_{n+1}^{(k)}(\widehat{\mathsf{w}})) = r_k \cdot \mathcal{SD}_n^{(k)}(\partial_{n+1}(\widehat{\mathsf{w}})) = r_k \cdot \mathcal{SD}_n^{(k)}(\mathsf{v}).$$

So we conclude that $r_k \cdot \mathcal{SD}_n^{(k)}(\mathsf{v})$ is a boundary, therefore $[r_k \cdot \mathcal{SD}_n^{(k)}(\mathsf{v})]_\sim = 0$, and since $[r_k \cdot \mathcal{SD}_n^{(k)}(\mathsf{v})]_\sim = [\mathsf{v}]_\sim \in {}_{\vec{m}}\mathcal{H}_n(X, \mathcal{U})$ it follows $[\mathsf{v}]_\sim = 0 = \mathbf{x}$! □

Hence $j_* : {}_{\vec{m}}\mathcal{H}_n(X, \mathcal{U}) \xrightarrow{\cong} {}_{\vec{m}}\mathcal{H}_n(X)$ is an isomorphism, and the proof of Proposition 1 is finished. This proposition leads directly to the excision axiom, see again [**9**, p.30,31].

7. Computing Homology Groups. Dividing by the Degenerate Maps

In the last section we proved the excision axiom for a 'weight' $\vec{m} = (m_0, m_1)$ which has the property $\mathcal{NCD}$. In this way we constructed an extraordinary homology theory. When you read the construction of this homology theory for the first time it seems to be very difficult to compute the homology modules of any space, except for a point. But fortunately there is an old (1968) paper [**2**], which helps us by using the ordinary singular homology theory.

Theorem 4. *For abelian groups $\mathcal{A}$ and all $n \in \mathbb{N}_0$ and all pairs of finite CW-complexes (X, B) let ${}_{S}H_n[(X,B);\mathcal{A}]$ be the nth ordinary singular homology group (with coefficient group ${}_{S}H_0[(point)] \cong \mathcal{A}$). Let $\mathcal{R} := \mathbb{Z}$. Let $\vec{m} = (a,b) \in \mathbb{Z}^2$ with $gcd\{a,b\} = 1$. Then we have for each pair of finite CW-complexes (X, B) and for all $n \in \mathbb{N}_0$:*
If $\{a,b\} = \{1,-1\}$ then there is an isomorphism

$$_{\vec{m}}\mathcal{H}_n(X,B) \cong \sum_{k=0}^{n} {}_{S}H_k[(X,B);\mathbb{Z}].$$

If $\{a,b\} \neq \{1,-1\}$, so that the index $\sigma = a+b \neq 0$, then

$$_{\vec{m}}\mathcal{H}_n(X,B) \cong \begin{cases} \sum_{k\in\{0,1,2,\ldots\}\,\wedge\, 2k\leqslant n} \ {}_{S}H_{2k}[(X,B);\mathbb{Z}_\sigma] & \text{if } n \text{ is even },\\ \sum_{k\in\{0,1,2,\ldots\}\,\wedge\, 2k+1\leqslant n} \ {}_{S}H_{2k+1}[(X,B);\mathbb{Z}_\sigma] & \text{if } n \text{ is odd.}\end{cases}$$

Proof. See [**2**], and use the homology groups of a point, computed in section 3. □

Recall that in section 3 we calculated the homology groups of a point for an arbitrary $\mathcal{R}$, and note that 'our' homology theory $_{\vec{m}}\mathcal{H}_n$ differs from the usual singular homology theory. But, as we announced in the abstract, we can divide the chain modules $\mathcal{K}_n(X)$ by suitable submodules, and in the case of $\vec{m} = (m_0, m_1)$ where (m_0, m_1) has the property $\mathcal{NCD}$, we shall obtain the usual singular homology theory with the coefficient module $\mathcal{R}/(\sigma\mathcal{R})$. Compare [**9**, p.12,13], or [**8**, p.236 ff], where this process is called a *normalization*.

As always, a new part begins with definitions, see the definitions 6 - 10.

Lemma 6. *For all weights $\vec{m} = (m_0, m_1, \ldots, m_L) \in \mathcal{R}^{L+1}$ (hence its index is $\sigma = \sum_{i=0}^{L} m_i$), for $n \in \mathbb{N}$ the boundary operator $\partial_n : \mathcal{K}_n(X) \longrightarrow \mathcal{K}_{n-1}(X)$ yields a map*

$$\partial_{n|\Gamma_{\sigma,n}(X)} : \Gamma_{\sigma,n}(X) \longrightarrow \Gamma_{\sigma,n-1}(X).$$

Proof. Let $\mathsf{u} \in \Gamma_{\sigma,n}(X)$. We know that

$$\mathsf{u} = \mathsf{u}_1 + \mathsf{u}_2, \ \text{ with } \mathsf{u}_1 \in \mathrm{Ideal}_{\sigma,n}(X),\ \mathsf{u}_2 \in \mathcal{K}_{\mathcal{D},n}(X).$$

Hence it follows that $\partial_n(\mathsf{u}_1) \in \mathrm{Ideal}_{\sigma,n-1}(X)$, since ∂_n is linear. And we have $\mathsf{u}_2 = \sum_{k=1}^{p} r_k \cdot T_k$, and all the $T_k's$ are degenerate. Take $T := T_k$, and assume that T is degenerate at the $\widehat{j}^{th}$ component, $\widehat{j} \in \{1,2,\ldots,n\}$, i.e. for all $y, z \in [0,1]$ we have (see Definition 6)

$$T(x_1, x_2, \ldots, x_{\widehat{j}-1}, y, x_{\widehat{j}+1}, \ldots, x_n) = T(x_1, x_2, \ldots, x_{\widehat{j}-1}, z, x_{\widehat{j}+1}, \ldots, x_n).$$

By the definition of $\partial_n(T)$ it follows that

$$\partial_n(T) = \sum_{j\in\{1,2,\ldots n\}\wedge j\neq\widehat{j}} (-1)^{j+1}\cdot\sum_{i=0}^{L} m_i\cdot\langle T\rangle_{n,\,i,\,j} \ + \ (-1)^{\widehat{j}+1}\cdot\sum_{i=0}^{L} m_i\cdot\langle T\rangle_{n,\,i,\,\widehat{j}}\ .$$

The first summand is a linear combination of degenerate maps. We compute the second summand. For a point $(x_1, x_2, \ldots, x_{n-1}) \in \mathbf{I}^{n-1}$ we have:

$$\begin{aligned}
&\sum_{i=0}^{L} m_i \cdot \langle T\rangle_{n,\,i,\,\hat{j}}\,(x_1, x_2, \ldots, x_{\hat{j}-1}, x_{\hat{j}}, \ldots, x_{n-1}) \\
&= \sum_{i=0}^{L} m_i \cdot\, T\left(x_1, x_2, \ldots, x_{\hat{j}-1}, \frac{i}{L}, x_{\hat{j}}, \ldots, x_{n-1}\right) \\
&= T(x_1, \ldots, x_{\hat{j}-1}, *, x_{\hat{j}}, \ldots, x_{n-1}) \cdot \sum_{i=0}^{L} m_i \\
&= T(x_1, \ldots, x_{\hat{j}-1}, *, x_{\hat{j}}, \ldots, x_{n-1}) \cdot \sigma.
\end{aligned}$$

Thus, the second summand is an element of $\mathrm{Ideal}_{\sigma,n-1}(X)$. □

In Definition 10 we defined the quotient $\mathcal{R}$-module $\mathcal{K}_n(X,A)_{\sim\Gamma,\sigma}$. By the previous Lemma 6 the boundary operators $(\partial_n)_{n\geqslant 0}$ yield a chain complex

$$\cdots\cdots \xrightarrow{\partial_{n+1}} \mathcal{K}_n(X,A)_{\sim\Gamma,\sigma} \xrightarrow{\partial_n} \mathcal{K}_{n-1}(X,A)_{\sim\Gamma,\sigma} \xrightarrow{\partial_{n-1}} \cdots \xrightarrow{\partial_0} \{0\}.$$

This leads to homology $\mathcal{R}$-modules as usual, ${}_{\vec{m}}\mathcal{H}_{n/\Gamma}(X,A) := \frac{kernel(\partial_n)}{image(\partial_{n+1})}$, for $n \in \mathbb{N}_0$.

Example: In section 3 we calculated the homology groups for a one-point space $\{p\}$. We had for $\sigma = 0$ that ${}_{\vec{m}}\mathcal{H}_n(p) \cong \mathcal{R}$ for all $n \in \mathbb{N}_0$, and for arbitrary indexes σ we got:

$${}_{\vec{m}}\mathcal{H}_n(p) \cong \begin{cases} \{x \in \mathcal{R} \,|\, \sigma \cdot x = 0\} & \text{if } n \text{ is odd} \\ \mathcal{R}/(\sigma \cdot \mathcal{R}) & \text{if } n \text{ is even.} \end{cases}$$

For the space $\{p\}$ and for $n \in \mathbb{N}$ the single map $T : \mathbf{I}^n \to \{p\}$ is degenerate, but $T : \mathbf{I}^0 \to \{p\}$ is not. Hence $\Gamma_{\sigma,0}(p) = \mathrm{Ideal}_{\sigma,0}(p) \cong \sigma \cdot \mathcal{R}$, thus the generating chain complex $\quad {}_{\vec{m}}\mathcal{K}_*(p) \;=\; \cdots \xrightarrow{\partial_4} \mathcal{K}_3(p) \xrightarrow{\partial_3} \mathcal{K}_2(p) \xrightarrow{\partial_2} \mathcal{K}_1(p) \xrightarrow{\partial_1} \mathcal{K}_0(p) \xrightarrow{\partial_0} \{0\},$

$$\text{i. e.} \quad {}_{\vec{m}}\mathcal{K}_*(p) \;\cong\; \cdots \xrightarrow{\partial_4} \mathcal{R} \xrightarrow{\partial_3} \mathcal{R} \xrightarrow{\partial_2} \mathcal{R} \xrightarrow{\partial_1} \mathcal{R} \xrightarrow{\partial_0} \{0\},$$

turns, by dividing for each $n \in \mathbb{N}_0$ by $\Gamma_{\sigma,n}(p)$, into

$$\begin{aligned}
&\cdots \xrightarrow{\partial_4} \mathcal{K}_3(p)_{\sim\Gamma,\sigma} \xrightarrow{\partial_3} \mathcal{K}_2(p)_{\sim\Gamma,\sigma} \xrightarrow{\partial_2} \mathcal{K}_1(p)_{\sim\Gamma,\sigma} \xrightarrow{\partial_1} \mathcal{K}_0(p)_{\sim\Gamma,\sigma} \xrightarrow{\partial_0} \{0\} \\
\cong\; &\cdots \xrightarrow{\partial_4} \{0\} \xrightarrow{\partial_3} \{0\} \xrightarrow{\partial_2} \{0\} \xrightarrow{\partial_1} \mathcal{R}/(\sigma \cdot \mathcal{R}) \xrightarrow{\partial_0} \{0\}.
\end{aligned}$$

$$\text{Hence it follows that} \qquad {}_{\vec{m}}\mathcal{H}_{n/\Gamma}(p) \cong \begin{cases} \{0\} & \text{for } n \in \mathbb{N} \\ \mathcal{R}/(\sigma \cdot \mathcal{R}) & \text{for } n = 0. \end{cases}$$

Corollary 1. *If we take a weight $\vec{m} = (a,b) \in \mathcal{R}^2$, and if (a,b) has the property $\mathcal{NCD}$, the homology theory ${}_{\vec{m}}\mathcal{H}_{/\Gamma} := ({}_{\vec{m}}\mathcal{H}_{n/\Gamma})_{n\geqslant 0}$ is isomorphic to the ordinary singular homology theory on all pairs of finite CW-complexes, and we have a coefficient module ${}_{\vec{m}}\mathcal{H}_{0/\Gamma}(p) \cong \mathcal{R}/(\sigma \cdot \mathcal{R})$, with $\sigma = a + b$.*

Proof. As we proved above, the homology theory ${}_{\vec{m}}\mathcal{H}$ fulfils all of the Eilenberg-Steenrod axioms except one. That means the axioms of exactness, homotopy and excision are satisfied. By dividing the chain modules $\mathcal{K}_n(X, A)$ by $\Gamma_{\sigma,n}(X, A)$ and using the boundary operator ∂_n we get the homology modules ${}_{\vec{m}}\mathcal{H}_{n/\Gamma}(X, A)$, and the dimension axiom will be added, while the other three axioms remain. Thus, with the uniqueness theorem proved by Eilenberg and Steenrod, we have the uniqueness of the homology groups for all finite CW-complexes (X, A). See [**6**, p.51 ff], or [**3**, p.100 ff]. □

Corollary 2. *The usual singular homology theory is a special case of the class which is developed here. If we take the weight* $\vec{m} := (1, -1) \in \mathbb{Z}^2$, *the homology theory* ${}_{\vec{m}}\mathcal{H}_{/\Gamma}$ *is isomorphic to the usual singular homology theory on all pairs of finite CW-complexes, and the coefficient group is* $\mathbb{Z}$.

Proof. By the previous Corollary 1. Or see for the last time [**9**, p.11-37]. □

8. Final Suggestion

We cannot decide whether this new homology theory has any important application. Perhaps it might be an interesting tool for other mathematicians. It is easy to see one difficulty: The computation of the homology modules of the one-point space is very simple. But to do the same for other topological spaces might be more complicate (except for finite CW-complexes, see Theorem 4), although the homotopy axiom and the excision axiom and the exactness axiom hold. There are not enough $\{0\}$'s in the homology modules of a point, e.g. for $\mathcal{R} := \mathbb{Z}$ and $\sigma \neq 0$ only every second is the trivial group $\{0\}$. So it would be an improvement for a better application if we can increase the number of $\{0\}$'s.

If we consider for all $n \in \mathbb{N}_0$ the canonical quotient map t_n,

$$t_n : \mathcal{K}_n(X) \longrightarrow \frac{\mathcal{K}_n(X)}{\Gamma_{\sigma,n}(X)}, \quad \text{then the following diagram commutes :}$$

$$\begin{array}{ccccccccc}
\cdots & \xrightarrow{\partial_3} & \mathcal{K}_2(X) & \xrightarrow{\partial_2} & \mathcal{K}_1(X) & \xrightarrow{\partial_1} & \mathcal{K}_0(X) & \xrightarrow{\partial_0} & \{0\} \\
 & & \downarrow t_2 & & \downarrow t_1 & & \downarrow t_0 & & \downarrow 0 \\
\cdots & \xrightarrow[\partial_3]{} & \mathcal{K}_2(X)_{\sim\Gamma,\sigma} & \xrightarrow[\partial_2]{} & \mathcal{K}_1(X)_{\sim\Gamma,\sigma} & \xrightarrow[\partial_1]{} & \mathcal{K}_0(X)_{\sim\Gamma,\sigma} & \xrightarrow[\partial_0]{} & \{0\}
\end{array}$$

Let β be a fixed element from the set $\mathbb{N}_0 \cup \{\infty\}$. Then we generate a chain complex

$$\cdots\cdots \xrightarrow{\overline{\partial_{n+1}}} \overline{{}_\beta\mathcal{K}_n(X)} \xrightarrow{\overline{\partial_n}} \overline{{}_\beta\mathcal{K}_{n-1}(X)} \xrightarrow{\overline{\partial_{n-1}}} \cdots \xrightarrow{\overline{\partial_1}} \overline{{}_\beta\mathcal{K}_0(X)} \xrightarrow{\overline{\partial_0}} \{0\}$$

if we define for all $n \in \mathbb{N}_0$:

$$\overline{{}_\beta\mathcal{K}_n(X)} := \begin{cases} \mathcal{K}_n(X) & \text{if } n \geqslant \beta \\ \mathcal{K}_n(X)_{\sim\Gamma,\sigma} & \text{if } 0 \leqslant n < \beta. \end{cases}$$

If we have chosen a number $\beta \in \mathbb{N}$, then let $\overline{\partial_\beta} : \mathcal{K}_\beta(X) \to \mathcal{K}_{\beta-1}(X)_{\sim\Gamma,\sigma}$ be the βth boundary operator by defining $\overline{\partial_\beta} := \partial_\beta \circ t_\beta = t_{\beta-1} \circ \partial_\beta$. This means that for the special cases $\beta = 0$ and $\beta = \infty$ we have for all $n \in \mathbb{N}_0$:

$$\overline{{}_0\mathcal{K}_n(X)} = \mathcal{K}_n(X) \quad \text{and} \quad \overline{{}_\infty\mathcal{K}_n(X)} = \mathcal{K}_n(X)_{\sim\Gamma,\sigma}, \quad \text{respectively.}$$

For arbitrary weights $\vec{m}$ we define for $n \in \mathbb{N}_0$: ${}_{\vec{m},\beta}\mathcal{H}_n(X) := \frac{kernel(\overline{\partial_n})}{image(\overline{\partial_{n+1}})}$, and we get sequences ${}_{\vec{m},\beta}\mathcal{H} := ({}_{\vec{m},\beta}\mathcal{H}_n)_{n\geqslant 0}$ of homology modules; with the two we have developed here as special cases, i.e. for $\beta = 0$ and $\beta = \infty$ we get

$${}_{\vec{m},0}\mathcal{H} = {}_{\vec{m}}\mathcal{H} \quad \text{and} \quad {}_{\vec{m},\infty}\mathcal{H} = {}_{\vec{m}}\mathcal{H}_{/\Gamma}.$$

Example: For the one-point space $\{p\}$ (which is our favourite topological space obviously) and $\mathcal{R} := \mathbb{Z}$ and for the weight $\vec{m} := (1,4)$ and $\beta := 7$ or $\beta := 8$ we obtain for $n \in \mathbb{N}_0$:

$${}_{(1,4),7}\mathcal{H}_n(p) \cong \begin{cases} \mathbb{Z}_5 & \text{for} \quad n \in \{0,8,10,12,14,\ldots\} \\ \mathbb{Z} & \text{for} \quad n = 7 \\ \{0\} & \text{for} \quad n \in \{1,\ldots,6,9,11,13,15,\ldots\}, \end{cases}$$

$${}_{(1,4),8}\mathcal{H}_n(p) \cong \begin{cases} \mathbb{Z}_5 & \text{for} \quad n \in \{0,8,10,12,14,\ldots\} \\ \{0\} & \text{for} \quad n \in \{1,\ldots,7,9,11,13,15,\ldots\}. \end{cases}$$

References

[1] R. Brown, P. J. Higgins and R. Sivera, *Nonabelian algebraic topology: filtered spaces, crossed complexes, cubical homotopy groupoids.* EMS Tracts in Mathematics Vol. 15 (to appear 2011).

[2] R. O. Burdick, P. E. Conner and E. E. Floyd, *Chain Theories and Their Derived Homology*, Proc. Amer. Math. Soc., Vol. 19, No. 5 (1968), 1115-1118.

[3] S. Eilenberg and N. Steenrod, *Foundations of Algebraic Topology*, Princeton, 1952.

[4] A. Hatcher, *Algebraic Topology*, Cambridge University Press, 2001.

[5] P. J. Hilton and S. Wylie, *Homology theory: An introduction to algebraic topology*, Cambridge University Press, New York, 1960.

[6] S. T. Hu, *Homology Theory*, Holden-Day, 1966.

[7] T. Kaczynski, K. Mischaikow and M. Mrozek, *Computational Homology*, Springer, 2004.

[8] S. Mac Lane, *Homology*, Springer, 1963, Fourth Printing 1994.

[9] W. S. Massey, *Singular Homology Theory*, Springer, 1980.

[10] J. Rotman, *An Introduction to Algebraic Topology*, Springer, 1988.

[11] R. Schön, *Acyclic Models and Excision*, Proc. Amer. Math. Soc., Vol. 59, No. 1 (1976), 167-168.

[12] E. H. Spanier, *Algebraic Topology*, Springer, 1966.

[13] J. W. Vick, *Homology Theory*, Springer, 1994.

This article may be accessed via WWW at `http://tcms.org.ge/Journals/JHRS/`

Volker W. Thürey
volker@thuerey.de

Germany
28199 Bremen
Rheinstr. 91
T: 49 (0)421/591777

Journal of Homotopy and Related Structures, vol. 6(1), 2011, pp.103–112

VANISHING OF UNIVERSAL CHARACTERISTIC CLASSES FOR HANDLEBODY GROUPS AND BOUNDARY BUNDLES

JEFFREY GIANSIRACUSA AND ULRIKE TILLMANN

(*communicated by Michael Weiss*)

Abstract

Using certain Thom spectra appearing in the study of cobordism categories, we show that the odd half of the Miller-Morita-Mumford classes on the mappping class group of a surface with negative Euler characteristic vanish in integral cohomology when restricted to the handlebody subgroup. This is a special case of a more general theorem valid in all dimensions: universal characteristic classes made from monomials in the Pontrjagin classes (and even powers of the Euler class) vanish when pulled back from $B\mathrm{Diff}(\partial W)$ to $B\mathrm{Diff}(W)$.

1. Introduction

Let Σ_g denote a closed oriented surface of genus g. Its mapping class group $\Gamma_g := \pi_0\mathrm{Diff}(\Sigma_g)$ is the group of connected components of its group of orientation preserving diffeomorphisms $\mathrm{Diff}(\Sigma_g)$. Miller, Morita, and Mumford [**Mil86, Mor87, Mum83**] defined characteristic classes, known as the MMM classes, $\kappa_i \in H^{2i}(\Gamma_g;\mathbb{Z})$. By the proof of the Mumford conjecture [**MW07**] these classes freely generate the rational cohomology ring in degrees increasing with g:

$$\lim_{g\to\infty} H^*(\Gamma_g;\mathbb{Q}) \simeq \mathbb{Q}[\kappa_1,\kappa_2,\ldots].$$

The mapping class group of a surface has various interesting subgroups, and it is a natural question to ask how the MMM-classes restrict to these subgroups. Here we will be interested in the handlebody subgroup H_g. To define it, fix a handlebody W with boundary $\partial W = \Sigma_g$. H_g contains those mapping classes of Σ_g that can be extended across the interior of W.

Theorem A. *For $g \geqslant 2$, the odd MMM-classes $\kappa_{2i+1} \in H^{4i+2}(\Gamma_g;\mathbb{Z})$ vanish when restricted to the handlebody subgroup $H_g \subset \Gamma_g$.*

Remark 1.1. It is well-known that the analogue of Theorem A holds rationally for the Torelli group $I_g := \ker(\Gamma_g \to \mathrm{Aut}(H_1(\Sigma_g;\mathbb{Z})))$. This can be proved by index theory, see [**Mor87, Mum83**]. It remains a significant open problem whether the even kappa classes restrict non-trivially to the Torelli group.

The first author thanks Oscar Randal-Williams for helpful discussions.
Received August 12, 2010, revised February 17, 2011; published on March 27, 2011.
2000 Mathematics Subject Classification: 57R20, 55R40, 57R90.
Key words and phrases: mapping class group, cobordism theory, generalized MMM-classes.

Motivated by these questions, Sakasai [**Sak09**] has recently proved a result closely related to Theorem A by rather different methods. He shows that in a stable range the odd kappa classes rationally vanish when restricted to the Lagrangian mapping class subgroup $L_g := H_g I_g$. As our result holds without restriction to the stable range and integrally, the question arises whether the same holds also for I_g and L_g.

Remark 1.2. Recently (and after the completion of this work), Hatcher has announced an analogue of the Madsen-Weiss theorem [**MW07**] for the handlebody mapping class group. The proof is an adaptation of the Galatius's proof [**Gal**] of the analogue of the Madsen-Weiss theorem for automorphism groups of free groups. Hatcher determines the cohomology of H_g in the stable range, which by [**HW**] is $(g-4)/2$, as that of a component of $QBSO(3)_+$. In view of Proposition 2.2 below, Hatcher's result implies Theorem A for the stable range and also implies that the even MMM classes freely generate the cohomology ring of H_g in the stable range.

Theorem A is a special case of the more general Theorem B below which is a statement about the diffeomorphism groups of manifolds of any dimension. Recall that for $g \geqslant 2$, the mapping class group Γ_g is homotopy equivalent to the diffeomorphism group $\mathrm{Diff}(\Sigma_g)$ [**EE69**], and the handlebody subgroup is homotopy equivalent to the diffeomorphism group of a 3-dimensional handlebody of genus g [**Hat76, Hat99**]. Thus the discrete mapping class groups may be replaced by the diffeomorphism groups. The more general result is about how generalizations of the MMM-classes are pulled back in cohomology under the restriction-to-the-boundary map,

$$r : B\mathrm{Diff}(W) \to B\mathrm{Diff}(\partial W),$$

where W is a $(d+1)$-dimensional manifold with boundary $\partial W = M$.

More precisely, let $\pi : E \to B$ be an oriented fibre bundle with closed fibres M of dimension d, and let $T^{\pi}E \to E$ denote the fibrewise tangent bundle. The generalized MMM classes (or universal tangential classes) are defined by taking a monomial X in the Euler class e and the Pontrjagin classes p_i of $T^{\pi}E$ and then forming the pushforward

$$\widehat{X}(E) := \pi_! X(T^{\pi}E) \in H^*(B;\mathbb{Z}),$$

where $\pi_! : H^*(E) \to H^{*-d}(B)$ is the Gysin map of π, also known as the integration over the fibre map. In particular, one obtains universal characteristic classes $\widehat{X} \in H^*(B\mathrm{Diff}(M);\mathbb{Z})$ by taking $E \to B$ to be the universal M-bundle over $B\mathrm{Diff}(M)$. In this notation $\kappa_i = \widehat{e^{i+1}}$ for $M = \Sigma_g$.

These generalized MMM classes have been studied intensively. Sadykov [**Sad**] shows that for d even they are the only rational characteristic classes of d-dimensional manifolds that are stable in an appropriate sense. Ebert [**Ebe1**] furthermore shows that for each of these classes there is a bundle of d-manifolds on which it does not vanish (though this is not quite the case when d is odd [**Ebe2**]).

Theorem B. *Suppose W is an oriented manifold with boundary. Then $r^*\widehat{X} \in H^*(B\mathrm{Diff}(W);\mathbb{Z})$ vanishes whenever the dimension of W is even, or whenever it is odd and X can be written as a monomial just in the Pontrjagin classes.*

It is worth stating an immediate corollary of the above theorem.

Corollary C. *Given an oriented bundle $E \to B$ of closed manifolds, the classes $\widehat{X}(E)$ coming from monomials X in Pontrjagin classes give obstructions to fibrewise oriented null-bordism of the bundle.*

An analogue of Theorem B for not necessarily orientable manifolds states that in cohomology with $\mathbb{Z}/2\mathbb{Z}$ coefficients $r^*\widehat{X}$ is trivial for any monomial X in the Stiefel-Whitney classes.

We shall take a geometric approach to the mapping class groups that was first introduced in [**MT01**]. From this point of view the universal MMM-classes can be interpreted as elements in the (stable) cohomology of the infinite loop space associated to a certain Thom spectrum denoted by $\mathbf{MTSO}(2)$. More generally, the proof of Theorem B comes out of the theory of the Thom spectra $\mathbf{MTSO}(d)$ (defined below in section 2) and is related to the theory of "spaces of manifolds" or cobordism categories as in [**GMTW09, Gen**], although we do not actually rely on their results.

Recall, there is a homotopy fibre sequence of infinite loop spaces

$$\Omega^\infty \mathbf{MTSO}(d+1) \to QBSO(d+1)_+ \xrightarrow{\delta} \Omega^\infty \mathbf{MTSO}(d). \tag{1}$$

A bundle of oriented d-manifolds over a base B has a classifying map $B \to \Omega^\infty \mathbf{MTSO}(d)$, and the generalized MMM-classes are pulled back from universal classes in the cohomology of this infinite loop space. A simple calculation in section 2.3 shows that $\delta^*\widehat{X} = 0$ if and only if either d is odd or d is even and X can be written as a product of Pontrjagin classes (i.e. using only even powers of the Euler class). The proof of Theorem B consists of observing, see section 3, that if a bundle of d-manifolds is the fibrewise boundary of a bundle of $(d+1)$-manifolds with boundary then its classifying map factors up-to-homotopy through $QBSO(d+1)_+$. This factorization trick is motivated by the philosophy that the homotopy fibre sequence (1) corresponds to the exact sequence

$$\{\text{closed } (d+1)\text{-manifolds}\} \hookrightarrow \{(d+1)\text{-manifolds with boundary}\} \xrightarrow{\partial} \{\text{closed } d\text{-manifolds}\}.$$

2. A cofibre sequence of Thom spectra

For the reader's convenience we will recall the definition and construction of the fibre sequence (1) and compute the map δ in cohomology.

2.1. Definition of the spectra

Let γ_d denote the tautological bundle of oriented d-planes over $BSO(d)$, and let $\mathbf{MTSO}(d)$ denote the Thom spectrum, $\mathbf{Th}(-\gamma_d)$, of the virtual bundle $-\gamma_d$. Explicitly, let $G_{d,n}$ denote the Grassmannian of oriented d-planes in $\mathbb{R}^{d+n}$, let $\gamma_{d,n}$ denote the tautological d-plane bundle over it, and let $\gamma_{d,n}^\perp$ denote the complementary n-plane bundle. The $(d+n)^{th}$ space of the spectrum $\mathbf{MTSO}(d)$ is the Thom space

$$\mathrm{Th}(\gamma_{d,n}^\perp).$$

The space $G_{d,n}$ sits inside $G_{d,n+1}$ and the restriction of $\gamma_{d,n+1}^\perp$ to $G_{d,n}$ is canonically $\gamma_{d,n}^\perp \oplus \mathbb{R}$. The structure maps of the spectrum are defined by the composition

$$\Sigma\mathrm{Th}(\gamma_{d,n}^\perp) \simeq \mathrm{Th}(\gamma_{d,n}^\perp \oplus \mathbb{R}) \simeq \mathrm{Th}(\gamma_{d,n+1}^\perp|_{G_{d,n}}) \hookrightarrow \mathrm{Th}(\gamma_{d,n+1}^\perp).$$

2.2. A homotopy cofibre sequence of Thom spectra

The suspension spectrum $\Sigma^\infty BSO(d+1)_+$ can be regarded as the Thom spectrum of the trivial bundle of rank 0. In explicit terms, the $(d+1+n)^{th}$ space is $\mathrm{Th}(\gamma_{d+1,n} \oplus \gamma_{d+1,n}^\perp)$ and the structure maps are as above. The inclusion

$$\mathrm{Th}(\gamma_{d+1,n}^\perp) \hookrightarrow \mathrm{Th}(\gamma_{d+1,n} \oplus \gamma_{d+1,n}^\perp) \tag{2}$$

induces a map of spectra

$$\mathbf{MTSO}(d+1) \to \Sigma^\infty BSO(d+1)_+. \tag{3}$$

The cofibre of (3) is known to be homotopy equivalent to $\mathbf{MTSO}(d)$; for convenience we include a proof here.

Lemma 2.1. *Let E and F be vector bundles over a base B, let $p : S(F) \to B$ be the unit sphere bundle of F, and let L denote the tautological line bundle on $S(F)$. There is a cofibre sequence*

$$Th(E) \hookrightarrow Th(E \oplus F) \xrightarrow{\delta} Th(p^*E \oplus L).$$

Proof. Observe that the quotient space $Th(E \oplus F)/Th(E)$ consists of a basepoint together with the space of all triples $(b \in B, u \in E_b, v \in F_b \setminus \{0\})$, suitably topologised. Sending

$$(b,u,v) \mapsto \left(\frac{v}{|v|} \in S(F_b),\ u \in (p^*E)_{v/|v|},\ \log(|v|) \cdot \frac{v}{|v|} \in L_{v/|v|} \right)$$

defines a homeomorphism $Th(E \oplus F)/Th(E) \cong Th(p^*E \oplus L)$. □

Under the identification of the above lemma, one can see that δ corresponds to the map defined by collapsing the complement of an appropriate tubular neighbourhood of the embedding $j : S(F) \hookrightarrow E \oplus F$ and using the canonical identification of the normal bundle of j with $p^*E \oplus L$. In particular, if $E \oplus F$ is isomorphic to a trivial bundle $\mathbb{R}^n$ then δ is the pre-transfer for the projection p, and hence the Gysin map $p_!$ on cohomology is given by δ^* composed with the Thom isomorphism.

We are concerned with the case when B is the Grassmannian $G_{d+1,n}$ of oriented $(d+1)$-planes in $\mathbb{R}^{d+1+n}$, F is the tautological $(d+1)$-plane bundle $\gamma_{d+1,n}$, and E is the complementary n-plane bundle $\gamma_{d+1,n}^\perp$. In this case there is a map

$$q : S(\gamma_{d+1,n}) \to G_{d,n+1}$$

given by sending $(M \in G_{d+1,n}, v \in S(M))$ to the d-plane $M \cap v^\perp$. This map is a fibration, and the fibre over a d-plane N is the n-sphere $S(N^\perp)$. Hence q is n-connected. Observe that $q^*\gamma_{d,n+1}^\perp$ is canonically isomorphic to $p^*\gamma_{d+1,n}^\perp \oplus L$, where $p : S(\gamma_{d+1,n}) \to G_{d+1,n}$ is the projection and L is the tautological line bundle over $S(\gamma_{d+1,n})$. Hence there is a map of Thom spaces,

$$Th(p^*\gamma_{d+1,n}^\perp \oplus L) \to Th(\gamma_{d,n+1}^\perp)$$

that is $(2n+1)$-connected. Combining this with the above cofibre sequence and passing to spectra indexed by n now gives the desired homotopy cofibre sequence of spectra,

$$\mathbf{MTSO}(d+1) \to \Sigma^\infty BSO(d+1)_+ \xrightarrow{\tilde{\delta}} \mathbf{MTSO}(d) \tag{4}$$

and hence a homotopy fibre sequence of infinite loop spaces

$$\Omega^\infty \mathbf{MTSO}(d+1) \to QBSO(d+1)_+ \xrightarrow{\delta} \Omega^\infty \mathbf{MTSO}(d).$$

2.3. Cohomology of Thom spectra and universal tangential classes

For any spectrum E there is a map

$$\sigma^* : H^*(E) \to \widetilde{H}^*(\Omega_0^\infty E)$$

from the spectrum cohomology of E to the reduced cohomology of the basepoint component $\Omega_0^\infty E$ of the associated infinite loop space. This map is induced by the evaluation map

$$\sigma : \Sigma^n \Omega^n E_n \to E_n$$

that takes (t, f) to $f(t)$ for $t \in S^n$ and $f : S^n \to E_n$. Thus σ commutes with maps of spectra.

Let V be a virtual vector bundle of virtual dimension $-d$ over a space B. There is a Thom class, u, in the degree $-d$ cohomology of the associated Thom spectrum $\mathbf{Th}(V)$ (with arbitrary coefficients if V is orientable and with $\mathbb{Z}/2\mathbb{Z}$ coefficients otherwise) and by the Thom isomorphism, the spectrum cohomology $H^*(\mathbf{Th}(V))$ is a free $H^*(B)$-module of rank one generated by the Thom class u. For the Thom spectrum $\mathbf{MTSO}(d) = \mathrm{Th}(-\gamma_d)$ we thus have

$$H^*(\mathbf{MTSO}(d);\mathbb{Z}) \cong u \cdot H^*(BSO(d);\mathbb{Z}),$$

with $\deg u = -d$.

Now, let X be a monomial in the Euler class e and the Pontrjagin classes p_i. We define the associated *universal tangential class* as

$$\widehat{X} := \sigma^*(uX) \in \widetilde{H}^*(\Omega_0^\infty \mathbf{MTSO}(d);\mathbb{Z}).$$

Note that by definition all universal tangential classes are stable in the sense that they come from spectrum cohomology. Rationally, these classes (as X ranges over a basis for the degree $> d$ monomials) freely generate the cohomology ring of $\Omega_0^\infty \mathbf{MTSO}(d)$.

Proposition 2.2. *Let $r = \lfloor d/2 \rfloor$ and $X = p_1^{k_1} \dots p_r^{k_r} e^s \in H^*(BSO(d);\mathbb{Z})$. Consider the image of $\widehat{X}$ under $\delta^* : H^*(\Omega^\infty \mathbf{MTSO}(d);\mathbb{Z}) \to H^*(QBSO(d+1)_+;\mathbb{Z})$.*
(i.) For d odd, $\delta^* \widehat{X} = 0$;
(ii.) For d even, $\delta^* \widehat{X} = 0$ when s is even, and $\delta^* \widehat{X} = 2\sigma^*(X/e)$ when s is odd.

Proof. Identify the inclusion $BSO(d) \hookrightarrow BSO(d+1)$ with the projection

$$\pi : S(\gamma_{d+1}) \to BSO(d+1)$$

of the unit sphere bundle of γ_{d+1}. The map $\widetilde{\delta}^* : H^*(\mathbf{MTSO}(d);\mathbb{Z}) \to H^*(BSO(d+1);\mathbb{Z})$ can then be identified, via the Thom isomorphism, with the Gysin map $\pi_!$. The image of the Euler class e under the Gysin map is the Euler characteristic of the fiber. Thus $\pi_! e = 0$ when d is odd and $\pi_! e = 2$ when d is even. The Pontrjagin classes on $BSO(d) = S(\gamma_{d+1})$ are the pullbacks of the Pontrjagin classes on $BSO(d+1)$. The statement now follows from the formula $\pi_!(\pi^*\alpha \cdot \beta) = \alpha \cdot \pi_! \beta$. Indeed, as $e^2 = 0$ for d odd and $e^2 = p_{d/2}$ for d even, we may assume that $s = 0$ or $s = 1$ in the definition of X, and compute

$$\delta^* \widehat{X} = \delta^* \sigma^*(uX) = \sigma^* \widetilde{\delta}^*(uX) = \sigma^*(\pi_! X) = \sigma^*(p_1^{k_1} \dots p_r^{k_r} \pi_! e^s),$$

which gives the desired result. □

To illustrate the above result consider the case when $d = 2$. In that case we have

$$H^*(\Omega_0^\infty \mathbf{MTSO}(2); \mathbb{Q}) = \mathbb{Q}[\kappa_1, \kappa_2, \ldots]$$

with $\kappa_i = \widehat{e^{i+1}}$ of degree $2i$, while

$$H^*(QBSO(3); \mathbb{Q}) = \mathbb{Q}[\rho_1, \rho_2, \ldots]$$

with $\rho_i = \sigma^* p_1^i$ of degree $4i$. Then $\delta^* \kappa_{2i+1} = 0$ while $\delta^* \kappa_{2i} = 2\rho_i$.

Remark 2.2: When working over $\mathbb{Z}/2\mathbb{Z}$ (in the orientable as well as non-orientable case) a similar computation yields that for any monomial X in the Stiefel-Whitney classes δ^* maps $\widehat{X}$ to zero.

3. Classifying maps

We show here that δ is the universal restriction-to-the-boundary map $r : B\mathrm{Diff}(W) \to B\mathrm{Diff}(\partial W)$.

3.1. Bundles of closed manifolds

Pontrjagin-Thom theory allows one to show that the infinite loop space $\Omega^\infty \mathbf{MTSO}(d)$ classifies concordance classes of oriented d-dimensional *formal bundles*, which are objects slightly more general than fibre bundles of closed oriented d-manifolds. Such an object over a smooth base B consists of a smooth proper map $\pi : E \to B$ of codimension $-d$ and a bundle epimorphism $\delta\pi : TE \to TB$ (which need not be the differential of π) with an orientation of $ker(\delta\pi)$, cf. [**MW07**], [**EG06**].

For a bundle $\pi : E \to B$, the classifying map

$$\alpha_\pi : B \to \Omega^\infty \mathbf{MTSO}(d)$$

is defined concisely as follows. Let $T^\pi E$ denote the fibrewise tangent bundle. The classifying map is the pre-transfer,

$$\text{pre-trf} : B \to \Omega^\infty \mathbf{Th}(-T^\pi E)$$

followed by the map $\Omega^\infty \mathbf{Th}(-T^\pi E) \to \Omega^\infty \mathbf{Th}(-\gamma_d) = \Omega^\infty \mathbf{MTSO}(d)$ induced by the classifying map for $T^\pi E$. To construct this map α_π explicitly, choose a lift of π to an embedding

$$\widetilde{\pi} : E \hookrightarrow B \times \mathbb{R}^{d+n}$$

for some sufficiently large n, and choose a fibrewise tubular neighborhood $U \subset B \times \mathbb{R}^{d+n}$. Let $N^{\widetilde{\pi}} E$ denote the fibrewise normal bundle of $\widetilde{\pi}$. We obtain a map

$$\Sigma^{d+n} B_+ \to \mathrm{Th}(N^{\widetilde{\pi}} E) \tag{5}$$

by identifying U with the normal bundle and collapsing the complement of U to the basepoint. Classifying the fibrewise normal bundle gives a map

$$\mathrm{Th}(N^{\widetilde{\pi}} E) \to \mathrm{Th}(\gamma_{d+n}^\perp). \tag{6}$$

The adjoint of the composition of (5) and (6) is a map $B \to \Omega^{d+n} \mathrm{Th}(\gamma_{d,n}^\perp)$ which gives the classifying map α_π upon composing with the map to $\Omega^\infty \mathbf{MTSO}(d)$. One can check that the homotopy class of this map is independent of the choice of embedding and tubular neighborhood.

The following propositions follow immediately from the description of the classifying map in terms of the pre-transfer, and are well-known.

Proposition 3.1. *The classifying map α_π is natural (up to homotopy) with respect to pull-backs.*

Proposition 3.2. *Given a bundle $\pi : E \to B$ and a class $\widehat{X} \in H^*(\Omega_0^\infty \mathbf{MTSO}(d);\mathbb{Z})$ defined by a monomial X in the Pontrjagin classes and Euler class on $BSO(d)$,*

$$\alpha_\pi^* \widehat{X} = \pi_! X(T^\pi E).$$

3.2. Bundles of manifolds with boundary

Given a bundle, $\pi : E \to B$, of oriented $(d+1)$-manifolds with boundary, let

$$\beta_\pi : B \to QBSO(d+1)_+$$

denote the composition of the transfer, $\mathrm{trf} : B \to QE_+$, followed by the map $QE_+ \to QBSO(d+1)_+$ induced by classifying $T^\pi E$. To construct this map β_π concretely, choose an embedding $\widetilde{\pi}$ of E into $E \times \mathbb{R}^{d+1+n}$ over π. A tubular neighborhood U of E can then be identified with the subspace of $T^\pi E \oplus N^{\widetilde{\pi}} E \cong E \times \mathbb{R}^{d+1+n}$ consisting of those vectors for which the tangential component is zero if they sit over the interior of a fibre and over the boundary the tangential component is outward pointing normal to the boundary.

Hence collapsing the complement of U and classifying the fibrewise tangent bundle gives maps

$$\Sigma^{d+1+n} B_+ \to U^+ \to \mathrm{Th}(T^\pi E \oplus N^{\widetilde{\pi}} E) \to \mathrm{Th}(\gamma_{d+1,n} \oplus \gamma_{d+1,n}^\perp),$$

where $()^+$ denotes the one-point compactification. Taking the adjoint of this composition and then mapping into the colimit as n goes to infinity gives the desired map

$$\beta_\pi : B \to QBSO(d+1)_+.$$

Again, one can check that the homotopy class of this map is independent of the choices made in the construction. Analogous to α_π, β_π can be interpreted as the classifying map of formal bundles of $(d+1)$-dimensional manifolds with boundary.

Proposition 3.3. *The classifying map β_π is natural (up to homotopy) with respect to pull-backs.*

3.3. Restricting to the boundary

The classifying maps α and β constructed above for bundles of closed manifolds and manifolds with boundary are compatible in two ways. First, it is easy to see that regarding a bundle of closed manifolds as a bundle of manifolds with (empty) boundary is compatible with the map $\Omega^\infty \mathbf{MTSO}(d+1) \to QBSO(d+1)_+$. More importantly for us, the fibrewise boundary of a bundle of $(d+1)$-manifolds is a bundle of d-manifolds and the classifying maps for these two bundles are compatible in the following sense.

Proposition 3.4. *Given a bundle of oriented $(d+1)$-manifolds $\pi : E \to B$ with fibrewise boundary bundle $\partial\pi : \partial E \to B$, the diagram*

$$\begin{array}{ccc} B & & \\ {\scriptstyle \beta_\pi}\downarrow & \searrow^{\alpha_{\partial\pi}} & \\ QBSO(d+1)_+ & \xrightarrow[\delta]{} & \Omega^\infty \mathbf{MTSO}(d) \end{array}$$

commutes up to homotopy.

Proof. Fix an embedding $\widetilde{\pi} : E \hookrightarrow B \times \mathbb{R}^{d+n}$ over π and a tubular neighborhood U. Let $U_\partial \subset U$ be the subspace sitting over the fibrewise boundary of E. The lower composition in the diagram comes from the adjoint of a map

$$\Sigma^{d+1+n} B_+ \to \mathrm{Th}(\gamma_{d+1,n} \oplus \gamma_{d+1,n}^\perp) \to \mathrm{Th}(\gamma_{d,n+1}^\perp) \tag{7}$$

which collapses the complement of U_∂ to the basepoint. The space U_∂ is identified with the subspace of the vector bundle $(T^\pi E \oplus N^{\widetilde{\pi}} E)|_{\partial E}$ consisting of vectors for which the tangential component is outward pointing normal to ∂E. In the fibre over any point $p \in \partial E$ there is a unique point v_p which is sent by the map (7) to the zero section in $\mathrm{Th}(\gamma_{d,n+1}^\perp)$. Explicitly, the component of v_p that is normal to E is zero and the tangential component is outward pointing unit normal to ∂E. The map

$$\rho : p \mapsto v_p \in U_\partial \subset B \times \mathbb{R}^{d+1+n}$$

gives an embedding of ∂E into $B \times \mathbb{R}^{d+1+n}$ over $\partial\pi$. Observe that U_∂ is a tubular neighborhood of the embedding ρ, and the composition (7) collapses the complement of U_∂, identifies it with the normal bundle of ρ and classifies this bundle of oriented $(n+1)$-planes in $\mathbb{R}^{d+1+n}$. Hence the lower composition in the diagram in the statement of the proposition is a map $\alpha'_{\partial\pi}$ constructed exactly as $\alpha_{\partial\pi}$ but with a different choice of embedding and tubular neighborhood. Since different choices lead to homotopic maps, the diagram commutes up to a homotopy. □

4. Proofs of the theorems

Theorem B follows directly from Proposition 3.4 and Proposition 2.2. Theorem A is the special case when W is a 3-dimensional oriented handlebody of genus $g \geqslant 2$.

References

[EE69] C. J. Earle and J. Eells, *A fibre bundle description of Teichmüller theory*, J. Differential Geometry **3** (1969), 19–43.

[Ebe1] J. Ebert, *Algebraic independence of generalized Morita-Miller-Mumford classes*, arXiv:0910.1030.

[Ebe2] J. Ebert, *A vanishing theorem for characteristic classes of odd-dimensional manifold bundles*, arXiv:0902.4719.

[EG06] Y. Eliashberg and S. Galatius, *Homotopy theory of compactified moduli spaces*, Oberwolfach Reports (2006), 761–766.

[Gal] S. Galatius, *Stable homology of automorphism groups of free groups*, to appear in Ann. of Math.

[Gen] J. Genauer, *Cobordism categories of manifolds with corners*, arXiv:0810.0581 (2008).

[GMTW09] S. Galatius, I. Madsen, U. Tillmann and M. Weiss, *The homotopy type of the cobordism category*, Acta Math. **202** (2009), no. 2, 195–239.

[Hat76] A. Hatcher, *Homeomorphisms of sufficiently large P^2-irreducible 3-manifolds*, Topology **15** (1976), no. 4, 343–347.

[Hat99] A. Hatcher, *Spaces of imcompressible surfaces*, arXiv:math/9906074 (1999).

[HW] A. Hatcher, N. Wahl, *Stabilization for mapping class groups of 3-manifolds*, to appear in Duke Math. J.; arXiv:0709.2173.

[Mil86] E. Y. Miller, *The homology of the mapping class group*, J. Differential Geom. **24** (1986), no. 1, 1–14.

[Mor87] S. Morita, *Characteristic classes of surface bundles*, Invent. Math. **90** (1987), no. 3, 551–577.

[MT01] I. Madsen and U. Tillmann, *The stable mapping class group and $Q(CP^\infty_+)$*, Invent. Math. **145** (2001), no. 3, 509–544.

[Mum83] D. Mumford, *Towards an enumerative geometry of the moduli space of curves*, Arithmetic and geometry, Vol. II, Progr. Math., vol. 36, Birkhäuser Boston, Boston, MA, 1983, pp. 271–328.

[MW07] I. Madsen and M. Weiss, *The stable moduli space of Riemann surfaces: Mumford's conjecture*, Ann. of Math. **165** (2007), 843–941.

[Sad] R. Sadykov, *Stable characteristic classes of smooth manifold bundles*, arXiv:0910.4770.

[Sak09] T. Sakasai, *Lagrangian mapping class groups from a group homological point of view*, arXiv:0910.5262 (2009).

This article may be accessed via WWW at `http://tcms.org.ge/Journals/JHRS/`

Jeffrey Giansiracusa
j.h.giansiracusa@gmail.com

Department of Mathematical Sciences, University of Bath
Claverton Down
Bath, BA2 7AY
United Kingdom

Ulrike Tillmann
tillmann@maths.ox.ac.uk

Mathematical Institute, Oxford University
24-29 St. Giles'
Oxford, OX1 3LB
United Kingdom

Journal of Homotopy and Related Structures, vol. 6(1), 2011, pp.113–114

ERRATUM TO "INFINITY-INNER-PRODUCTS ON A-INFINITY-ALGEBRAS", JOURNAL OF HOMOTOPY AND RELATED STRUCTURES, VOL. 3(1), 2008, PP.245–271

THOMAS TRADLER

(*communicated by James Stasheff*)

The following two items should be corrected in [**T**].

- The sign $(-1)^{\varepsilon}$ in Lemma 2.9 was falsely stated. Correctly, it should read

$$\varepsilon := (|a_1| + \ldots + |a_k|) \cdot (|m^*| + |a_{k+1}| + \ldots + |a_{k+l}| + |m|) \\ + |m^*| \cdot (k + l + 1) + (k + 1) \cdot (l + 1).$$

- An edge in the pairahedron for $k = 2, l = 1$ (or for $k = 1, l = 2$) was misplaced on page 270. The pairahedron associated to $< \cdots >_{2,1}$ (and also associated to $< \cdots >_{1,2}$) consists of 4 square, 4 pentagons, and 3 hexagons. It is depicted below:

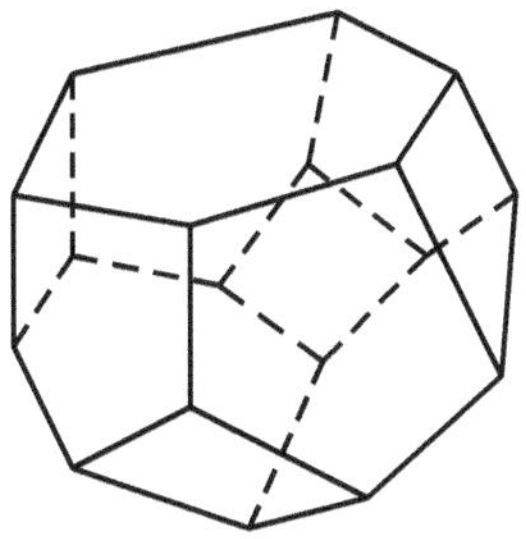

References

[T] T. Tradler, *Infinity-Inner-Products on A-Infinity-Algebras*, Journal of Homotopy and Related Structures, vol. 3(1), 2008, pp.245–271

This article may be accessed via WWW at `http://tcms.org.ge/Journals/JHRS/`

I would like to thank Andrea Ferrario for pointing out the sign issue in Lemma 2.9, and Stefan Forcey and Jim Stasheff for bringing the misplaced edge to my attention.
Received March 08, 2011; published on March 27, 2011.

Thomas Tradler
`ttradler@citytech.cuny.edu`

Department of Mathematics
College of Technology of the City University of New York
300 Jay Street, Brooklyn, NY 11201, USA

Journal of Homotopy and Related Structures, vol. 6(1), 2011, pp.115–118

THE HELP-LEMMA AND ITS CONVERSE IN QUILLEN MODEL CATEGORIES

R.M. VOGT

(*communicated by Ross Staffeldt*)

Abstract

We show that a map $p : X \to Y$ between fibrant objects in a closed model category is a weak equivalence if and only if it has the right homotopy extension lifting property with respect to all cofibrations. The dual statement holds for maps between cofibrant objects.

The HELP-Lemma states that a homotopy equivalence $p : X \to Y$ of topological spaces has the homotopy extension lifting property, HELP for short, for all closed cofibrations [**2**, Appendix Thm. 3.5]. The lemma, variants of it and their Eckmann-Hilton duals (e.g. see [**1**, II.1.11], [**4**, Thm. 4, Thm. 4*]) have proven to be very useful tools in homotopy theory.

The main purpose of this paper is to make this tool and its Eckmann-Hilton dual available in arbitrary closed model categories in the sense of Quillen [**7**], (see also [**3**]). In addition, we prove a converse.

Surprisingly, this converse has never been explicitely stated in the past with the exception of the very classical case of weak homotopy equivalences of topological spaces (e.g. see [**5**, p. 68]; we refer to it as May's lemma), but it follows from a lemma due to Reedy [**8**, Lemma 2.1]. Our proof is a bit more elementary than Reedy's. Applying the theorem below to the category $\mathcal{T}op$ of topological spaces with the Strøm model structure [**9**] we for example obtain

Proposition 1. *A map of topological spaces $X \to Y$ is a homotopy equivalence if and only if it has the HELP for all closed cofibrations.*

Of course, its Eckmann-Hilton dual also holds.

Our theorem covers Reedy's lemma and May's lemma for weak equivalences of topological spaces, but not for n-equivalences, because n-equivalences in his sense do not satisfy the two-out-of-three axiom for weak equivalences in a model category.

Throughout the paper let $\mathcal{M}$ denote a closed model category in the sense of Quillen.

I am indebted to P. May for drawing my attention to [**5**, p. 68] and to M. Stelzer for helpful discussions.

Received April 29, 2010, revised March 03, 2011; published on March 27, 2011.

2000 Mathematics Subject Classification: 55P05, 55P30

Key words and phrases: Homotopy-extension-lifting-property, weak equivalences, model categories

Definition 2. Let $i : A \to B$ and $p : X \to Y$ be maps in $\mathcal{M}$.

(1) We say that p has the right HELP with respect to i, if for each not necessarily commutative square

$$\begin{array}{ccc} A & \xrightarrow{f_A} & X \\ {\scriptstyle i}\downarrow & & \downarrow{\scriptstyle p} \\ B & \xrightarrow{g} & Y \end{array} \qquad (*)$$

and each right homotopy $h_A : A \to Y^I$ from $p \circ f_A$ to $g \circ i$, where Y^I is a path object $Y \xrightarrow[\sim]{j} Y^I \xrightarrow{\pi} Y \times Y$ for Y, there is a map $f : B \to X$ and a right homotopy $h : B \to Y^I$ from $p \circ f$ to g such that $f \circ i = f_A$ and $h \circ i = h_A$.

(2) We say that i has the left HELP with respect to p, if for each not necessarily commutative square

$$\begin{array}{ccc} A & \xrightarrow{f} & X \\ {\scriptstyle i}\downarrow & & \downarrow{\scriptstyle p} \\ B & \xrightarrow{g_Y} & Y \end{array}$$

and each left homotopy $h_Y : Z_A \to Y$ from $g_Y \circ i$ to $p \circ f$, where Z_A is a cylinder object $A \sqcup A \xrightarrow{j} Z_A \xrightarrow[\sim]{\sigma} A$ for A, there is a map $g : B \to X$ and a left homotopy $h : Z_A \to X$ from $g \circ i$ to f such that $p \circ g = g_Y$ and $p \circ h = h_Y$.

Theorem 3. *(1) A map $p : X \to Y$ of fibrant objects is a weak equivalence in $\mathcal{M}$ if and only if it has the right HELP with respect to all cofibrations.*

(2) A map $i : A \to B$ of cofibrant objects is a weak equivalence in $\mathcal{M}$ if and only if it has the left HELP with respect to all fibrations.

Proof. The two statements are dual so we just prove the first one.

Since X and Y are fibrant the projections

$$X \xleftarrow{p_X} X \times Y \xrightarrow{p_Y} Y \qquad \text{and} \qquad Y \xleftarrow{p_1} Y \times Y \xrightarrow{p_2} Y$$

are fibrations.

Suppose that p is a weak equivalence and that we are given a square $(*)$. Consider the commutative diagram

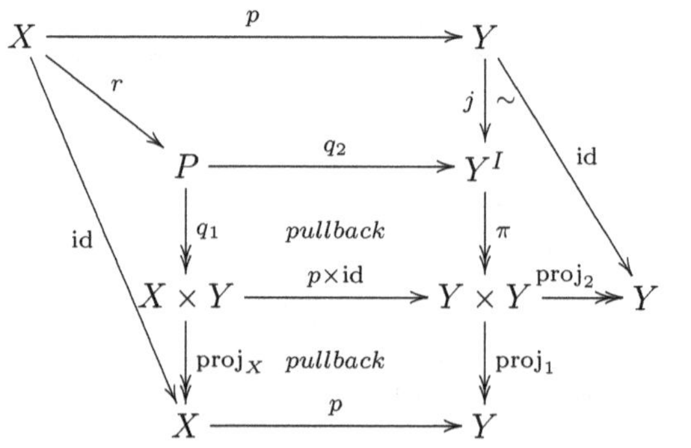

where $r = (\mathrm{id}, j \circ p)$. Since $\mathrm{proj}_1 \circ \pi$ is a weak equivalence, so is $\mathrm{proj}_X \circ q_1$ and hence r. It follows that q_2 and $\mathrm{proj}_2 \circ \pi \circ q_2$ are weak equivalences. Hence $q_3 = \mathrm{proj}_Y \circ q_1 = \mathrm{proj}_2 \circ \pi \circ q_2 : P \twoheadrightarrow Y$ is a weak equivalence. Now consider

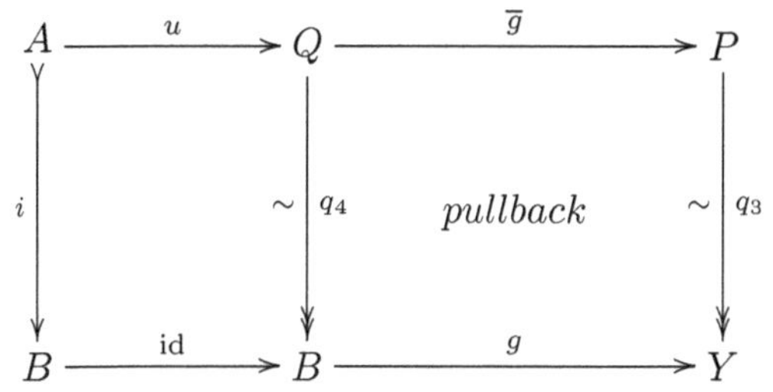

where $u = (i, v)$ and $v : A \to P$ is induced by $(f_A, g \circ i) : A \to X \times Y$ and $h_A : A \to Y^I$. Since q_4 is a trivial fibration there is a section $s : B \to Q$ such that $s \circ i = u$. Define

$$f = \mathrm{proj}_X \circ q_1 \circ \overline{g} \circ s : B \to X$$
$$h = q_2 \circ \overline{g} \circ s : B \to Y^I$$

Conversely, suppose that p has the right HELP. Consider the diagram

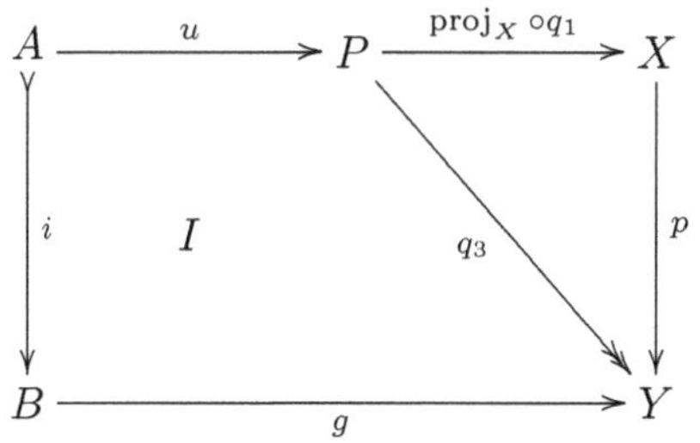

where square I is supposed to commute. We define

$$f_A = \mathrm{proj}_X \circ q_1 \circ u : A \to X \quad \text{and} \quad h_A = q_2 \circ u : A \to Y^I.$$

Then h_A is a right homotopy from $p \circ f_A$ to $g \circ i$. Hence there exist

$$f : B \to X \quad \text{and} \quad h : B \to Y^I$$

such that h is a right homotopy from $p \circ f$ to g and $f \circ i = f_A$ and $h \circ i = h_A$. Then f and h induce a map

$$k : B \to P$$

such that $k \circ i = u$ and $q_3 \circ k = \mathrm{proj}_2 \circ \pi \circ k = g$.

Hence q_3 has the right lifting property with respect to all cofibration and has to be a trivial fibration. Since $q_2 : P \to Y^I$ is a right homotopy from $p \circ \mathrm{proj}_X \circ q_1$ to q_3 and since a map right homotopic to a weak equivalence is itself a weak equivalence, $p \circ \mathrm{proj}_X \circ q_1$ is a weak equivalence. Since $\mathrm{proj}_X \circ q_1$ is a weak equivalence, so is p. □

References

[1] H.J. Baues, *Algebraic homotopy*, Cambridge University Press 1989.

[2] J.M. Boardman, R.M. Vogt, *Homotopy invariant structures on topological spaces*, Lecture Notes in Math. 347, Springer Verlag, Berlin 1973.

[3] W.G. Dwyer, J. Spalinski, *Homotopy theories and model categories*, Handbook of Algebraic Topology (I.M. James, ed.), Elsevier Science B.V., 1995.

[4] J.P. May, *The dual Whitehead theorems*, Topological Topics (I.M. James, ed.), London Math. Soc. Lecture Notes Ser. 86, Cambridge University Press 1983.

[5] J.P. May, *A concise course in algebraic topology*, Chicago Lecture Notes in Math., The University of Chicago Press, 1999.

[6] D.G. Quillen, *Homotopical algebra*, Springer Lecture Notes in Math. 43 (1967).

[7] D.G. Quillen, *Rational homotopy theory*, Ann. Math. 90 (1969), 205-295.

[8] C.L. Reedy, *Homotopy theory of model categories*, Unpublished manuscript (1974), available at the Hopf Topology Archive as ftp:://hopf.math.purdue.edu/pub/Reedy/reedy.dvi.

[9] A. Strøm, *The homotopy category is a homotopy category*, Arch. Math. 23 (1972), 435-441.

This article may be accessed via WWW at `http://tcms.org.ge/Journals/JHRS/`

R.M. Vogt
`rvogt@uos.de`

Fachbereich Mathematik/Informatik
Universität Osnabrück
D-49069 Osnabrück
Germany

Journal of Homotopy and Related Structures, vol. 6(1), 2011, pp.119–159

CONTROLLED ALGEBRAIC G-THEORY, I

GUNNAR CARLSSON AND BORIS GOLDFARB

(*communicated by Ross Staffeldt*)

Abstract

This paper extends the notion of geometric control in algebraic K-theory from additive categories with split exact sequences to other exact structures. In particular, we construct exact categories of modules over a Noetherian ring filtered by subsets of a metric space and sensitive to the large scale properties of the space. The algebraic K-theory of these categories is related to the bounded K-theory of geometric modules of Pedersen and Weibel the way G-theory is classically related to K-theory. We recover familiar results in the new setting, including the nonconnective bounded excision and equivariant properties. We apply the results to the G-theoretic Novikov conjecture which is shown to be stronger than the usual K-theoretic conjecture.

Contents

1. Introduction

Since the invention of algebraic K-groups of a ring defined using the finitely generated projective R-modules, there existed a companion K-theory defined using arbitrary finitely generated R-modules, called G-theory. Its usefulness comes from the computational tool available in G-theory, the localization exact sequence, and

---

The authors acknowledge support from the National Science Foundation.
Received May 06, 2009, revised March 25, 2011; published on May 01, 2011.
2000 Mathematics Subject Classification: 18E10, 18E30, 18E35, 18F25, 19D35, 19J99
Key words and phrases: Algebraic K-theory, assembly map, Novikov Conjecture, Borel Conjecture, controlled algebra, algebraic G-theory

the close relation to K-theory via the Cartan map which becomes an isomorphism when R is a regular ring. The recent success of controlled K-theory in algebra and topology, where the ring involved is usually the regular ring of integers $\mathbb{Z}$, makes it natural to look for a similar controlled analogue of G-theory. This paper constructs and exploits such an analogue.

The bounded control is introduced by fixing a basis B in a free module M and defining a locally finite set function $s\colon B \to X$ into a metric space X. The control comes from restrictions on the maps one allows between the based modules. Since each element x in M is written uniquely as a sum $x = \sum_{b \in B} r_b b$, there is the notion of support, $\operatorname{supp}(x)$, which is the set of all points $s(b)$ in X with $b \in B$ such that $r_b \neq 0$.

For two sets of choices (M_i, B_i, s_i), $i = 1, 2$, an R-homomorphism $\phi\colon M_1 \to M_2$ is *bounded* if there is a number $D > 0$ such that for every $b \in B_1$ the support $\operatorname{supp} s_2(\phi(b))$ is contained in the metric ball of radius D centered at $s_1(b)$.

The triples (M, B, s) and the bounded homomorphisms form the *bounded category* $\mathcal{B}(X, R)$. It is in fact an additive category since the direct products can be defined in the evident way. To each additive category $\mathcal{A}$, one associates a sequence of groups $K_i(\mathcal{A})$, $i \in \mathbb{Z}$, or rather a nonconnective spectrum $\operatorname{Spt}(\mathcal{A})$ whose stable homotopy groups are $K_i(\mathcal{A})$, as in [**18**]. This construction applied to the bounded category $\mathcal{B}(X, R)$ gives the *bounded algebraic K-theory* $K_i(X, R)$.

The general goal of this paper is to construct larger categories associated to a metric space X and a Noetherian ring R and to recover in this context the basic results from bounded K-theory. We are mostly concerned with controlled excision established in section 3. In many ways these categories are more flexible than the bounded categories and allow application of recent powerful results in algebraic K-theory. Their properties are essential for our study of the Borel isomorphism conjecture continued elsewhere but indicated in section 4. In the same section, we prove the integral Novikov conjecture in this context. It turns out that earlier results asserting the split injectivity of the assembly map follow from this result, due to the fact that the natural transformation from K-theory to G-theory is an equivalence for a regular Noetherian ring.

The following is a sketch of the constructions and results of the paper.

First notice that, given a triple (M, B, s) in $\mathcal{B}(X, R)$, to every subset $S \subset X$ there is associated a free submodule $M(S)$ generated by those $b \in B$ with the property $s(b) \in S$. The restriction to bounded homomorphisms can be described entirely in terms of these submodules. We generalize this as follows. The objects of the new category $\mathbf{U}^{\mathrm{b}}(X, R)$ are left R-modules M filtered by the subsets of X in the sense that they are functors from the category of subsets of X and inclusions to the category of submodules of M and inclusions for which the value on the whole space X is the whole module M and the value on the empty set $\varnothing$ is the zero submodule. By abuse of notation we usually denote the functor by the same letter M. We also make several additional assumptions spelled out in Definition 2.16, in particular, that the values on the bounded subsets are finitely generated submodules.

The morphisms in $\mathbf{U}^{\mathrm{b}}(X, R)$ are the R-homomorphisms $\phi\colon M_1 \to M_2$ for which there exists a number $D \geqslant 0$ such that the image $\phi(M_1(S))$ is contained in the submodule $M_2(S[D])$ for all subsets $S \subset X$. Here $S[D]$ stands for the metric D-enlargement of S in X. In this context we say a submodule $N \subset M$ is *supported on a subset* $S \subset X$ if $N \subset M(S)$.

The *boundedly controlled* category $\mathbf{B}(X, R)$ is the full subcategory of $\mathbf{U}^{\mathrm{b}}(X, R)$ on filtered modules M generated by elements supported on subsets of diameter less than d for some number $d > 0$ specific to M.

The additive structure on $\mathbf{B}(X, R)$ gives it the *split exact structure* where the admissible monomorphisms are all split monics and admissible epimorphisms are all split epis. In order to describe a different Quillen exact structure on $\mathbf{B}(X, R)$, we define an additional property a boundedly controlled homomorphism $\phi \colon M_1 \to M_2$ in $\mathbf{U}^{\mathrm{b}}(X, R)$ may or may not have: ϕ is *boundedly bicontrolled* if there exists a number $D \geqslant 0$ such that

$$\phi(M_1(S)) \subset M_2(S[D])$$

and

$$\phi(M_1) \cap M_2(S) \subset \phi M_1(S[D])$$

for all subsets S of X. We define the admissible monomorphisms in the new Quillen exact structure to be the boundedly bicontrolled injections of modules, both for the case of $\mathbf{B}(X, R)$ and $\mathbf{U}^{\mathrm{b}}(X, R)$. We define the admissible epimorphisms in $\mathbf{U}^{\mathrm{b}}(X, R)$ to be the boundedly bicontrolled surjections. The admissible epimorphisms in $\mathbf{B}(X, R)$ are the boundedly bicontrolled surjections with kernels in $\mathbf{B}(X, R)$. In both cases the exact sequences are simply the short exact sequences when viewed as sequences in $\mathbf{U}^{\mathrm{b}}(X, R)$ so that all kernels and cokernels are well-defined filtered submodules in the respective category. Notice that split injections and surjections are boundedly bicontrolled, so the split exact structure is an exact subcategory of the new one.

Recall that a map $f \colon X \to Y$ of metric spaces is *bi-Lipschitz* if there is a number $k \geqslant 1$ such that

$$k^{-1} \operatorname{dist}(x_1, x_2) \leqslant \operatorname{dist}(f(x_1), f(x_2)) \leqslant k \operatorname{dist}(x_1, x_2)$$

for all x_1, $x_2 \in X$. More generally, f is *quasi-bi-Lipschitz* if there is a real positive function l such that

$$\begin{aligned} \operatorname{dist}(x_1, x_2) \leqslant r &\implies \operatorname{dist}(f(x_1), f(x_2)) \leqslant l(r), \\ \operatorname{dist}(f(x_1), f(x_2)) \leqslant r &\implies \operatorname{dist}(x_1, x_2) \leqslant l(r). \end{aligned}$$

For example, a bounded function $f \colon X \to X$, with the property $\operatorname{dist}(x, f(x)) \leqslant D$ for all $x \in X$ and a fixed number $D \geqslant 0$, is quasi-bi-Lipschitz with $l(r) = r + 2D$. An isometry $g \colon X \to Y$ is quasi-bi-Lipschitz with $l(r) = r$.

Both constructions, $\mathbf{U}^{\mathrm{b}}(X, R)$ and $\mathbf{B}(X, R)$, are functorial in the metric space variable X with respect to quasi-bi-Lipschitz maps, as should be expected from [**18**]. Recall that an exact functor between Quillen exact categories is an additive functor which sends exact sequences to exact sequences. So to each quasi-bi-Lipschitz map $f \colon X \to Y$ one associates an exact functor $f_* \colon \mathbf{B}(X, R) \to \mathbf{B}(Y, R)$. For example, if Z is a metric subspace of X then the isometric inclusion $i \colon Z \to X$ induces an exact functor $i_* \colon \mathbf{B}(Z, R) \to \mathbf{B}(X, R)$.

To an exact category $\mathbf{E}$, one associates a sequence of groups $K_i(\mathbf{E})$, $i \geqslant 0$, as in Quillen [**20**] or a connective spectrum $K(\mathbf{E})$ whose stable homotopy groups are $K_i(\mathbf{E})$. If the exact structure is split, these groups are the same as the K-groups of $\mathbf{E}$ as an additive category. When this construction is applied to the exact category $\mathbf{B}(X, R)$, we call the result the *connective controlled G-theory* of X and denote the spectrum by $G(X, R)$.

A proper metric space is a metric space where all closed metric balls in are compact. Let X be a proper metric space. Suppose Z is a metric subspace of X. There is a construction of an exact category $\mathbf{B}/\mathbf{Z}$ associated to Z and an exact functor $\mathbf{B}(X, R) \to \mathbf{B}/\mathbf{Z}$ such that the following is true.

Theorem (Localization, Corollary 4.16). *The sequence*

$$G(Z, R) \longrightarrow G(X, R) \longrightarrow K(\mathbf{B}/\mathbf{Z})$$

is a homotopy fibration.

The Localization Theorem can be used to construct nonconnective deloopings of $G(X, R)$. We will indicate the corresponding nonconnective spectra with superscripts "$-\infty$". The construction is similar to the K-theory delooping using bounded K-theory due to Pedersen–Weibel [**17**]. Therefore, there is a natural transformation $K^{-\infty}(X, R) \to G^{-\infty}(X, R)$.

The following is the analogue of a major tool in many proofs of the Novikov conjecture. If a proper metric space X is the union of proper metric subspaces X_1 and X_2, let $\mathbf{B}(X_1, X_2; R)$ stand for the full subcategory of $\mathbf{B}(X, R)$ on the modules supported on the intersection of bounded enlargements of X_1 and X_2 and let $G(X_1, X_2; R)$ denote its K-theory.

Theorem (Nonconnective controlled excision, Theorem 4.25). *There is a homotopy pushout*

$$\begin{array}{ccc} G^{-\infty}(X_1, X_2; R) & \longrightarrow & G^{-\infty}(X_1, R) \\ \downarrow & & \downarrow \\ G^{-\infty}(X_2, R) & \longrightarrow & G^{-\infty}(X, R) \end{array}$$

Finally, we describe the application to splitting integral G-theoretic assembly maps. There is a close relation to the same problem in K-theory.

Earlier applications of bounded K-theory to conjectures of Novikov type use in a critical way the existence of equivariant versions of the bounded K-theory functors attached to actions of discrete groups of isometries. The applications we envision of the present theory will also require such a theory, and we develop it in the last section of the paper. It turns out that we will need to develop the equivariant theory for a more general class of actions than isometric actions, namely the class of actions by discrete groups on metric spaces by quasi-bi-Lipschitz equivalences. In carrying this out, we find that we obtain a novel exact structure $\mathbf{B}(R[\Gamma])$ on the a category of (not necessarily projective) finitely generated modules over a group ring $R[\Gamma]$, where R is a Noetherian ring and Γ is a discrete group.

Recall that the integral assembly map in algebraic K-theory

$$A_K \colon B\Gamma_+ \wedge K^{-\infty}(R) \longrightarrow K^{-\infty}(R[\Gamma])$$

is defined for any group Γ and any ring R and relates the homology of Γ with coefficients in the K-theory of R to the K-theory of the group ring. The *integral Novikov conjecture* for Γ is the statement that A_K is a split injection of spectra. It is speculated to be true whenever Γ is a discrete torsion-free group.

For a Noetherian ring R and the spectrum $G^{-\infty}(R[\Gamma])$ defined as $K^{-\infty}\mathbf{B}(R[\Gamma])$ there is a similar map

$$A_G \colon B\Gamma_+ \wedge G^{-\infty}(R) \longrightarrow G^{-\infty}(R[\Gamma])$$

which we call the *integral assembly map* in algebraic G-theory. In this paper we show that it is a split injection for many geometric groups.

Theorem. *Let Γ be a discrete group of finite asymptotic dimension and a finite classifying space. Let R be a Noetherian ring. Then the assembly map A_G is a split injection.*

It turns out that earlier results asserting the split injectivity of the assembly map follow from this result, due to the fact that the natural transformation from K-theory to G-theory for rings is an equivalence when the ring is regular Noetherian, for example the integers $\mathbb{Z}$.

Indeed, notice that from the commutative square

$$\begin{array}{ccc} B\Gamma_+ \wedge K^{-\infty}(R) & \xrightarrow{A_K} & K^{-\infty}(R[\Gamma]) \\ \simeq \downarrow & & \downarrow C \\ B\Gamma_+ \wedge G^{-\infty}(R) & \xrightarrow{A_G} & G^{-\infty}(R[\Gamma]) \end{array}$$

the assembly map A_G is, up to homotopy, the composition of A_K followed by the Cartan map

$$C \colon K^{-\infty}(R[\Gamma]) \longrightarrow G^{-\infty}(R[\Gamma]).$$

If A_G is a split injection, it follows that A_K is a split injection.

We are grateful to Marco Schlichting for showing us the preliminary version of his thesis [**21**] and several very fruitful discussions. We are grateful for the critique and suggestions of the referee and the editor which have improved the paper.

The authors acknowledge support from the National Science Foundation.

2. Controlled categories of filtered objects

This work is motivated by the delooping of algebraic K-theory of a small additive category in [**17**] and, in particular, the introduction of bounded control in a cocomplete additive category $\mathbf{A}$ which we briefly recall.

A category is *cocomplete* if it contains colimits of arbitrary small diagrams, cf. Mac Lane [**16**], chapter V.

Definition 2.1 (Pedersen–Weibel). Let X be a proper metric space, in the sense that all closed metric balls in X are compact. An X-graded object is a function F from the set X to the set of objects of $\mathbf{A}$ such that the set $\{x \in S \mid F(x) \neq 0\}$ is finite for every bounded $S \subset X$. We will also refer to the object

$$F = \bigoplus_{x \in X} F(x)$$

in $\mathbf{A}$ as an X-graded object.

The X-graded objects form a new category $\mathcal{B}(X, \mathbf{A})$. The morphisms are collections of $\mathbf{A}$-morphisms $f(x, y) \colon F(x) \to G(y)$ with the property that there is a number $D > 0$ such that $f(x, y) = 0$ if $\operatorname{dist}(x, y) > D$.

If $\mathbf{B}$ is a subcategory of $\mathbf{A}$ closed under the direct sum, one obtains the additive bounded category $\mathcal{B}(X, \mathbf{B})$ as the full subcategory of $\mathcal{B}(X, \mathbf{A})$ on objects F with $F(x) \in \mathbf{B}$ for all $x \in X$. Notice that $\mathbf{B}$ does not need to be cocomplete.

The *bounded algebraic K-theory* $K(X, \mathbf{B})$ is the K-theory spectrum associated to the additive category $\mathcal{B}(X, \mathbf{B})$.

To generalize this construction from additive to more general exact categories $\mathbf{E}$, first notice the following. Given an object F in $\mathcal{B}(X, \mathbf{B})$, to every subset $S \subset X$ there is associated a direct sum

$$F(S) = \bigoplus_{x \in S} F(x).$$

Since the condition $f(x, y) = 0$ if $\operatorname{dist}(x, y) > D$ is equivalent to the condition that $f(F(x)) \subset F(x[D])$ or, more generally,

$$f(F(S)) \subset F(S[D]),$$

the restriction from arbitrary to bounded morphisms can be described entirely in terms of the subobjects $F(S)$.

We start by recalling definitions and several standard facts about exact and abelian categories.

Definition 2.2 (Quillen exact categories). Let $\mathbf{C}$ be an additive category. Suppose $\mathbf{C}$ has two classes of morphisms $\mathbf{m}(\mathbf{C})$, called *admissible monomorphisms*, and $\mathbf{e}(\mathbf{C})$, called *admissible epimorphisms*, and a class $\mathcal{E}$ of *exact* sequences, or extensions, of the form

$$C^{\cdot} \colon \quad C' \xrightarrow{i} C \xrightarrow{j} C''$$

with $i \in \mathbf{m}(\mathbf{C})$ and $j \in \mathbf{e}(\mathbf{C})$ which satisfy the three axioms:

a) any sequence in $\mathbf{C}$ isomorphic to a sequence in $\mathcal{E}$ is in $\mathcal{E}$; the canonical sequence

$$C' \xrightarrow{\text{incl}_1} C' \oplus C'' \xrightarrow{\text{proj}_2} C''$$

is in $\mathcal{E}$; for any sequence $C^{\cdot}$, i is a kernel of j and j is a cokernel of i in $\mathbf{C}$,

b) both classes $\mathbf{m}(\mathbf{C})$ and $\mathbf{e}(\mathbf{C})$ are subcategories of $\mathbf{C}$; $\mathbf{e}(\mathbf{C})$ is closed under base-changes along arbitrary morphisms in $\mathbf{C}$ in the sense that for every exact sequence $C' \to C \to C''$ and any morphism $f \colon D'' \to C''$ in $\mathbf{C}$, there is a pullback commutative diagram

$$\begin{array}{ccccc} C' & \longrightarrow & D & \xrightarrow{j'} & D'' \\ {\scriptstyle =}\downarrow & & \downarrow{\scriptstyle f'} & & \downarrow{\scriptstyle f} \\ C' & \longrightarrow & C & \xrightarrow{j} & C'' \end{array}$$

where $j' \colon D \to D''$ is an admissible epimorphism; $\mathbf{m}(\mathbf{C})$ is closed under cobase-changes along arbitrary morphisms in $\mathbf{C}$ in the (dual) sense that for every exact sequence $C' \to C \to C''$ and any morphism $g \colon C' \to D'$ in $\mathbf{C}$, there is a pushout diagram

$$\begin{array}{ccccc} C' & \xrightarrow{i} & C & \longrightarrow & C'' \\ {\scriptstyle g}\downarrow & & {\scriptstyle g'}\downarrow & & \downarrow{\scriptstyle =} \\ D' & \xrightarrow{i'} & D & \longrightarrow & C'' \end{array}$$

where $i' \colon D' \to D$ is an admissible monomorphism,

c) if $f \colon C \to C''$ is a morphism with a kernel in $\mathbf{C}$, and there is a morphism $D \to C$ so that the composition $D \to C \to C''$ is an admissible epimorphism, then f is an admissible epimorphism; dually for admissible monomorphisms.

According to Keller [**13**], axiom (c) follows from the other two. We will use the standard notation $\rightarrowtail$ for admissible monomorphisms and $\twoheadrightarrow$ for admissible epimorphisms.

A *preabelian category* is an additive category in which every morphism has a kernel and a cokernel [**10**]. Every morphism $f \colon F \to G$ in a preabelian category has a canonical decomposition

$$f \colon X \xrightarrow{\operatorname{coim}(f)} \operatorname{coim}(f) \xrightarrow{\bar{f}} \operatorname{im}(f) \xrightarrow{\operatorname{im}(f)} Y$$

where $\operatorname{coim}(f) = \operatorname{coker}(\ker f)$ is the coimage of f and $\operatorname{im}(f) = \ker(\operatorname{coker} f)$ is the image of f. Recall that an *abelian category* is a preabelian category such that every morphism f is *balanced*, that is, the canonical map $\bar{f} \colon \operatorname{coim}(f) \to \operatorname{im}(f)$ is an isomorphism. An abelian category has the canonical exact structure where all kernels and cokernels are respectively admissible monomorphisms and admissible epimorphisms.

A *subobject* of a fixed object F is a monic $m \colon F' \to F$. The collection of all subobjects of F forms a category where morphisms are morphisms $j \colon F' \to F''$ between two subobjects of F such that $m''j = m'$. Notice that such j are also monic. If the category is an exact category, there is the subcategory of *admissible subobjects* of F represented by admissible monomorphisms. If both m' and m'' are admissible, it follows from exactness axiom 3 that j is also an admissible monomorphism.

Given two subobjects $m' \colon F' \to F$, $m'' \colon F'' \to F$, the *intersection* $F' \cap F''$, which is the pullback of m' along m'', is a subobject of F and can be written as the kernel of a morphism. If F' and F'' are admissible subobjects then the intersection $F' \cap F''$ is an admissible subobject.

Now let $\mathbf{A}$ be a cocomplete abelian category. The power set $\mathcal{P}(X)$ of a proper metric space X is partially ordered by inclusion which makes it into the category with subsets of X as objects and unique morphisms (S,T) when $S \subset T$. A *presheaf* of $\mathbf{A}$-objects on X is a functor $F \colon \mathcal{P}(X) \to \mathbf{A}$. This corresponds to the usual notion of presheaf of $\mathbf{A}$-objects on the discrete topological space X^δ if the chosen Grothendieck topology on $\mathcal{P}(X)$ is the partial order given by inclusion, cf. section II.1 of [**12**]. We will use terms which are standard in sheaf theory such as *structure maps*, when referring to the morphisms $F(S,T)$.

Definition 2.3. A presheaf of objects in $\mathbf{A}$ on X is an *X-filtered object in* $\mathbf{A}$ if all the structure maps of F are monomorphisms. We will often suppress the reference to $\mathbf{A}$ when the meaning is clear from the context.

For each presheaf F there is an associated X-filtered object given by

$$F_X(S) = \operatorname{im} F(S,X).$$

Suppose F is an X-filtered object. Given a subobject $F' \subset F(X)$ in $\mathbf{A}$, define the *standard filtration* of F' induced from F by the formula

$$F'(S) = F(S) \cap F'.$$

In other words, $F'(S)$ is the image of the pullback

$$\begin{array}{ccc} P & \longrightarrow & F(S) \\ \downarrow & & \downarrow \\ F' & \longrightarrow & F(X) \end{array}$$

Definition 2.4. The *uncontrolled* category $\mathbf{U}(X,\mathbf{A})$ is the category of X-filtered objects in $\mathbf{A}$. The morphisms $F \to G$ in $\mathbf{U}(X,\mathbf{A})$ are the morphisms $F(X) \to G(X)$ in $\mathbf{A}$.

Let $S[D]$ denote the subset $\{x \in X \mid \operatorname{dist}(x,S) \leqslant D\}$. A morphism $f \colon F \to G$ in $\mathbf{U}(X,\mathbf{A})$ is *boundedly controlled* if there is a number $D \geqslant 0$ such that the image of f restricted to $F(S)$ is a subobject of $G(S[D])$ for every subset $S \subset X$.

The category $\mathbf{U}^{\mathrm{b}}(X,\mathbf{A})$ is the full subcategory of $\mathbf{U}(X,\mathbf{A})$ on the objects with the property $F(\varnothing) = 0$ and the boundedly controlled morphisms.

If f in addition has the property that for all subsets $S \subset X$ the pullback $\operatorname{im}(f) \cap G(S)$ is a subobject of $fF(S[D])$, then f is called *boundedly bicontrolled.* In this case we say that f has filtration degree D and write $\operatorname{fil}(f) \leqslant D$.

Lemma 2.5. *Let $f_1 \colon F \to G$, $f_2 \colon G \to H$ be in $\mathbf{U}^{\mathrm{b}}(X,\mathbf{A})$ and $f_3 = f_2 f_1$.*

1. *If f_1, f_2 are boundedly bicontrolled morphisms and either $f_1 \colon F(X) \to G(X)$ is an epi or $f_2 \colon G(X) \to H(X)$ is a monic, then f_3 is also boundedly bicontrolled.*
2. *If f_1, f_3 are boundedly bicontrolled and f_1 is epic then f_2 is also boundedly bicontrolled; if f_3 is only boundedly controlled then f_2 is also boundedly controlled.*
3. *If f_2, f_3 are boundedly bicontrolled and f_2 is monic then f_1 is also boundedly bicontrolled; if f_3 is only boundedly controlled then f_1 is also boundedly controlled.*

Proof. Suppose $\mathrm{fil}(f_i) \leqslant D$ and $\mathrm{fil}(f_j) \leqslant D'$ for $\{i,j\} \subset \{1,2,3\}$, then in fact $\mathrm{fil}(f_{6-i-j}) \leqslant D+D'$ in each of the three cases. For example, there are factorizations

$$f_2G(S) \subset f_2f_1F(S[D]) = f_3F(S[D]) \subset H(S[D+D'])$$
$$f_2G(X) \cap H(S) \subset f_3F(S[D']) = f_2f_1F(S[D']) \subset f_2G(S[D+D'])$$

which verify part 2 with $i = 1$, $j = 3$. □

Proposition 2.6. $\mathbf{U}^{\mathrm{b}}(X, \mathbf{A})$ *is an additive category with kernels and cokernels.*

Proof. Additive properties are inherited from $\mathbf{A}$. In particular, the biproduct is given by the filtration-wise operation

$$(F \oplus G)(S) = F(S) \oplus G(S)$$

in $\mathbf{A}$. For any boundedly controlled morphism $f\colon F \to G$, the kernel of f in $\mathbf{A}$ has the standard X-filtration K where

$$K(S) = \ker(f) \cap F(S)$$

which gives the kernel of f in $\mathbf{U}^{\mathrm{b}}(X, \mathbf{A})$. The canonical monic $\kappa\colon K \to F$ has filtration degree 0 and is therefore boundedly bicontrolled. It follows from part 3 of Lemma 2.5 that K has the universal properties of the kernel in $\mathbf{U}^{\mathrm{b}}(X, \mathbf{A})$.

Similarly, let I be the standard X-filtration of the image of f in $\mathbf{A}$ by

$$I(S) = \mathrm{im}(f) \cap G(S).$$

Then there is a presheaf C over X with

$$C(S) = G(S)/I(S)$$

for $S \subset X$. Of course $C(X)$ is the cokernel of f in $\mathbf{A}$. Recall fron Definition 2.3 that there is an X-filtered object C_X associated to C given by

$$C_X(S) = \mathrm{im}\, C(S, X).$$

The canonical morphism $\pi\colon G(X) \to C(X)$ gives a morphism of filtration 0 (and which is therefore boundedly bicontrolled) $\pi\colon G \to C_X$ since

$$\mathrm{im}(\pi G(S, X)) = \mathrm{im}\, C(S, X) = C_X(S).$$

This in conjunction with part 2 of Lemma 2.5 also verifies the universal cokernel properties of C_X and π in $\mathbf{U}^{\mathrm{b}}(X, \mathbf{A})$. □

Remark 2.7. If $\mathbf{A}$ is an abelian category and X is unbounded then $\mathbf{U}^{\mathrm{b}}(X, \mathbf{A})$ is not necessarily an abelian category.

For an explicit description of a boundedly controlled morphism in $\mathbf{U}(\mathbb{Z}, \mathbf{Mod}(R))$ which is an isomorphism of left R-modules but whose inverse is not boundedly controlled, we refer to Example 1.5 of [**17**].

This indicates that under any embedding of $\mathbf{U}^{\mathrm{b}}(X, \mathbf{A})$ in an abelian category $\mathbf{F}$ the kernels and cokernels of some morphisms in $\mathbf{F}$ will be different from those in $\mathbf{U}^{\mathrm{b}}(X, \mathbf{A})$.

One consequence of Remark 2.7 is that $\mathbf{U}^{\mathrm{b}}(X, \mathbf{A})$ is not a balanced category.

Proposition 2.8. *A morphism in* $\mathbf{U}^{\mathrm{b}}(X, \mathbf{A})$ *is balanced if and only if it is boundedly bicontrolled.*

Lemma 2.9. *An isomorphism in* $\mathbf{U}^{\mathrm{b}}(X, \mathbf{A})$ *is boundedly bicontrolled.*

Proof. Suppose f is an isomorphism in $\mathbf{U}^{\mathrm{b}}(X, \mathbf{A})$ bounded by $D(f)$ and let f^{-1} be the inverse bounded by $D(f^{-1})$. Then f is boundedly bicontrolled with filtration degree $\mathrm{fil}(f) \leqslant \max\{D(f^{-1}), D(f)\}$. □

Lemma 2.10. *In any additive category, a morphism* h *is monic if and only if* $\ker(h)$ *exists and is the* 0 *object. Similarly,* h *is epic if and only if* $\mathrm{coker}(h)$ *exists and is the* 0 *object.*

Proof. Suppose $h_1, h_2 \colon F \to G$ and $h \colon G \to H$ are such that $hh_1 = hh_2$, then $h(h_1 - h_2) = 0$. So there is a morphism $F \to \ker(h) = 0$ such that $F \to \ker(h) \to G$ is precisely $h_1 - h_2$. Hence $h_1 - h_2 = 0$ and $h_1 = h_2$. Conversely, if h is monic in a category with a zero object, it is clear that $\ker(h) = 0$. The fact about epics is similar. □

Proof of Proposition 2.8. Let $f \colon F \to G$ be a morphism in $\mathbf{U}^{\mathrm{b}}(X, \mathbf{A})$, and let J be the coimage and I be the image of f. The standard filtration $I(S) = I \cap G(S)$ makes the inclusion $i \colon I \to G$ boundedly bicontrolled of filtration 0. Similarly, J is the cokernel of the inclusion of $\ker(f)$ in F, so J has the filtration described in the proof of Proposition 2.6 which makes the projection $p \colon F \to J$ boundedly bicontrolled of filtration 0.

Necessity of the condition follows from Lemma 2.9.

Now f factors as the composition

$$F \xrightarrow{p} J \xrightarrow{\theta} I \xrightarrow{i} G,$$

where θ is the canonical map. If f is bounded by D then clearly θ is bounded by D and has the 0 object for the kernel and the cokernel. This shows that θ is an isomorphism in $\mathbf{A}$ by Lemma 2.10. In particular, there is an inverse $\theta^{-1} \colon J \to I$ in $\mathbf{A}$. Now the condition that

$$I \cap G(S) \subset fF(S[b])$$

for some number b and a subset $S \subset X$ is equivalent to the condition

$$\theta^{-1}(I(S)) \subset J(S[b]).$$

So f is boundedly bicontrolled if and only if θ^{-1} is bounded and, therefore, is an isomorphism in $\mathbf{U}^{\mathrm{b}}(X, \mathbf{A})$. □

Corollary 2.11. *An isomorphism in* $\mathbf{U}^{\mathrm{b}}(X, \mathbf{A})$ *is a morphism which is an isomorphism in* $\mathbf{A}$ *and is boundedly bicontrolled.*

Definition 2.12. The *admissible monomorphisms* $\mathbf{mU}^{\mathrm{b}}(X, \mathbf{A})$ in $\mathbf{U}^{\mathrm{b}}(X, \mathbf{A})$ consist of boundedly bicontrolled morphisms $m \colon F_1 \to F_2$ such that $m \colon F_1(X) \to F_2(X)$ is a monic in $\mathbf{A}$. The *admissible epimorphisms* $\mathbf{eU}^{\mathrm{b}}(X, \mathbf{A})$ are the boundedly bicontrolled morphisms $e\colon F_1 \to F_2$ such that $e\colon F_1(X) \to F_2(X)$ is an epi in $\mathbf{A}$. The class $\mathcal{E}$ of exact sequences consists of the sequences

$$E^{\cdot}: \quad E' \xrightarrow{i} E \xrightarrow{j} E''$$

with $i \in \mathbf{mU}^{\mathrm{b}}(X, \mathbf{A})$ and $j \in \mathbf{eU}^{\mathrm{b}}(X, \mathbf{A})$ which are exact at E in the sense that $\mathrm{im}(i)$ and $\ker(j)$ represent the same subobject of E.

Theorem 2.13. $\mathbf{U}^{\mathrm{b}}(X, \mathbf{A})$ *is an exact category.*

Proof. (a) It follows from Lemma 2.5 that any short exact sequence $F^{\cdot}$ isomorphic to some $E^{\cdot} \in \mathcal{E}$ is also in $\mathcal{E}$, that

$$F' \xrightarrow{[\mathrm{id},0]} F' \oplus F'' \xrightarrow{[0,\mathrm{id}]^T} F''$$

is in $\mathcal{E}$, and that $i = \ker(j)$, $j = \mathrm{coker}(i)$ in any $E^{\cdot} \in \mathcal{E}$.

(b) The collections of morphisms $\mathbf{mU}^{\mathrm{b}}(X, \mathbf{A})$ and $\mathbf{eU}^{\mathrm{b}}(X, \mathbf{A})$ are closed under composition by part 1 of Lemma 2.5.

Now suppose we are given an exact sequence

$$E^{\cdot}: \quad E' \xrightarrow{i} E \xrightarrow{j} E''$$

in $\mathcal{E}$ and a morphism $f\colon A \to E'' \in \mathbf{U}^{\mathrm{b}}(X, \mathbf{A})$. Let $D(j)$ be a filtration constant for j as a boundedly controlled epi and let $D(f)$ be a bound for f as a boundedly controlled map. There is a base change diagram

$$\begin{array}{ccccc} E' & \longrightarrow & E \times_f A & \xrightarrow{j'} & A \\ {\scriptstyle =}\downarrow & & \downarrow{\scriptstyle f'} & & \downarrow{\scriptstyle f} \\ E' & \longrightarrow & E & \xrightarrow{j} & E'' \end{array}$$

where $m\colon E \times_f A \to E \oplus A$ is the kernel of the epi

$$j \circ \mathrm{pr}_1 - f \circ \mathrm{pr}_2 \colon E \oplus A \longrightarrow E''$$

and $f' = \mathrm{pr}_1 \circ m$, $j' = \mathrm{pr}_2 \circ m$. The X-filtration on $E \times_f A$ is the standard filtration

$$(E \times_f A)(S) = E \times_f A \cap (E(S) \times A(S)).$$

The induced map j' has the same kernel as j and is bounded by 0 since

$$j'((E \times_f A)(S)) \subset A(S).$$

In fact,

$$fA(S) \subset E''(S[D(f)]),$$

so

$$fA(S) \subset jE(S[D(f) + D(j)]),$$

and

$$\mathrm{im}(j') \cap A(S) \subset j'(E \times_f A)(S[D(f) + D(j)]).$$

This shows that j' is boundedly bicontrolled of filtration degree $D(f) + D(j)$.

Therefore, the class of admissible epimorphisms is closed under base change by arbitrary morphisms in $\mathbf{U}^{\mathrm{b}}(X, \mathbf{A})$. Cobase changes by admissible monomorphisms are similar. □

The following Proposition is an organic characterization of the exact structure in $\mathbf{U}^{\mathrm{b}}(X, \mathbf{A})$.

Proposition 2.14. *The exact structure $\mathcal{E}$ in $\mathbf{U}^{\mathrm{b}}(X, \mathbf{A})$ consists of sequences isomorphic to those*

$$E^{\cdot}\colon \quad E' \xrightarrow{i} E \xrightarrow{j} E''$$

which possess restrictions

$$E^{\cdot}(S)\colon \quad E'(S) \xrightarrow{i} E(S) \xrightarrow{j} E''(S)$$

for all subsets $S \subset X$, and each $E^{\cdot}(S)$ is an exact sequence in $\mathbf{A}$.

Proof. Clearly, each of the sequences $E^{\cdot}$ described in the statement is an exact sequence in $\mathbf{U}^{\mathrm{b}}(X, \mathbf{A})$. Indeed, the restriction $i\colon E'(S) \to E(S)$ is monic and $j\colon E(S) \to E''(S)$ is epic, so $i\colon E' \to E$ and $j\colon E \to E''$ are both bicontrolled of filtration 0.

Suppose $F^{\cdot}$ is a sequence isomorphic to such $E^{\cdot}$. There is a commutative diagram

$$\begin{array}{ccccc} F' & \xrightarrow{f} & F & \xrightarrow{g} & F'' \\ \cong\downarrow & & \downarrow\cong & & \downarrow\cong \\ E' & \xrightarrow{i} & E & \xrightarrow{j} & E'' \end{array}$$

Then f and g are compositions of two isomorphisms (which are boundedly bicontrolled by Lemma 2.9) which are either preceded by a boundedly bicontrolled monic or followed by a boundedly bicontrolled epi. By Lemma 2.5, part (1), both f and g are boundedly bicontrolled.

Now suppose $F^{\cdot}$ is an exact sequence in $\mathcal{E}$. Let $K = \ker(g)$ and $C = \operatorname{coker}(f)$, then we obtain a commutative diagram

$$\begin{array}{ccccc} F' & \xrightarrow{f} & F & \xrightarrow{g} & F'' \\ \cong\downarrow & & \downarrow = & & \uparrow\cong \\ K & \xrightarrow{i} & F & \xrightarrow{j} & C \end{array}$$

where the vertical maps are the canonical isomorphisms. By the construction of kernels and cokernels in Proposition 2.6, there are exact sequences

$$K(S) \xrightarrow{i} F(S) \xrightarrow{j} C(S)$$

for all subsets $S \subset X$. □

Definition 2.15. A full subcategory $\mathbf{H}$ of a small exact category $\mathbf{C}$ is said to be *closed under extensions* in $\mathbf{C}$ if $\mathbf{H}$ contains a zero object and for any exact sequence $C' \to C \to C''$ in $\mathbf{C}$, if C' and C'' are isomorphic to objects from $\mathbf{H}$ then so is C.

A *Grothendieck subcategory* of an exact category is a subcategory which is closed under isomorphisms, exact extensions, admissible subobjects, and admissible quotients.

It is known [20] that a subcategory closed under extensions inherits an exact structure from $\mathbf{C}$.

Now let $\mathbf{E}$ be a Grothendieck subcategory of $\mathbf{A}$ and let F be an object of $\mathbf{U}^{\mathrm{b}}(X, \mathbf{A})$.

Definition 2.16. (1) F is $\mathbf{E}$-*local* if $F(V)$ is an object of $\mathbf{E}$ for every bounded subset $V \subset X$.

(2) F is *lean* or *D-lean* if there is a number $D \geqslant 0$ such that for every subset S of X

$$F(S) \subset \sum_{x \in S} F(B_D(x)),$$

where $B_D(x)$ is the metric ball of radius D centered at x.

(3) F is *insular* or *d-insular* if there is a number $d \geqslant 0$ such that

$$F(T) \cap F(U) \subset F(T[d] \cap U[d])$$

for every pair of subsets T, U of X.

Notice that a d-insular object has the property that for any subset $T \subset X$,

$$F(T) \cap F(U) = 0$$

whenever $T \cap U[2d] = \varnothing$.

Remark 2.17. It is clear that properties (1), (2), and (3) are preserved under isomorphisms in $\mathbf{U}^{\mathrm{b}}(X, \mathbf{A})$.

Proposition 2.18. (1) *Lean objects are closed under exact extensions in* $\mathbf{U}^{\mathrm{b}}(X, \mathbf{A})$, *that is, if*

$$E' \longrightarrow E \longrightarrow E''$$

is an exact sequence in $\mathbf{U}^{\mathrm{b}}(X, \mathbf{A})$, *and* E', E'' *are lean, then* E *is lean.*

(2) *Insular objects are closed under exact extensions in* $\mathbf{U}^{\mathrm{b}}(X, \mathbf{A})$.

(3) *If in the exact sequence above the object* E *is lean and insular then*

(a) E' *is insular,*

(b) E'' *is lean,*

(c) E'' *is insular if and only if* E' *is lean.*

Proof. Let

$$E' \xrightarrow{f} E \xrightarrow{g} E''$$

be an exact sequence in $\mathbf{U}^{\mathrm{b}}(X, \mathbf{A})$ and let $b \geqslant 0$ be a common filtration degree for both f and g as boundedly bicontrolled maps.

(1) Assume that both E' and E'' are D-lean as objects of $\mathbf{U}^{\mathrm{b}}(X, \mathbf{A})$. Consider $E(S)$, then

$$gE(S) \subset E''(S[b])$$

and so

$$gE(S) \subset \sum_{x \in S[b]} E''(B_D(x)).$$

For each $x \in S[b]$,

$$E''(B_D(x)) \subset gE(B_{D+b}(x)),$$

so

$$E(S) \subset \sum_{x \in S[b]} E(B_{D+2b}(x)) + \sum_{x \in S[b]} fE'(B_{D+2b}(x)).$$

Therefore

$$E(S) \subset \sum_{x \in S[b]} E(B_{D+3b}(x)) \subset \sum_{x \in S} E(B_{D+4b}(x)),$$

so E is $(D+4b)$-lean.

(2) Assuming that both E' and E'' are d-insular, for any pair of subsets T and U of X,

$$\begin{aligned} & g(E(T) \cap E(U)) \\ \subset\ & E''(T[b]) \cap E''(U[b]) \\ \subset\ & E''(T[b+d] \cap U[b+d]). \end{aligned}$$

Now we have

$$\begin{aligned} & E(T) \cap E(U) \\ \subset\ & E(T[2b+d] \cap U[2b+d]) + fE' \cap E(T[2b+d]) \cap E(U[2b+d]) \\ \subset\ & E(T[2b+d] \cap U[2b+d]) + f(E'(T[3b+d]) \cap E'(U[3b+d]) \\ \subset\ & E(T[2b+d] \cap U[2b+d]) + fE'(T[3b+2d] \cap U[3b+2d]) \\ \subset\ & E(T[4b+2d] \cap U[4b+2d]). \end{aligned}$$

So E is $(4b+2d)$-insular.

(3a) Suppose E is d-insular. Given subsets T and U of X,

$$\begin{aligned} & f(E'(T) \cap E'(U)) \\ \subset\ & fE'(T) \cap fE'(U) \\ \subset\ & E(T[b]) \cap E(U[b]) \\ \subset\ & E(T[b+d]) \cap E(U[b+d]), \end{aligned}$$

so

$$E'(T) \cap E'(U) \subset E'(T[2b+d] \cap U[2b+d]).$$

Thus E' is $(2b+d)$-insular.

(3b) If E is D-lean then for any $S \subset X$, $E''(S) \subset gE(S[b])$. Since

$$E(S[b]) \subset \sum_{x \in S[b]} E(B_D(x)),$$

$$E''(S) \subset \sum_{x \in S[b]} E''(B_{D+b}(x)).$$

So

$$E''(S) \subset \sum_{x \in S} E''(B_{D+2b}(x)),$$

and E'' is $(D+2b)$-lean.

(3c) Suppose E' is D-lean and E is d-insular. For any pair of subsets T, $U \subset X$,

$$E''(T) \cap E''(U) \subset gE(T[b]) \cap gE(U[b]).$$

Given

$$z \in E''(T) \cap E''(U),$$

let $y' \in E(T[b])$ and $y'' \in E(U[b])$ so that

$$g(y') = g(y'') = z.$$

Now

$$k = y' - y'' \in \big(E(T[b]) + E(U[b])\big) \cap \ker(g),$$

so there is

$$\overline{k} \in E'(T[2b]) + E'(U[2b]) \subset E'(T[2b] \cup U[2b])$$

with $f(\overline{k}) = k$. Since E' is D-lean,

$$\overline{k} \in \sum_{x \in T[2b] \cup U[2b]} E'(B_D(x)) = \sum_{x \in T[2b]} E'(B_D(x)) + \sum_{y \in U[2b]} E'(B_D(y)).$$

Hence,

$$\overline{k} \in E'(T[2b+D]) + E'(U[2b+D]).$$

This allows us to write $\overline{k} = \overline{k}_1 + \overline{k}_2$, where $\overline{k}_1 \in E'(T[2b+D])$ and $\overline{k}_2 \in E'(U[2b+D])$. Now

$$k = f\overline{k}_1 + f\overline{k}_2.$$

Notice that

$$y' = y'' + k = y'' + f\overline{k}_1 + f\overline{k}_2.$$

So

$$y = y' - f\overline{k}_1 = y'' + f\overline{k}_2$$

has the property

$$y \in E(T[3b+D]) \cap E(U[3b+D]) \subset E(T[3b+D+d] \cap U[3b+D+d]),$$

and $g(y) = z$. Hence

$$z \in E''(T[4b+D+d] \cap U[4b+D+d]).$$

We conclude that E'' is $(4b+D+d)$-insular. The converse is proved similarly; it is not used in this paper. □

Definition 2.19. An object F of $\mathbf{U}^{\mathrm{b}}(X, \mathbf{A})$ is called *ℓ-strict* or simply *strict* if there exists an order preserving function

$$\ell \colon \mathcal{P}(X) \longrightarrow [0, \infty)$$

such that for every subset S of X the subobject $F_S = F(S)$ is $\mathbf{E}$-local, ℓ_S-lean and ℓ_S-insular with respect to the standard filtration $F_S(U) = F_S \cap F(U)$.

Unlike the subcategory of lean and insular objects, the subcategory of strict objects is not necessarily closed under isomorphisms.

Definition 2.20. The *boundedly controlled* category $\mathbf{B}(X, \mathbf{E})$ is the full subcategory of $\mathbf{U}^{\mathrm{b}}(X, \mathbf{A})$ on objects which are isomorphic to strict objects.

The terminology adopted here is convenient and should not suggest relations to boundedly controlled spaces and maps introduced earlier by Anderson and Munkholm [**1**].

Remark 2.21. The exact subcategory $\mathbf{E}$ is not assumed to be cocomplete. In fact, the construction is most interesting when it is not. Notice also that the notation $\mathbf{B}(X, \mathbf{E})$ does not suggest that the objects F have the terminal piece $F(X)$ in $\mathbf{E}$, unlike the notation for $\mathbf{U}^{\mathrm{b}}(X, \mathbf{A})$ where $F(X)$ are in $\mathbf{A}$. The object $F(X)$ is contained in the cocompletion of $\mathbf{E}$ in $\mathbf{A}$.

Theorem 2.22. $\mathbf{B}(X, \mathbf{E})$ *is closed under extensions in* $\mathbf{U}^{\mathrm{b}}(X, \mathbf{A})$.

Proof. Let

$$E' \xrightarrow{f} E \xrightarrow{g} E''$$

be an exact sequence in $\mathbf{U}^{\mathrm{b}}(X, \mathbf{E})$ and let $b \geqslant 0$ be a common filtration degree for both f and g as boundedly bicontrolled maps. We will also assume, without loss of generality, that both E' and E'' are ℓ-strict for some function $\ell \geqslant 0$. We need to check that E is isomorphic to a strict object.

Since $\mathbf{E}$ is a Grothendieck subcategory of $\mathbf{A}$, for every bounded subset $V \subset X$ the restriction

$$g|E(V) \colon E(V) \longrightarrow gE(V)$$

is an admissible epimorphism onto an admissible subobject of $E''(V[D])$, which is in $\mathbf{E}$. The kernel of $g|E(V)$ is the admissible subobject $\ker(g) \cap E(V)$ of $E(V)$, which is also in $\mathbf{E}$. So $E(V)$ is in $\mathbf{E}$ by closure under extensions in $\mathbf{A}$.

To see that E is isomorphic to a strict object, consider $S \subset X$ so that $E''(S[b])$ is $\ell_{S[b]}$-lean and $\ell_{S[b]}$-insular. The induced epi

$$g \colon E(S[2b]) \cap g^{-1}E''(S[b]) \longrightarrow E''(S[b])$$

extends to another epi

$$g' \colon fE'(S[3b]) + E(S[2b]) \cap g^{-1}E''(S[b]) \longrightarrow E''(S[b])$$

with $\ker(g') = E'(S[3b])$.

Since both $E'(S[3b]$ and $E''(S[b])$ are $\mathbf{E}$-local, $\ell_{S[3b]}$-lean, and $\ell_{S[3b]}$-insular, parts (1) and (2) of Proposition 2.18 show that

$$\widehat{E}(S) = fE'(S[3b]) + E(S[2b]) \cap g^{-1}E''(S[b])$$

is $(3b + 4b)$-lean and $(6b + 4b)$-insular. This makes the filtration $\widehat{E}$ ϕ-strict for the function $\phi_S = \ell_{S[10b]}$.

Clearly, the identity map $\mathrm{id}\colon E(X) \to E(X)$ gives an isomorphism $\mathrm{id}\colon \widehat{E} \to E$ with $\mathrm{fil}(\mathrm{id}) \leqslant 4b$. □

Corollary 2.23. *$\mathbf{B}(X, \mathbf{E})$ is an exact category in the sense of Quillen.*

The bounded theory of geometric free modules described in the Introduction can be generalized to arbitrary additive categories.

Definition 2.24. Given a proper metric space M and an additive category $\mathcal{A}$, Pedersen–Weibel [**17**] define the category of *geometric objects* $\mathcal{B}(M, \mathcal{A})$ as follows. The objects are functions F from M to the objects of $\mathcal{A}$ which satisfy the local finiteness condition: a bounded subset of M contains only finitely many points $x \in M$ such that the values $F(x)$ are nonzero. A morphism $\phi\colon F \to G$ of degree $D \geqslant 0$ is a collection of $\mathcal{A}$-morphisms

$$\phi(x, y)\colon F(x) \to G(y)$$

such that $\phi(x, y)$ is zero unless $d(x, y) \leqslant D$.

The category $\mathcal{B}(M, \mathcal{A})$ is an additive category with the biproduct

$$(F \oplus G)(x) = F(x) \oplus G(x).$$

The associated bounded K-theory spectrum is denoted by $K(M, \mathcal{A})$.

Given an exact category $\mathbf{E}$, one can view $\mathbf{E}$ as an additive category with the underlying split exact structure. When we use $\mathbf{E}$ as coefficients in a category of geometric objects $\mathcal{B}(M, \mathbf{E})$, this is the structure that is implicitly understood.

Corollary 2.25. *The additive category $\mathcal{B}(X, \mathbf{E})$ of geometric objects with the split exact structure is an exact subcategory of $\mathbf{B}(X, \mathbf{E})$.*

Proof. The X-filtration of the geometric objects in $\mathcal{B}(X, \mathbf{E})$ is given by

$$F(S) = \bigoplus_{x \in S} F(x),$$

and the structure maps are the boundedly controlled inclusions and projections onto direct summands. □

Recall that a morphism $e\colon F \to F$ is an idempotent if $e^2 = e$. Categories in which every idempotent is the projection onto a direct summand of F are called *idempotent complete.* Abelian categories are clearly idempotent complete. Thus $\mathbf{A}$ and its Grothendieck subcategories, which are abelian, are idempotent complete.

Corollary 2.26. *The subcategory $\mathbf{B}(X, \mathbf{E})$ is idempotent complete.*

Proof. Since the restriction of an idempotent e to the image of e is the identity, every idempotent is boundedly bicontrolled of filtration 0. It follows easily that the splitting of e in $\mathbf{A}$ is in fact a splitting in $\mathbf{B}(X, \mathbf{E})$. □

Proposition 2.27. *The subcategory $\mathbf{B}(X, \mathbf{E})$ is closed under admissible quotients of strict objects. Precisely, for a given boundedly bicontrolled epi $f\colon F \to G$ in $\mathbf{U}^{\mathrm{b}}(X, \mathbf{A})$ where both F and the kernel $k\colon K \to F$ with the standard filtration $K(S) = K \cap F(S)$ are strict, the cokernel G is isomorphic to a strict object.*

Proof. Suppose $\mathrm{fil}(f) \leqslant b$, then from the assumptions

$$K(S[b]) \longrightarrow F(S[b]) \longrightarrow fF(S[b])$$

is an exact sequence in $\mathbf{U}^{\mathrm{b}}(X, \mathbf{A})$ for any subset $S \subset X$. Since $F(S[b])$ is lean and insular and $K(S[b])$ is lean, the quotient $fF(S[b])$ is lean and insular by Proposition 2.18. It is clear that $fF(S[b])$ is also $\mathbf{E}$-local. Thus the object $\widehat{G}$ with filtration

$$\widehat{G}(S) = fF(S[b])$$

is strict. The identity map induces an isomorphism $\mathrm{id}\colon G \to \widehat{G}$ with $\mathrm{fil}(\mathrm{id}) \leqslant 2b$ because for all $S \subset X$ we have $G(S) \subset \widehat{G}(S)$ and $\widehat{G}(S) \subset G(S[2b])$. □

Remark 2.28. For additional flexibility, one may want to impose weaker requirements on objects in $\mathbf{U}^{\mathrm{b}}(X, \mathbf{A})$. Restricting as in Definition 2.20 to objects F with a fixed locally finite covering $\mathcal{U} \subset \mathcal{P}(X)$ by bounded subsets $U \in \mathcal{B}(X)$ such that

$$F(X) = \sum_{U \in \mathcal{U}} F(U)$$

gives another exact category. In this case, one may also relax the bounded control conditions on the maps to those of Lipschitz type. Similar modifications have become useful in recent work of Hambleton–Pedersen [**11**] and Pedersen–Weibel [**19**] in controlled K-theory.

3. Localization in controlled categories

Definition 3.1. Let F be an object of $\mathbf{B}(X, \mathbf{E})$ and Z be a subset of X. We say F is *supported near* Z if there is a number $d \geqslant 0$ such that $F(X) \subset F(Z[d])$.

Let $\mathbf{B}(X, \mathbf{E})_{<Z}$ be the full subcategory of $\mathbf{B}(X, \mathbf{E})$ on objects supported near Z. If $\mathbf{B}_d(X, \mathbf{E})_{<Z}$ denotes the full subcategory of $\mathbf{B}(X, \mathbf{E})$ with objects F as above then

$$\mathbf{B}(X, \mathbf{E})_{<Z} = \underset{\overrightarrow{d}}{\mathrm{colim}}\, \mathbf{B}_d(X, \mathbf{E})_{<Z}.$$

Proposition 3.2. *$\mathbf{B}(X, \mathbf{E})_{<Z}$ is a Grothendieck subcategory of $\mathbf{B}(X, \mathbf{E})$.*

Proof. First we show closure under exact extensions. Let

$$F' \xrightarrow{\ f\ } F \xrightarrow{\ g\ } F''$$

be an exact sequence in $\mathbf{B}(X, \mathbf{E})$. Let b be a common filtration degree of f and g and let d', $d'' \geqslant 0$ be numbers with $F' = F'(Z[d'])$ and $F'' = F''(Z[d''])$. Since $F = I + M$, where $I = \mathrm{im}(f)$ and M is any subobject $M \subset F$ with $g(M) = F''$, it suffices to show that for some $d \geqslant 0$

$$I(X) = I(Z[d]) \subset F(Z[d]),$$

and that M can be chosen to be a subobject of $F(Z[d])$. Indeed,

$$I(X) = fF'(X) = fF'(Z[d']) \subset F(Z[d' + b]),$$
$$F''(X) = gF(X) \cap F''(Z[d'']) \subset gF(Z[d'' + b]).$$

Let $M = F(Z[d''+b])$. If we choose $d = \max\{d'+b, d''+b\}$ then $F = F(Z[d])$ is in $\mathbf{B}(X,R)_{<Z}$.

Now suppose $f\colon F' \to F$ is an admissible subobject in $\mathbf{B}(X,\mathbf{E})$, which is a boundedly bicontrolled monic with $\mathrm{fil}(f) \leqslant b$, $F = F(Z[d])$, and F is c-insular. Also suppose F and F' are respectively D- and D'-lean, then notice that

$$fF'(B_{D'}(x)) \subset F(B_{D'+b}(x)),$$

while $F(X) \subset F(Z[d])$. Therefore,

$$F'(B_{D'}(x)) = 0$$

for

$$x \in X - Z[d+D+D'+b+2c].$$

This means that

$$F' = F'(Z[d+D+2D'+b+2c]).$$

On the other hand, if $g\colon F \to F''$ is an admissible quotient of filtration b then it is easy to see that

$$F'' = F''(Z[d+D+b])$$

is also in $\mathbf{B}(X,\mathbf{E})_{<Z}$. Since $\mathbf{B}(X,\mathbf{E})_{<Z}$ is clearly closed under isomorphisms, this proves the assertion. □

Given an object $F \in \mathbf{B}(X,\mathbf{E})$ and a subset $T \subset X$, we will need a construction of an admissible subobject $\widetilde{F}$ of F in $\mathbf{B}(X,\mathbf{E})$ such that

$$F(T) \subset \widetilde{F} \subset F(T[D])$$

for some $D \geqslant 0$.

Choose a strict F' isomorphic to F in $\mathbf{B}(X,\mathbf{E})$ and assume the chosen isomorphism and its inverse are of filtration b.

Lemma 3.3. *The subobject $\widetilde{F} = F'(T[b])$ is an admissible subobject of F in $\mathbf{B}(X,\mathbf{E})$ and satisfies*

$$F(T) \subset \widetilde{F} \subset F(T[2b]).$$

Proof. The cokernel G' of the inclusion $k\colon F'(T[b]) \to F$ is in $\mathbf{B}(X,\mathbf{E})$ by Proposition 2.27. We can view $F'(T[b])$ as an admissible subobject of F with the cokernel G isomorphic to G'. □

Definition 3.4. A class of morphisms Σ in an additive category $\mathbf{C}$ *admits a calculus of right fractions* if

1. the identity of each object is in Σ,
2. Σ is closed under composition,

3. each diagram $F \xrightarrow{f} G \xleftarrow{s} G'$ with $s \in \Sigma$ can be completed to a commutative square

$$\begin{array}{ccc} F' & \xrightarrow{f'} & G' \\ \downarrow{\scriptstyle t} & & \downarrow{\scriptstyle s} \\ F & \xrightarrow{f} & G \end{array}$$

with $t \in \Sigma$, and

4. if f is a morphism in $\mathbf{C}$ and $s \in \Sigma$ such that $sf = 0$ then there exists $t \in \Sigma$ such that $ft = 0$.

In this case there is a construction of the *localization* $\mathbf{C}[\Sigma^{-1}]$ which has the same objects as $\mathbf{C}$. The morphism sets $\mathrm{Hom}(F, G)$ in $\mathbf{C}[\Sigma^{-1}]$ consist of equivalence classes of diagrams

$$(s, f)\colon \quad F \xleftarrow{s} F' \xrightarrow{f} G$$

with the equivalence relation generated by $(s_1, f_1) \sim (s_2, f_2)$ if there is a map $h\colon F_1' \to F_2'$ so that $f_1 = f_2 h$ and $s_1 = s_2 h$. Let $(s|f)$ denote the equivalence class of (s, f).

The composition of morphisms in $\mathbf{C}[\Sigma^{-1}]$ is defined by

$$(s|f) \circ (t|g) = (st'|gf')$$

where g' and s' fit in the commutative square

$$\begin{array}{ccc} F'' & \xrightarrow{f'} & G' \\ \downarrow{\scriptstyle t'} & & \downarrow{\scriptstyle t} \\ F & \xrightarrow{f} & G \end{array}$$

from axiom 3.

Proposition 3.5. *The localization* $\mathbf{C}[\Sigma^{-1}]$ *is a category. The morphisms of the form* $(\mathrm{id}\,|s)$ *where* $s \in \Sigma$ *are isomorphisms in* $\mathbf{C}[\Sigma^{-1}]$*. The rule* $P_\Sigma(f) = (\mathrm{id}\,|f)$ *gives a functor*

$$P_\Sigma\colon \mathbf{C} \longrightarrow \mathbf{C}[\Sigma^{-1}]$$

which is universal among the functors making the morphisms Σ *invertible.*

Proof. The proofs of these facts can be found in Chapter I of [**9**]. The inverse of $(\mathrm{id}\,|s)$ is $(s|\,\mathrm{id})$. □

Suppose $\mathbf{E}$ is a Grothendieck subcategory of a cocomplete abelian category $\mathbf{A}$, and let $\mathbf{Z}$ be the subcategory $\mathbf{B}(X, \mathbf{E})_{<Z}$ of $\mathbf{B} = \mathbf{B}(X, \mathbf{E})$ for a fixed choice of $Z \subset X$. Let the class of *weak equivalences* Σ consist of all finite compositions of admissible monomorphisms with cokernels in $\mathbf{Z}$ and admissible epimorphisms with kernels in $\mathbf{Z}$. We will show that the class Σ admits a calculus of right fractions.

Definition 3.6. A Grothendieck subcategory $\mathbf{Z} \subset \mathbf{B}$ is *right filtering* if each morphism $f\colon F \to G$ in $\mathbf{B}$, where G is an object of $\mathbf{Z}$, factors through an admissible epimorphism $e\colon F \to \overline{G}$, where $\overline{G}$ is in $\mathbf{Z}$.

Lemma 3.7. *The subcategory* $\mathbf{Z} = \mathbf{B}(X, \mathbf{E})_{<Z}$ *of* $\mathbf{B} = \mathbf{B}(X, \mathbf{E})$ *is right filtering.*

Proof. Suppose first that F and G are strict with the characteristic functions ℓ_F and ℓ_G respectively. Let $L_F = \ell_F(X)$ and $L_G = \ell_G(X)$. If

$$G = G(Z[d_G])$$

and the given morphism $f \colon F \to G$ is bounded by d then we have

$$fF(B_{L_G}(x)) \subset G(B_{L_G+b}(x)) = 0$$

for all

$$x \in X - Z[d_G + L_G + 2L_G + d + L_F].$$

Let

$$E = F(X - Z[d_G + L_G + 2L_G + d + L_F]),$$

then $fE = 0$. Now E is an admissible subobject of F by Lemma 3.3; let $\overline{G}$ be the cokernel of the inclusion. Since

$$\overline{G}(B_{L_F}(x)) = 0$$

for all

$$x \in X - Z[d_G + L_G + 2L_G + d + L_F],$$

we have

$$\overline{G} = \overline{G}(Z[d_G + 2L_G + 2L_G + d + L_F])$$

as an object of $\mathbf{Z}$. The required factorization is the right square in the map between the two exact sequences

$$\begin{array}{ccccc} E & \longrightarrow & F & \xrightarrow{j'} & \overline{G} \\ {\scriptstyle i}\downarrow & & \downarrow{\scriptstyle =} & & \downarrow \\ K & \xrightarrow{k} & F & \xrightarrow{f} & G \end{array}$$

If F and G are not strict, one considers a map $f' \colon F' \to G'$ between strict objects isomorphic to F and G and chooses the subobject

$$E = F'(X - Z[d_G + L_G + 2L_G + d + L_F + 4b])$$

of F' for an appropriate value of b, as in the proof of Lemma 3.3. □

Corollary 3.8. *The class* Σ *admits a calculus of right fractions.*

Proof. This follows from Lemma 3.7, see Lemma 1.13 of [**22**]. □

Definition 3.9. The *quotient category* $\mathbf{B}/\mathbf{Z}$ is the localization $\mathbf{B}[\Sigma^{-1}]$.

It is clear that the quotient $\mathbf{B}/\mathbf{Z}$ is an additive category, and P_Σ is an additive functor. In fact, we have the following.

Theorem 3.10. *The short sequences in* $\mathbf{B}/\mathbf{Z}$ *which are isomorphic to images of exact sequences from* $\mathbf{B}$ *form a Quillen exact structure.*

The proof uses the following fact.

Definition 3.11. An extension closed full subcategory $\mathbf{Z}$ of an exact category $\mathbf{B}$ is called *right s-filtering* if whenever $f\colon F \to G$ is an admissible monomorphism with F in $\mathbf{Z}$ then there exist an object E in $\mathbf{Z}$ and an admissible epimorphism $e\colon G \to E$ such that the composition ef is an admissible monomorphism.

Lemma 3.12 (Schlichting). *Let $\mathbf{Z}$ be a right filtering and right s-filtering subcategory in $\mathbf{B}$. Then the quotient category $\mathbf{B}/\mathbf{Z}$, equipped with the images of the exact sequences from $\mathbf{B}$, is an exact category. Moreover, exact functors from $\mathbf{B}$ vanishing on $\mathbf{Z}$ are in bijective correspondence with exact functors from $\mathbf{B}/\mathbf{Z}$.*

Proof. See Proposition 1.16 of [**22**]. □

Proof of Theorem 3.10. Since $\mathbf{Z}$ is right filtering by Lemma 3.7, it remains to check that $\mathbf{Z}$ is right s-filtering.

Again, suppose that F and G are strict with the characteristic functions ℓ_F and ℓ_G and let $L_F = \ell_F(X)$ and $L_G = \ell_G(X)$. Assume that $F = F(Z[d_F])$, $\mathrm{fil}(f) \leqslant d$, and let

$$G' = G(X - Z[d_F + 2L_F + 2L_G + d + L_G]).$$

Let $e\colon G \to E$ be the cokernel of the inclusion, then $f(F) \cap G' = 0$, so that ef is an admissible monomorphism with $\mathrm{fil}(ef) = \mathrm{fil}(f) \leqslant d$. If F and G are not strict but are isomorphic to strict objects, one makes obvious adjustments. □

The main tool in proving controlled excision in the boundedly controlled K-theory will be the following localization theorem.

Theorem 3.13 (Theorem 2.1 of Schlichting [**22**]). *Let $\mathbf{Z}$ be an idempotent complete right s-filtering subcategory of an exact category $\mathbf{B}$. Then the sequence of exact categories $\mathbf{Z} \to \mathbf{B} \to \mathbf{B}/\mathbf{Z}$ induces a homotopy fibration of K-theory spectra*

$$K(\mathbf{Z}) \longrightarrow K(\mathbf{B}) \longrightarrow K(\mathbf{B}/\mathbf{Z}).$$

4. Bounded excision theorem

The proof of controlled excision in the boundedly controlled G-theory requires the context of Waldhausen K-theory of derived categories.

Definition 4.1 (Waldhausen categories). A *Waldhausen category* is a category $\mathbf{D}$ with a zero object 0 together with two chosen subcategories of *cofibrations* $\mathrm{co}(\mathbf{D})$ and *weak equivalences* $\boldsymbol{w}(\mathbf{D})$ satisfying the four axioms:

1. every isomorphism in $\mathbf{D}$ is in both $\mathrm{co}(\mathbf{D})$ and $\boldsymbol{w}(\mathbf{D})$,
2. every map $0 \to D$ in $\mathbf{D}$ is in $\mathrm{co}(\mathbf{D})$,
3. if $A \to B \in \mathrm{co}(\mathbf{D})$ and $A \to C \in \mathbf{D}$ then the pushout $B \cup_A C$ exists in $\mathbf{D}$, and the canonical map $C \to B \cup_A C$ is in $\mathrm{co}(\mathbf{D})$,

4. ("gluing lemma") given a commutative diagram

$$\begin{array}{ccccc} B & \xleftarrow{a} & A & \longrightarrow & C \\ \downarrow & & \downarrow & & \downarrow \\ B' & \xleftarrow{a'} & A' & \longrightarrow & C' \end{array}$$

in $\mathbf{D}$, where the morphisms a and a' are in $\mathrm{co}(\mathbf{D})$ and the vertical maps are in $\boldsymbol{w}(\mathbf{D})$, the induced map $B \cup_A C \to B' \cup_{A'} C'$ is also in $\boldsymbol{w}(\mathbf{D})$.

A Waldhausen category $\mathbf{D}$ with weak equivalences $\boldsymbol{w}(\mathbf{D})$ is often denoted by $\boldsymbol{w}\mathbf{D}$ as a reminder of the choice. A functor between Waldhausen categories is exact if it preserves cofibrations and weak equivalences.

A Waldhausen category may or may not satisfy the following additional axioms.

Saturation axiom 4.2. Given two morphisms $\phi\colon F \to G$ and $\psi\colon G \to H$ in $\mathbf{D}$, if any two of ϕ, ψ, or $\psi\phi$, are in $\boldsymbol{w}(\mathbf{D})$ then so is the third.

Extension axiom 4.3. Given a commutative diagram

$$\begin{array}{ccccc} F & \longrightarrow & G & \longrightarrow & H \\ \downarrow{\scriptstyle\phi} & & \downarrow{\scriptstyle\psi} & & \downarrow{\scriptstyle\mu} \\ F' & \longrightarrow & G' & \longrightarrow & H' \end{array}$$

with exact rows, if both ϕ and μ are in $\boldsymbol{w}(\mathbf{D})$ then so is ψ.

A *cylinder functor* on $\mathbf{D}$ is a functor C from the category of morphisms $f\colon F \to G$ in $\mathbf{D}$ to $\mathbf{D}$ together with three natural transformations $j_1\colon F \to C(f)$, $j_2\colon G \to C(f)$, and $p\colon C(f) \to G$ which satisfy $pj_2 = \mathrm{id}_G$ and $pj_1 = f$ for all f. The cylinder functor has to satisfy a number of properties listed in point 1.3.1 of [**25**]. They will be rather automatic for the functors we construct later.

Cylinder axiom 4.4. A cylinder functor C satisfies this axiom if for all morphisms $f\colon F \to G$ the required map p is in $\boldsymbol{w}(\mathbf{D})$.

Let $\mathbf{D}$ be a small Waldhausen category with respect to two categories of weak equivalences $\boldsymbol{v}(\mathbf{D}) \subset \boldsymbol{w}(\mathbf{D})$ with a cylinder functor T both for $\boldsymbol{v}\mathbf{D}$ and for $\boldsymbol{w}\mathbf{D}$ satisfying the cylinder axiom for $\boldsymbol{w}\mathbf{D}$. Suppose also that $\boldsymbol{w}(\mathbf{D})$ satisfies the extension and saturation axioms.

Define $\boldsymbol{v}\mathbf{D}^{\boldsymbol{w}}$ to be the full subcategory of $\boldsymbol{v}\mathbf{D}$ whose objects are F such that $0 \to F \in \boldsymbol{w}(\mathbf{D})$. Then $\boldsymbol{v}\mathbf{D}^{\boldsymbol{w}}$ is a small Waldhausen category with cofibrations $\mathrm{co}(\mathbf{D}^{\boldsymbol{w}}) = \mathrm{co}(\mathbf{D}) \cap \mathbf{D}^{\boldsymbol{w}}$ and weak equivalences $\boldsymbol{v}(\mathbf{D}^{\boldsymbol{w}}) = \boldsymbol{v}(\mathbf{D}) \cap \mathbf{D}^{\boldsymbol{w}}$. The cylinder functor T for $\boldsymbol{v}\mathbf{D}$ induces a cylinder functor for $\boldsymbol{v}\mathbf{D}^{\boldsymbol{w}}$. If T satisfies the cylinder axiom then the induced functor does so too.

Theorem 4.5 (Approximation theorem). *Let $E\colon \mathbf{D}_1 \to \mathbf{D}_2$ be an exact functor between two small saturated Waldhausen categories. It induces a map of K-theory spectra*

$$K(E)\colon K(\mathbf{D}_1) \longrightarrow K(\mathbf{D}_2).$$

Assume that $\mathbf{D}_1$ has a cylinder functor satisfying the cylinder axiom. If E satisfies two conditions:

1. *a morphism $f \in \mathbf{D}_1$ is in $\boldsymbol{w}(\mathbf{D}_1)$ if and only if $E(f) \in \mathbf{D}_2$ is in $\boldsymbol{w}(\mathbf{D}_2)$,*
2. *for any object $D_1 \in \mathbf{D}_1$ and any morphism $g \colon E(D_1) \to D_2$ in $\mathbf{D}_2$, there is an object $D_1' \in \mathbf{D}_1$, a morphism $f \colon D_1 \to D_1'$ in $\mathbf{D}_1$, and a weak equivalence $g' \colon E(D_1') \to D_2 \in \boldsymbol{w}(\mathbf{D}_2)$ such that $g = g'E(f)$,*

then $K(E)$ is a homotopy equivalence.

Proof. This is Theorem 1.6.7 of [**26**]. The presence of the cylinder functor with the cylinder axiom allows to make condition 2 weaker than that of Waldhausen, see point 1.9.1 in [**25**]. □

Definition 4.6. In any additive category, a sequence of morphisms

$$E^{\cdot}\colon \quad 0 \longrightarrow E^1 \xrightarrow{d_1} E^2 \xrightarrow{d_2} \ \ldots \ \xrightarrow{d_{n-1}} E^n \longrightarrow 0$$

is called a *(bounded) chain complex* if the compositions $d_{i+1}d_i$ are the zero maps for all $i = 1, \ldots, n-1$. A *chain map* $f \colon F^{\cdot} \to E^{\cdot}$ is a collection of morphisms $f^i \colon F^i \to E^i$ such that $f^i d_i = d_i f^i$.

A chain map f is *null-homotopic* if there are morphisms $s_i \colon F^{i+1} \to E^i$ such that $f = ds + sd$. Two chain maps f, $g \colon F^{\cdot} \to E^{\cdot}$ are *chain homotopic* if $f - g$ is null-homotopic. Now f is a *chain homotopy equivalence* if there is a chain map $h \colon E^i \to F^i$ such that the compositions fh and hf are chain homotopic to the respective identity maps.

The Waldhausen structures on categories of bounded chain complexes are based on homotopy equivalence as a weakening of the notion of isomorphism of chain complexes.

Definition 4.7. A sequence of maps in an exact category is called *acyclic* if it is assembled out of short exact sequences in the sense that each map factors as the composition of the cokernel of the preceding map and the kernel of the succeeding map.

It is known that the class of acyclic complexes in an exact category is closed under isomorphisms in the homotopy category if and only if the category is idempotent complete, which is also equivalent to the property that each contractible chain complex is acyclic, cf. [**14**, sec. 11].

Definition 4.8. Given an exact category $\mathbf{E}$, there is a standard choice for the Waldhausen structure on the derived category $\mathbf{E}'$ of bounded chain complexes in $\mathbf{E}$. The cofibrations $\mathbf{co}(\mathbf{E}')$ are the degree-wise admissible monomorphisms. The weak equivalences $\boldsymbol{v}(\mathbf{E}')$ are the chain maps whose mapping cones are homotopy equivalent to acyclic complexes.

We will denote this Waldhausen structure by $\boldsymbol{v}\mathbf{E}'$.

Proposition 4.9. *The category $\boldsymbol{v}\mathbf{E}'$ is a Waldhausen category satisfying the extension and saturation axioms and has cylinder functor satisfying the cylinder axiom.*

Proof. The pushouts along cofibrations in $\mathbf{E}'$ are the complexes of pushouts in each degree. All standard Waldhausen axioms including the gluing lemma are clearly satisfied. The saturation and the extension axioms are also clear.

The cylinder functor C for $\boldsymbol{v}\mathbf{E}'$ is defined using the canonical homotopy pushout as in point 1.1.2 in Thomason–Trobaugh [**25**]. Given a chain map $f\colon F \to G$, $C(f)$ is the canonical homotopy pushout of f and the identity $\mathrm{id}\colon F \to F$. With this construction, the map $p\colon C(f) \to G$ is a chain homotopy equivalence, so the cylinder axiom is also satisfied. □

Definition 4.10. There are two choices for the Waldhausen structure on the bounded derived category $\mathbf{B}'$ for the exact boundedly controlled category $\mathbf{B} = \mathbf{B}(X, \mathbf{E})$. One is $\boldsymbol{v}\mathbf{B}'$ as in Definition 4.8. Given a metric subspace Z in X, the other choice for the weak equivalences $\boldsymbol{w}(\mathbf{B}')$ is the chain maps whose mapping cones are homotopy equivalent to acyclic complexes in the quotient $\mathbf{B}/\mathbf{Z}$.

We will denote the second Waldhausen structure by $\boldsymbol{w}\mathbf{B}'$.

Corollary 4.11. *The categories $\boldsymbol{v}\mathbf{B}'$ and $\boldsymbol{w}\mathbf{B}'$ are Waldhausen categories satisfying the extension and saturation axioms and have cylinder functors satisfying the cylinder axiom.*

Proof. All axioms and constructions, including the cylinder functor, for $\boldsymbol{w}\mathbf{B}'$ are inherited from $\boldsymbol{v}\mathbf{B}'$. □

The K-theory functor from the category of small Waldhausen categories $\mathbf{D}$ and exact functors to connective spectra is defined in terms of $S.$-construction as in Waldhausen [**26**]. It extends to simplicial categories $\mathbf{D}$ with cofibrations and weak equivalences and inductively gives the connective spectrum

$$n \mapsto |\boldsymbol{w}S.^{(n)}\,\mathbf{D}\,|.$$

We obtain the functor assigning to $\mathbf{D}$ the connective Ω-spectrum

$$K(\mathbf{D}) = \Omega^{\infty}|\boldsymbol{w}S.^{(\infty)}\,\mathbf{D}\,| = \underset{\overrightarrow{n \geqslant 1}}{\mathrm{colim}}\ \Omega^{n}|\boldsymbol{w}S.^{(n)}\,\mathbf{D}\,|$$

representing the Waldhausen algebraic K-theory of $\mathbf{D}$. For example, if $\mathbf{D}$ is the additive category of free finitely generated R-modules with the canonical Waldhausen structure, then the stable homotopy groups of $K(\mathbf{D})$ are the usual K-groups of the ring R. In fact, there is a general identification of the two theories.

Recall that for any exact category $\mathbf{E}$, the derived category $\mathbf{E}'$ has the Waldhausen structure $\boldsymbol{v}\mathbf{E}'$ as in Definition 4.8.

Theorem 4.12. *The Quillen K-theory of an exact category $\mathbf{E}$ is equivalent to the Waldhausen K-theory of $\boldsymbol{v}\mathbf{E}'$.*

Proof. The proof is based on repeated applications of the additivity theorem, cf. Thomason's Theorem 1.11.7 [**25**]. Thomason's proof of his Theorem 1.11.7 can be repeated verbatim here. It is in fact simpler in this case since condition 1.11.3.1 is not required. □

Let $\mathbf{E}$ be a Grothendieck subcategory of a cocomplete abelian category $\mathbf{A}$ and let (Z, X) be a pair of proper metric spaces.

We will use the notation $\mathbf{B} = \mathbf{B}(X, \mathbf{E})$ and $\mathbf{Z} = \mathbf{B}(X, \mathbf{E})_{<Z}$.

Theorem 4.13 (Localization). *If $\mathbf{E}$ is idempotent complete, there is a homotopy fibration*

$$K(Z, \mathbf{E}) \longrightarrow K(X, \mathbf{E}) \longrightarrow K(\mathbf{B}/\mathbf{Z}).$$

This is a direct consequence of Theorem 3.13 as soon as we identify $K(Z, \mathbf{E})$ with $K(\mathbf{Z}) = K(X, \mathbf{E})_{<Z}$.

Recall that the *essential full image* of a functor $F\colon \mathbf{C} \to \mathbf{D}$ is the full subcategory of $\mathbf{D}$ whose objects are those D such that $D \cong F(C)$ for some C from $\mathbf{C}$.

There is a fully faithful embedding

$$\epsilon\colon \mathbf{B}(Z, \mathbf{E}) \longrightarrow \mathbf{B}(X, \mathbf{E})$$

defined by associating to each filtered object $F \in \mathbf{B}(Z, \mathbf{E})$ the extension $\epsilon(F) \in \mathbf{B}(X, \mathbf{E})$ given by

$$\epsilon(F)(S) = F(S \cap Z).$$

It is clear that $\epsilon(F)$ is strict for strict F.

Lemma 4.14. *The essential full image of $\mathbf{B}(Z, \mathbf{E})$ in $\mathbf{B}(X, \mathbf{E})$ is the Grothendieck subcategory $\mathbf{B}(X, \mathbf{E})_{<Z}$.*

Proof. Of course for each F in $\mathbf{B}(Z, \mathbf{E})$, the image $\epsilon(F)$ is in $\mathbf{B}(X, \mathbf{E})_{<Z}$. Now if $G(X) = G(Z[d])$ is an object of $\mathbf{B}(X, \mathbf{E})_{<Z}$ then there is a bounded function $r\colon Z[d] \to Z$, bounded by d, which gives an object $R = R(G)$ of $\mathbf{B}(Z, \mathbf{E})$ by the assignment $R(S) = G(r^{-1}(S))$.

If G is strict then the new object R is $\mathbf{E}$-local and strict with $\ell_R = \ell_G + d$. Since the identity map $\mathrm{id}\colon R \to G$ is boundedly bicontrolled with $\mathrm{fil}(\mathrm{id}) \leqslant 2d$, it is an isomorphism in $\mathbf{B}(X, \mathbf{E})$. □

Corollary 4.15. *For any pair of proper metric spaces $Z \subset X$, there is a weak equivalence $K(Z, \mathbf{E}) \simeq K(X, \mathbf{E})_{<Z}$.*

Now Theorem 4.13 follows from the localization fibration in Theorem 3.13.

The results of Theorem 4.13 and Lemma 4.14 can be generalized to a more general and convenient geometric situation.

Suppose Z is an arbitrary subset of a proper metric space X. It is a metric subspace with the metric which is the restriction of the metric in X. When Z is a closed subset then the closed metric balls in Z are closed subsets of closed metric balls in X and, therefore, compact. If Z is an arbitrary subset of X, the closure $\overline{Z}$ is a proper metric subspace. There is an inclusion $\overline{Z} \subset Z[\epsilon]$ for any $\epsilon > 0$, so there is an ϵ-bounded retraction of $\overline{Z}$ onto Z. In addition to the obvious equivalence of categories $\mathbf{B}(X, \mathbf{E})_{<\overline{Z}} = \mathbf{B}(X, \mathbf{E})_{<Z}$, the retraction also induces an isomorphism $\mathbf{B}(\overline{Z}, \mathbf{E}) \cong \mathbf{B}(Z, \mathbf{E})$.

Corollary 4.16 (Localization). *Suppose X is a proper metric space and Z is a subset with the induced metric. There is a weak homotopy equivalence*

$$K(\overline{Z}, \mathbf{E}) \simeq K(Z, \mathbf{E}).$$

If $\mathbf{E}$ is idempotent complete, there is a homotopy fibration

$$K(Z, \mathbf{E}) \longrightarrow K(X, \mathbf{E}) \longrightarrow K(\mathbf{B}/\mathbf{Z}).$$

The computational tools from nonconnective bounded K-theory, the controlled excision theorems [**3, 17, 18**], can now be adapted to $\mathbf{B}(X, \mathbf{E})$. We will obtain a direct analogue, which is one of the main results of this paper.

Let $\mathbf{E}$ be a Grothendieck subcategory of a cocomplete abelian category $\mathbf{A}$ and let X be a proper metric space. Suppose X_1 and X_2 are subspaces in a proper metric space X, and $X = X_1 \cup X_2$.

We use the notation $\mathbf{B} = \mathbf{B}(X, \mathbf{E})$, $\mathbf{B}_i = \mathbf{B}(X, \mathbf{E})_{<X_i}$ for $i = 1$ or 2, and $\mathbf{B}_{12}$ for the intersection $\mathbf{B}_1 \cap \mathbf{B}_2$.

Now there is a commutative diagram

$$\begin{array}{ccccc} K(\mathbf{B}_{12}) & \longrightarrow & K(\mathbf{B}_1) & \longrightarrow & K(\mathbf{B}_1/\mathbf{B}_{12}) \\ \downarrow & & \downarrow & & \downarrow {\scriptstyle K(I)} \\ K(\mathbf{B}_2) & \longrightarrow & K(\mathbf{B}) & \longrightarrow & K(\mathbf{B}/\mathbf{B}_2) \end{array} \qquad (\dagger)$$

where the rows are homotopy fibrations from Theorem 3.13 and $I\colon \mathbf{B}_1/\mathbf{B}_{12} \to \mathbf{B}/\mathbf{B}_2$ is the exact functor induced from the exact inclusion $I\colon \mathbf{B}_1 \to \mathbf{B}$. We observe that I is not necessarily full and, therefore, not an isomorphism of categories as in similar applications in [**3**] and [**23**]. Nevertheless, we claim that $K(I)$ is a weak equivalence.

Lemma 4.17. *Let Z be a subset of X, so $\mathbf{Z} = \mathbf{B}(X, \mathbf{E})_{<Z}$ is a Grothendieck subcategory of $\mathbf{B}$. If $f^{\cdot}$ is a degreewise admissible monomorphism with cokernels in $\mathbf{Z}$ then $f^{\cdot}$ is a weak equivalence in $\boldsymbol{v}(\mathbf{B}/\mathbf{Z})'$.*

Proof. The mapping cone $Cf^{\cdot}$ of $f^{\cdot}$ is homotopy equivalent to the cokernel of $f^{\cdot}$, which is an acyclic complex in $\mathbf{B}/\mathbf{Z}$. □

Lemma 4.18. *The map*

$$K(I)\colon K(\mathbf{B}_1/\mathbf{B}_{12}) \longrightarrow K(\mathbf{B}/\mathbf{B}_2)$$

is a weak equivalence.

Proof. We will apply the Approximation Theorem to I. The first condition is clear. To check the second condition, consider

$$F^{\cdot}\colon \quad 0 \longrightarrow F^1 \xrightarrow{\phi_1} F^2 \xrightarrow{\phi_2} \ \ldots \ \xrightarrow{\phi_{n-1}} F^n \longrightarrow 0$$

in $\mathbf{B}_1$ and a chain map $g\colon F^{\cdot} \to G^{\cdot}$ to some complex

$$G^{\cdot}\colon \quad 0 \longrightarrow G^1 \xrightarrow{\psi_1} G^2 \xrightarrow{\psi_2} \ \ldots \ \xrightarrow{\psi_{n-1}} G^n \longrightarrow 0$$

in $\mathbf{B}$. Without loss of generality, suppose that each G^i is D-lean, suppose that $F^i = F^i(X_1[K])$ for all i, and suppose that b serves as a bound for all ϕ_i, ψ_i, and g_i. We define

$$F'^i = G^i(X_1[K + D + 3ib])$$

and define $\xi_i \colon F'^i \to F'^{i+1}$ to be the restrictions of ψ_i to F'^i. This gives a chain subcomplex (F'^i, ξ_i) of (G^i, ψ_i) in $\mathbf{B}_1$ with the inclusion $s \colon F'^i \to G^i$. Notice that the choices give the induced chain map $\overline{g} \colon F^{\cdot} \to F'^{\cdot}$ in $\mathbf{B}_1$ so that $g = s \circ I(\overline{g})$.

We will argue that $C^{\cdot} = \mathrm{coker}(s)$ is in $\mathbf{B}_2$. Given that, s is a weak equivalence in $\boldsymbol{v}(\mathbf{B}/\mathbf{B}_2)'$ by Lemma 4.17.

For each $x \in X_1$ and each i, we have $G^i(B_D(x)) \subset F'^i$, so $C^i(B_D(x)) = 0$. By Proposition 2.18, part (3b), C^i is D-lean, therefore

$$C^i(X) = \sum_{x \in X} C^i(B_D(x)) = \sum_{x \in X \setminus X_1} C^i(B_D(x)) \subset C^i((X \setminus X_1)[D]) \subset C^i(X_2).$$

So the complex $C^{\cdot}$ is indeed in $\mathbf{B}_2$. □

Let $\mathbb{Z}$, $\mathbb{Z}^{\geqslant 0}$, and $\mathbb{Z}^{\leqslant 0}$ denote the metric spaces of integers, nonnegative integers, and nonpositive integers with the restriction of the usual metric on the real line $\mathbb{R}$. Let $\mathbf{E}$ be an idempotent complete Grothendieck category of an abelian category $\mathbf{A}$. Then for any proper metric space X, we have the following instance of commutative diagram (†)

$$\begin{array}{ccccc}
K(X,\mathbf{E}) & \longrightarrow & K(X \times \mathbb{Z}^{\geqslant 0},\mathbf{E}) & \longrightarrow & K(\mathbf{B}_1/\mathbf{B}_{12}) \\
\downarrow & & \downarrow & & \downarrow {\scriptstyle K(I)} \\
K(X \times \mathbb{Z}^{\leqslant 0},\mathbf{E}) & \longrightarrow & K(X \times \mathbb{Z},\mathbf{E}) & \longrightarrow & K(\mathbf{B}/\mathbf{B}_2)
\end{array}$$

Lemma 4.19. *The spectra $K(X \times \mathbb{Z}^{\geqslant 0}, \mathbf{E})$ and $K(X \times \mathbb{Z}^{\leqslant 0}, \mathbf{E})$ are contractible.*

Proof. This follows from the fact that these controlled categories are flasque, that is, the usual shift functor T in the positive (respectively negative) direction along $\mathbb{Z}^{\geqslant 0}$ (respectively $\mathbb{Z}^{\leqslant 0}$) interpreted in the obvious way is an exact endofunctor, and there is a natural equivalence $1 \oplus \pm T \cong \pm T$. Contractibility follows from the additivity theorem, cf. Pedersen–Weibel [**17**]. □

In view of Lemma 4.18, we obtain a map

$$K(X,\mathbf{E}) \longrightarrow \Omega K(X \times \mathbb{Z},\mathbf{E})$$

which induces isomorphisms of K-groups in positive dimensions. Iterations of this construction give weak equivalences

$$\Omega^k K(X \times \mathbb{Z}^k,\mathbf{E}) \longrightarrow \Omega^{k+1} K(X \times \mathbb{Z}^{k+1},\mathbf{E})$$

for $k \geqslant 2$.

Definition 4.20 (Nonconnective controlled K-theory)**.** The *nonconnective controlled K-theory* of $\mathbf{E}$, relative to the embedding $\epsilon \colon \mathbf{E} \to \mathbf{A}$, over a proper metric space X is the spectrum

$$K_{\epsilon}^{-\infty}(X, \mathbf{E}) \overset{\text{def}}{=} \underset{\underset{k}{\longrightarrow}}{\text{hocolim}}\ \Omega^k K(X \times \mathbb{Z}^k, \mathbf{E}).$$

Since $\mathbf{B}(X, \mathbf{E})$ can be identified with $\mathbf{E}$ for a bounded metric space X, this definition gives the *nonconnective K-theory* of $\mathbf{E}$

$$K_{\epsilon}^{-\infty}(\mathbf{E}) \overset{\text{def}}{=} \underset{\underset{k>0}{\longrightarrow}}{\text{hocolim}}\ \Omega^k K(\mathbb{Z}^k, \mathbf{E}).$$

As $K_{\epsilon}^{-\infty}(\mathbf{E})$ is an Ω-spectrum in positive dimensions, the positive homotopy groups of $K^{-\infty}(\mathbf{E})$ coincide with those of $K(\mathbf{E})$, as desired. The class group $K_{\epsilon,0}(\mathbf{E})$ is the class group of the idempotent completion $K_0(\mathbf{E}^{\frown})$.

The first known delooping of the K-theory of a general exact category with these properties is due to M. Schlichting [**21**], however the construction here is different and is required in the excision theorem ahead.

Example 4.21. If $\mathbf{E}$ is an arbitrary small exact category, there is the full Gabriel–Quillen embedding of $\mathbf{E}$ in the cocomplete abelian category $\mathbf{A}$ of left exact functors $\mathbf{E}^{\text{op}} \to \mathbf{Mod}(\mathbb{Z})$ with the standard exact structure. The embedding is always closed under extensions in $\mathbf{A}$. It is not necessarily a Grothendieck subcategory, as when $\mathbf{E}$ is not balanced. But if $\mathbf{E}$ is abelian, for example, this gives a canonical delooping of $K(\mathbf{E})$.

Example 4.22. One may start with the cocomplete abelian category $\mathbf{Mod}(R)$ of modules over a ring R with the standard abelian exact structure where the admissible monomorphisms and epimorphisms are respectively all monics and epis. If R is a Noetherian ring, the subcategory $\mathbf{E}$ may be taken to be the noncocomplete abelian category of finitely generated R-modules $\mathbf{Modf}(R)$. Now $K^{-\infty}(\mathbf{E})$ gives the algebraic G-theory of R which we denote by $G^{-\infty}(R)$. We also use notation $\mathbf{B}(X, R)$ for $\mathbf{B}(X, \mathbf{E})$.

Definition 4.23 (Nonconnective controlled G-theory)**.** The *nonconnective controlled G-theory* of X-filtered modules over R is defined as

$$G^{-\infty}(X, R) = K^{-\infty}\mathbf{B}(X, \mathbf{E}).$$

Example 4.24. The negative K-theory of a regular ring R is trivial in the sense that $K_i(\mathbf{Modf}(R)) = 0$ for all $i < 0$. This is well-known in Bass' theory [**2**]. A proof that the negative K-theory is trivial for general abelian categories can be given using the same strategy as in chapter 9 of [**21**].

Of course, when the exact category $\mathbf{E}$ is itself cocomplete, its K-theory is contractible because of the Eilenberg swindle type argument.

We finally prove the main result of this section.

Theorem 4.25 (Nonconnective excision)**.** *Let $\mathbf{E}$ be a Grothendieck subcategory of a cocomplete abelian category $\mathbf{A}$ and let X be a proper metric space. Suppose X_1 and X_2 are subsets of X, and $X = X_1 \cup X_2$. Using the notation $\mathbf{B} = \mathbf{B}(X, \mathbf{E})$,*

$\mathbf{B}_i = \mathbf{B}(X, \mathbf{E})_{<X_i}$ *for* $i = 1$ *or* 2, *and* $\mathbf{B}_{12}$ *for the intersection* $\mathbf{B}_1 \cap \mathbf{B}_2$, *there is a homotopy pushout diagram of spectra*

$$\begin{array}{ccc} K^{-\infty}(\mathbf{B}_{12}) & \longrightarrow & K^{-\infty}(\mathbf{B}_1) \\ \downarrow & & \downarrow \\ K^{-\infty}(\mathbf{B}_2) & \longrightarrow & K^{-\infty}(\mathbf{B}) \end{array}$$

where the maps of spectra are induced from the exact inclusions.

Proof. Let us write $S^k\,\mathbf{B}$ for $\mathbf{B}(X \times \mathbb{Z}^k, \mathbf{E})$ whenever $\mathbf{B}$ is the boundedly controlled category for a general metric space X. If $\mathbf{Z}$ is a subset of X, consider the fibration

$$K(Z, \mathbf{E}) \longrightarrow K(X, \mathbf{E}) \longrightarrow K(\mathbf{B}/\mathbf{Z})$$

from Corollary 4.16. Notice that there is a map

$$K(\mathbf{B}/\mathbf{Z}) \longrightarrow \Omega K(S\,\mathbf{B}/S\,\mathbf{Z})$$

which is a weak equivalence in positive dimensions by the Five Lemma. If one defines

$$K^{-\infty}(\mathbf{B}/\mathbf{Z}) = \underset{\overrightarrow{k}}{\text{hocolim}}\ \Omega^k K(S^k\,\mathbf{B}/S^k\,\mathbf{Z}),$$

there is an induced fibration

$$K^{-\infty}(Z, \mathbf{E}) \longrightarrow K^{-\infty}(X, \mathbf{E}) \longrightarrow K^{-\infty}(\mathbf{B}/\mathbf{Z})$$

The theorem follows from the commutative diagram

$$\begin{array}{ccccc} K^{-\infty}(\mathbf{B}_{12}) & \longrightarrow & K^{-\infty}(\mathbf{B}_1) & \longrightarrow & K^{-\infty}(\mathbf{B}_1/\mathbf{B}_{12}) \\ \downarrow & & \downarrow & & \downarrow{\scriptstyle K^{-\infty}(I)} \\ K^{-\infty}(\mathbf{B}_2) & \longrightarrow & K^{-\infty}(\mathbf{B}) & \longrightarrow & K^{-\infty}(\mathbf{B}/\mathbf{B}_2) \end{array}$$

and the fact that now

$$K^{-\infty}(I)\colon\ K^{-\infty}(\mathbf{B}_1/\mathbf{B}_{12}) \longrightarrow K^{-\infty}(\mathbf{B}/\mathbf{B}_2)$$

is a weak equivalence. □

Remark 4.26. As in other versions of controlled K-theory, there is no excision theorem similar to Theorem 4.25 which employs the connective K-theory.

5. Equivariant theory and the Novikov conjecture

First we establish functoriality properties of the bounded G-theory.

Definition 5.1. A map $f\colon X \to Y$ of metric spaces f is *quasi-bi-Lipschitz* if there is a real positive function l such that

$$\text{dist}(x_1, x_2) \leqslant r \implies \text{dist}(f(x_1), f(x_2)) \leqslant l(r),$$
$$\text{dist}(f(x_1), f(x_2)) \leqslant r \implies \text{dist}(x_1, x_2) \leqslant l(r)$$

for all $x_1, x_2 \in X$.

Any bounded function $f\colon X \to X$, with the property that $\operatorname{dist}(x, f(x)) \leqslant D$ for all $x \in X$ and a fixed number $D \geqslant 0$, is quasi-bi-Lipschitz with $l(r) = r + 2D$. An isometry $g\colon X \to Y$ is quasi-bi-Lipschitz with $l(r) = r$. If only the first of the two conditions is satisfied, the map f is called *bornological.*

We will say that $f\colon X \to Y$ is a *quasi-bi-Lipschitz equivalence* if there is a map $g\colon Y \to X$ so that both f and g are quasi-bi-Lipschitz and the compositions $f \circ g$ and $g \circ f$ are bounded maps.

Proposition 5.2. *Consider the category of proper metric spaces X and quasi-bi-Lipschitz maps and the category of Noetherian rings R. Then $\mathbf{B}(X, R)$ is a bifunctor covariant in the first variable and contravariant in the second variable to exact categories and exact functors. Composing with the covariant functor $K^{-\infty}$ from Example 4.22 gives the spectrum-valued bifunctor $G^{-\infty}(X, R)$.*

Proof. If $f\colon X \to Y$ is a quasi-bi-Lipschitz map, the functor

$$f_*\colon \mathbf{B}(X, R) \longrightarrow \mathbf{B}(Y, R)$$

is given on objects by

$$f_* F(S) = F(f^{-1}(S)).$$

Using the containment

$$f^{-1}(S)[D] \subset f^{-1}(S[l(D)]),$$

one sees that if $\phi \in \mathbf{B}(X, R)$ is a boundedly bicontrolled morphism with $\operatorname{fil}(\phi) \leqslant D$ then $f_*\phi$ is boundedly bicontrolled with $\operatorname{fil}(f_*\phi) \leqslant l(D)$. □

A subset W of a metric space X is *boundedly dense* or *commensurable* if $W[D] = X$ for some $D \geqslant 0$.

Proposition 5.3. *For a commensurable metric subspace W of X, there is a natural exact equivalence of categories*

$$i\colon \mathbf{B}(W, R) \longrightarrow \mathbf{B}(X, R)$$

and the induced weak homotopy equivalence

$$i_*\colon G^{-\infty}(W, R) \longrightarrow G^{-\infty}(X, R).$$

Proof. Any surjective quasi-bi-Lipschitz equivalence $f\colon X \to Y$ induces two functors on filtered modules. One is contravariant

$$f^*\colon \mathbf{B}(Y, R) \longrightarrow \mathbf{B}(X, R)$$

given by $f^* F(S) = F(f(S))$; the other is covariant

$$f_*\colon \mathbf{B}(X, R) \longrightarrow \mathbf{B}(Y, R)$$

given by $f_* F(S) = F(f^{-1}(S))$, so that $f^* f_* = \mathrm{id}$.

Even when f is not surjective, there is the endofunctor $\omega = f^{-1} f$ of $\mathcal{P}(X)$ which induces an endofunctor ω_* of $\mathbf{B}(X, R)$. If $f\colon X \to X$ is bounded, that is $d(x, f(x)) \leqslant D$ for some $D \geqslant 0$ and all $x \in X$, there is always an isomorphism

$\omega_*(F) \cong F$ induced by the identity on $F(X)$. This shows that $f_*F \cong F$ for all $F \in \mathbf{B}(X, R)$.

If $W \subset X$ is commensurable, there is a bounded surjection $f\colon X \to W$, so f induces a natural transformation $\eta\colon \mathrm{id} \to f_*$ where all $\eta(F)$ are isomorphisms. □

Corollary 5.4. *If X is a bounded metric space then the natural equivalence*

$$\mathbf{B}(X, R) \cong \mathbf{B}(\text{point}, R) = \mathbf{Modf}(R)$$

induces a weak equivalence $G^{-\infty}(X, R) \simeq G^{-\infty}(R)$ on the level of K-theory.

Given a group Γ with a left action on X by quasi-bi-Lipschitz equivalences, there is a natural action of Γ on $\mathbf{B}(X, R)$ induced from the action on the power set $\mathcal{P}(X)$. However, this is not the correct choice for a useful equivariant controlled theory for essentially the same reasons as in the discussion of geometric modules in [**4**, ch. VI].

Definition 5.5. Let $\boldsymbol{E\Gamma}$ be the category with the object set Γ and the unique morphism $\mu\colon \gamma_1 \to \gamma_2$ for any pair γ_1, $\gamma_2 \in \Gamma$. There is a left Γ-action on $\boldsymbol{E\Gamma}$ induced by the left multiplication in Γ.

If $\mathcal{C}$ is a small category with left Γ-action, then the category of functors $\mathcal{C}_\Gamma = \mathrm{Fun}(\boldsymbol{E\Gamma}, \mathcal{C})$ is another category with the Γ-action given on objects by

$$\gamma(F)(\gamma') = \gamma F(\gamma^{-1}\gamma')$$

and

$$\gamma(F)(\mu) = \gamma F(\gamma^{-1}\mu).$$

It is always nonequivariantly equivalent to $\mathcal{C}$. The fixed subcategory $\mathrm{Fun}(\boldsymbol{E\Gamma}, \mathcal{C})^\Gamma \subset \mathcal{C}_\Gamma$ consists of equivariant functors and equivariant natural transformations.

Explicitly, when $C = \mathbf{B}(X, R)$ with the Γ-action described above, the objects of $\mathbf{B}_\Gamma(X, R)^\Gamma$ are the pairs (F, ψ) where $F \in \mathbf{B}(X, R)$ and ψ is a function on Γ with $\psi(\gamma) \in \mathrm{Hom}(F, \gamma F)$ such that

$$\psi(1) = 1 \quad \text{and} \quad \psi(\gamma_1\gamma_2) = \gamma_1\psi(\gamma_2)\psi(\gamma_1).$$

These conditions imply that $\psi(\gamma)$ is always an isomorphism as in [**24**]. The set of morphisms $(F, \psi) \to (F', \psi')$ consists of the morphisms $\phi\colon F \to F'$ in $\mathbf{B}(X, R)$ such that the squares

$$\begin{array}{ccc} F & \xrightarrow{\psi(\gamma)} & \gamma F \\ {\scriptstyle\phi}\downarrow & & \downarrow{\scriptstyle\gamma\phi} \\ F' & \xrightarrow{\psi'(\gamma)} & \gamma F' \end{array}$$

commute for all $\gamma \in \Gamma$. A slightly more refined theory is obtained by replacing $\mathbf{B}_\Gamma(X, R)$ with the full subcategory $\mathbf{B}_{\Gamma,0}(X, R)$ of functors sending all morphisms of $\boldsymbol{E\Gamma}$ to filtration 0 maps. So $\mathbf{B}_{\Gamma,0}(X, R)^\Gamma$ consists of (F, ψ) with $\mathrm{fil}\,\psi(\gamma) = 0$ for all $\gamma \in \Gamma$.

Proposition 5.6. *The fixed point category $\mathbf{B}_{\Gamma,0}(X, R)^\Gamma$ is exact.*

Proof. The exact structure is inherited from $\mathbf{B}(X,R)$ in the sense that a morphism $\phi\colon (F,\psi) \to (F',\psi')$ is an admissible monomorphism or epimorphism if the map $\phi\colon F \to F'$ is in $\mathbf{mB}(X,R)$ or $\mathbf{eB}(X,R)$ respectively. The fact that this is an exact structure follows from the proof of Theorem 2.13 by observing that all constructions in that proof produce equivariant objects and morphisms. □

Remark 5.7. One exact structure in the category of finitely generated $R[\Gamma]$-modules $\mathbf{Modf}(R[\Gamma])$ for a Noetherian ring R consists of short exact sequences of $R[\Gamma]$-modules with finitely generated kernels and quotients. When $R[\Gamma]$ is Noetherian, so that $\mathbf{Modf}(R[\Gamma])$ is an abelian category, this coincides with the conventional choice of all injections for admissible monomorphisms and all surjections for admissible epimorphisms.

However, there is a reasonable conjecture of P. Hall that only polycyclic-by-finite groups have Noetherian group rings, cf. Question 32 in Farkas [**8**].

We are going to define a new exact structure on a subcategory $\mathbf{B}(R[\Gamma])$ of $\mathbf{Modf}(R[\Gamma])$ and relate it to the exact category $\mathbf{B}_{\Gamma,0}(X,R)^{\Gamma}$.

Definition 5.8. The *word metric* on a finitely generated group Γ with a fixed generating set Ω is the path metric induced from the condition that $\mathrm{dist}(\gamma,\omega\gamma)=1$ whenever $\gamma\in\Gamma$ and $\omega\in\Omega$.

This metric clearly makes Γ a proper metric space. We will use the notation $B_d(\gamma)$ for the metric ball of radius d centered at γ.

Definition 5.9. Given a finitely generated $R[\Gamma]$-module F, fix a finite generating set Σ for F and define a Γ-filtration of the R-module F by

$$F(S) = \langle S\Sigma\rangle_R,$$

the R-submodule of F generated by $S\Sigma$. Let $s(F,\Sigma)$ stand for the resulting Γ-filtered R-module.

Lemma 5.10. *Every $R[\Gamma]$-homomorphism*

$$\phi\colon F \longrightarrow G$$

between finitely generated modules is boundedly controlled as an R-homomorphism between the filtered R-modules

$$\phi\colon s(F,\Sigma_F) \longrightarrow s(G,\Sigma_G)$$

with respect to any choice of the finite generating sets Σ_F and Σ_G.

Proof. Consider $x\in F(S)=\langle S\Sigma_F\rangle_R$, then

$$x=\sum_{s,\sigma} r_{s,\sigma}s\sigma$$

for a finite collection of pairs $s\in S$, $\sigma\in\Sigma_F$. Since $F(\{e\})=\langle\Sigma_F\rangle_R$ for the identity element e in Γ, there is a number $d\geqslant 0$ such that

$$\phi F(\{e\}) \subset G(B_d(e)).$$

Therefore,

$$\phi(x) = \sum_{s,\sigma} r_{s,\sigma}\phi(s\sigma) = \sum_{s,\sigma} r_{s,\sigma} s\phi(\sigma) \subset \sum_{s\in S} sG(B_d(e)) \subset G(S[d])$$

because the left translation action by any element $s \in S$ on $B_d(e)$ in Γ is an isometry onto $B_d(s)$. □

Corollary 5.11. *Given a finitely generated $R[\Gamma]$-module F and two choices of finite generating sets Σ_1 and Σ_2, the filtered R-modules $s(F, \Sigma_1)$ and $s(F, \Sigma_2)$ are isomorphic as Γ-filtered R-modules.*

Proof. The identity map and its inverse are boundedly controlled as maps between $s(F, \Sigma_1)$ and $s(F, \Sigma_2)$ by Lemma 5.10. □

Corollary 5.12. *Finitely generated $R[\Gamma]$-modules F with filtrations $s(F, \Sigma)$, with respect to arbitrary finite generating sets Σ, are locally finitely generated and lean. If $s(F, \Sigma)$ is insular and Σ' is another finite generating set then $s(F, \Sigma')$ is also insular.*

Proof. For a finite subset S, the submodule $F(S)$ is generated by the finite set $S\Sigma$. Since $F(x) = \langle x\Sigma\rangle_R$,

$$F(S) = \sum_{x\in\Sigma} \langle x\Sigma\rangle_R = \langle S\Sigma\rangle_R,$$

so $s(F, \Sigma)$ is 0-lean. The second claim follows from Corollary 5.11. □

Definition 5.13. Let $\mathbf{B}(R[\Gamma])$ be the full subcategory of $\mathbf{Modf}(R[\Gamma])$ on R-modules F which are *strict* as filtered modules $s(F, \Sigma)$ with respect to some choice of the finite generating set Σ.

Let $\mathbf{B}_\times(R[\Gamma])$ be the category of objects which are pairs (F, Σ) with F in $\mathbf{B}(R[\Gamma])$ and Σ a finite generating set for F. The morphisms are the $R[\Gamma]$-homomorphisms between the modules.

Lemma 5.10 shows that the map

$$s\colon \mathbf{B}_\times(R[\Gamma]) \longrightarrow \mathbf{B}(\Gamma, R)$$

described in Definition 5.9 is a functor. In fact, it is a functor

$$s_\Gamma\colon \mathbf{B}_\times(R[\Gamma]) \longrightarrow \mathbf{B}_{\Gamma,0}(\Gamma, R)^\Gamma$$

by interpreting $s_\Gamma(F, \Sigma) = (F, \psi)$ with $F = s(F, \Sigma)$ and $\psi(\gamma)\colon F \to \gamma F$ induced from $s\sigma \mapsto \gamma^{-1}s\sigma$. Since

$$(\gamma F)(S) = \langle\gamma^{-1}(S)\Sigma\rangle_R,$$

it follows that the object $s_\Gamma(F, \Sigma)$ lands in $\mathbf{B}_{\Gamma,0}(\Gamma, R)^\Gamma$, and s sends all $R[\Gamma]$-homomorphisms to Γ-equivariant homomorphisms.

Lemma 5.14. *Let $F \in \mathbf{B}_{\Gamma,0}(\Gamma, R)^\Gamma$ and let Σ be a finite generating set for the $R[\Gamma]$-module F. Then the identity homomorphism*

$$\mathrm{id}\colon s_\Gamma(F, \Sigma) \longrightarrow F$$

is boundedly controlled with respect to the induced and the original filtrations of F.

Proof. If Σ is contained in $F(B_d(e))$, where e is the identity element in Γ, then

$$\gamma\Sigma \subset F(B_d(\gamma))$$

for all $\gamma \in \Gamma$, and

$$s(F,\Sigma)(S) = \langle S\Sigma\rangle_R \subset F(S[d])$$

for all subsets $S \subset \Gamma$. □

Both functors s and s_Γ are additive with respect to the additive structure in $\mathbf{B}_\times(R[\Gamma])$ where the biproduct is given by

$$(F,\Sigma_F) \oplus (G,\Sigma_G) = (F \oplus G, \Sigma_F \times \Sigma_G).$$

Let the *admissible monomorphisms* $\phi\colon (F,\Sigma_F) \to (G,\Sigma_G)$ in $\mathbf{B}_\times(R[\Gamma])$ be the injections $\phi\colon F \to G$ of $R[\Gamma]$-modules ϕ such that

$$s(\phi)\colon s(F,\Sigma_F) \longrightarrow s(G,\Sigma_G)$$

is a boundedly bicontrolled homomorphism of Γ-filtered R-modules. This is equivalent to requiring that $s(\phi)$ be an admissible monomorphism in $\mathbf{B}(\Gamma,R)$. Let the *admissible epimorphisms* be the morphisms ϕ such that $s(\phi)$ are admissible epimorphisms in $\mathbf{B}(\Gamma,R)$.

Proposition 5.15. *The choice of admissible morphisms defines an exact structure on* $\mathbf{B}_\times(R[\Gamma])$ *such that both* s *and* s_Γ *are exact functors.*

Proof. When checking Quillen's axioms in $\mathbf{B}_\times(R[\Gamma])$, all required universal constructions are performed in $\mathbf{B}(R[\Gamma])$ with the canonical choices of finite generating sets. In particular, Σ in the pushout $B\cup_A C$ is the image of the product set $\Sigma_B \times \Sigma_C$ in $B \times C$.

The fact that all admissible morphisms are boundedly bicontrolled in $\mathbf{B}(\Gamma,R)$ or $\mathbf{B}_{\Gamma,0}(\Gamma,R)^\Gamma$ follows from the proof of Theorem 2.13. Exactness of s and s_Γ is immediate. □

Definition 5.16. We give $\mathbf{B}(R[\Gamma])$ the minimal exact structure that makes the forgetful functor

$$p\colon \mathbf{B}_\times(R[\Gamma]) \longrightarrow \mathbf{B}(R[\Gamma])$$

sending (F,Σ) to F an exact functor.

In other words, an $R[\Gamma]$-homomorphism $\phi\colon F \to G$ is an *admissible monomorphism* or *epimorphism* if for some choice of finite generating sets,

$$\phi\colon (F,\Sigma_F) \longrightarrow (G,\Sigma_G)$$

is respectively an admissible monomorphism or epimorphism in $\mathbf{B}_\times(R[\Gamma])$.

Corollary 5.11 shows that if $\phi\colon F \to G$ is boundedly bicontrolled as a map of filtered R-modules $s(F,\Sigma_F) \to s(G,\Sigma_G)$ then it is boundedly bicontrolled with respect to any other choice of finite generating sets, so this structure is well-defined.

Notation 5.17. The new exact category will be referred to as $\mathbf{B}(R[\Gamma])$, with the corresponding K-theory spectrum $G^{-\infty}(R[\Gamma])$.

Let (F, ψ) be an object of $\mathbf{B}_{\Gamma,0}(\Gamma, R)^{\Gamma}$. One may think of $\gamma F \in \mathbf{B}(\Gamma, R)$, $\gamma \in \Gamma$, as the module F with a new Γ-filtration. Now the R-module structure $\eta\colon R \to \operatorname{End} F$ induces an $R[\Gamma]$-module structure

$$\eta(\psi)\colon R[\Gamma] \longrightarrow \operatorname{End} F$$

given by

$$\sum_{\gamma} r_\gamma \gamma \mapsto \sum_{\gamma} \eta(r_\gamma)\psi(\gamma)$$

since the sums are taken over a finite subset of Γ. It is easy to see that this defines a map

$$\pi\colon \mathbf{B}_{\Gamma,0}(\Gamma, R)^{\Gamma} \longrightarrow \mathbf{B}(R[\Gamma])$$

by sending (F, ψ) to F, so that $p = \pi s_\Gamma$. Notice however that in general π is not exact as the identity homomorphism in Lemma 5.14 is not necessarily an isomorphism.

In the rest of the paper we assume Γ is torsion-free. The exact functors p and s_Γ induce maps in nonconnective K-theory

$$G^{-\infty}(R[\Gamma]) \xleftarrow{\;p\;} K^{-\infty}\mathbf{B}_\times(R[\Gamma]) \xrightarrow{\;s_\Gamma\;} G^{-\infty}_{\Gamma,0}(\Gamma, R)^{\Gamma}.$$

We claim that both of these maps are weak equivalences.

Proposition 5.18. *The functor f induces a weak equivalence*

$$K^{-\infty}\mathbf{B}_\times(R[\Gamma]) \simeq G^{-\infty}(R[\Gamma]).$$

Proof. This follows from the Approximation theorem applied to p'. The two categories are saturated, and $\mathbf{B}_\times(R[\Gamma])'$ has a cylinder functor satisfying the cylinder axiom which is constructed as the canonical homotopy pushout with the canonical product basis, see section 1 of [**25**].

The first condition of the Approximation theorem is clear. For the second condition, let $(F_1^{\cdot}, \Sigma_1^{\cdot})$ be a complex in $\mathbf{B}_\times(R[\Gamma])$ and let $g\colon F_1^{\cdot} \to F_2^{\cdot}$ be a chain map in $\mathbf{B}(R[\Gamma])'$. For each $R[\Gamma]$-module F_2^i choose any finite generating set Σ_2^i, then using $f = g$ and $g' = \mathrm{id}$, we have $g = g'p(f)$. □

Proposition 5.19. *The functor s_Γ induces a weak homotopy equivalence*

$$K^{-\infty}\mathbf{B}_\times(R[\Gamma]) \simeq G^{-\infty}_{\Gamma,0}(\Gamma, R)^{\Gamma}.$$

Proof. The target category is again saturated and has a cylinder functor satisfying the cylinder axiom. To check condition 2 of the approximation theorem, let

$$E^{\cdot}\colon \quad 0 \longrightarrow (E^1, \Sigma_1) \longrightarrow (E^2, \Sigma_2) \longrightarrow \;\ldots\; \longrightarrow (E^n, \Sigma_n) \longrightarrow 0$$

be a complex in $\mathbf{B}_\times(R[\Gamma])$,

$$(F^{\cdot}, \psi_{\cdot})\colon \quad 0 \longrightarrow (F^1, \psi_1) \xrightarrow{\;f_1\;} (F^2, \psi_2) \xrightarrow{\;f_2\;} \;\ldots\; \xrightarrow{\;f_{n-1}\;} (F^n, \psi_n) \longrightarrow 0$$

be a complex in $\mathbf{B}_{\Gamma,0}(\Gamma, R)^{\Gamma}$, and

$$g\colon s'_\Gamma(E^{\cdot}) \longrightarrow (F^{\cdot}, \psi_{\cdot})$$

be a chain map. Each F^i can be thought of as an $R[\Gamma]$-module, and there is a chain complex

$$F^{\cdot}\colon \quad 0 \longrightarrow F^1 \xrightarrow{f_1} F^2 \xrightarrow{f_2} \dots \xrightarrow{f_{n-1}} F^n \longrightarrow 0$$

in $\mathbf{Modf}(R[\Gamma])$. Choose arbitrary finite generating sets Ω_i in F^i for all $1 \leqslant i \leqslant n$. Now

$$\pi_\Omega F^{\cdot}\colon \quad 0 \longrightarrow (F^1, \Omega_1) \xrightarrow{f_1} (F^2, \Omega_2) \xrightarrow{f_2} \dots \xrightarrow{f_{n-1}} (F^n, \Omega_n) \longrightarrow 0$$

is a chain complex in $\mathbf{B}_\times(R[\Gamma])$.

The chain map g is degree-wise an $R[\Gamma]$-homomorphism, so there is a corresponding chain map

$$f\colon E^{\cdot} \longrightarrow \pi_\Omega F^{\cdot}$$

which coincides with g on modules. On the other hand, the degree-wise identity gives a chain map

$$g'\colon s'_\Gamma(\pi_\Omega F^{\cdot}) \longrightarrow F^{\cdot}$$

in $\mathbf{B}_{\Gamma,0}(\Gamma, R)^\Gamma$ by Lemma 5.14. This g' is a quasi-isomorphism, as required. □

Corollary 5.20. *Let Γ be a finitely generated torsion-free group and R be a Noetherian ring. There is a weak equivalence*

$$G^{-\infty}_{\Gamma,0}(\Gamma, R)^\Gamma \simeq G^{-\infty}(R[\Gamma]).$$

Corollary 5.21. *Let Γ be a finitely generated torsion-free group acting freely, properly discontinuously and cocompactly on a proper metric space X and let R be a Noetherian ring. There is a weak homotopy equivalence*

$$G^{-\infty}_{\Gamma,0}(X, R)^\Gamma \simeq G^{-\infty}(R[\Gamma]).$$

Proof. Let $p\colon X \to$ point be the geometric collapse. For any $x \in X$ such that the embedding i of the orbit Γx with the word metric is commensurable in X, there is a commutative diagram

$$\begin{array}{ccc} G^{-\infty}_{\Gamma,0}(\Gamma x, R)^\Gamma & \xrightarrow{\pi_*} & G^{-\infty}(R[\Gamma]) \\ \downarrow{\scriptstyle i_*} & & \downarrow{\scriptstyle =} \\ G^{-\infty}_{\Gamma,0}(X, R)^\Gamma & \xrightarrow{\pi_*} & G^{-\infty}(R[\Gamma]) \end{array}$$

The top π_* is a weak equivalence by Corollary 5.20. The vertical map i_* is a weak equivalence as in Proposition 5.3, so the lower map π_* is a weak equivalence. □

For a discrete group Γ and a ring R there is an assembly map

$$A_K\colon B\Gamma_+ \wedge K^{-\infty}(R) \longrightarrow K^{-\infty}(R[\Gamma]).$$

When the ring R is regular Noetherian, for example the integers $\mathbb{Z}$, the spectra $G^{-\infty}(R)$ and $K^{-\infty}(R)$ and, therefore, $B\Gamma_+ \wedge G^{-\infty}(R)$ and $B\Gamma_+ \wedge K^{-\infty}(R)$ can be naturally identified.

Definition 5.22. Let Γ be a finitely generated group and R be a regular Noetherian ring. The *assembly map in G-theory*

$$A_G \colon B\Gamma_+ \wedge G^{-\infty}(R) \longrightarrow G^{-\infty}(R[\Gamma])$$

is the composition of A_K and the canonical Cartan map

$$C \colon K^{-\infty}(R[\Gamma]) \longrightarrow G^{-\infty}(R[\Gamma])$$

induced by inclusion of categories.

The *integral Novikov conjecture* in algebraic G-theory is the statement that this is a split injection of spectra.

Remark 5.23. Notice that whenever the assembly map A_G is split injective, the map A_K is also split injective, so this conjecture is stronger than the K-theoretic conjecture when the ring R is regular.

Remark 5.24. The standard exact structure on $\mathbf{Modf}(R[\Gamma])$ has all injective and surjective $R[\Gamma]$-homomorphisms with finitely generated cokernels and kernels as admissible morphisms so that the exact sequences are the traditional short exact sequences. Let the corresponding K-theory spectrum be $G_m^{-\infty}(R[\Gamma])$. One might attempt to replace $G_\times^{-\infty}(R[\Gamma])$ with $G_m^{-\infty}(R[\Gamma])$ as the target of the assembly A_G. However, W. Lück has pointed out that this map would not be weakly injective even in the simple case when R is a commutative ring and Γ is the free group on two generators, cf. Remark 2.23 in [**15**].

This underscores the importance of choosing the coarse version $G^{-\infty}(R[\Gamma])$ as our approximation of $K^{-\infty}(R[\Gamma])$.

The equivariant *assembly map in boundedly controlled G-theory* can be defined as in [**4**]. For any Noetherian ring R, this is an equivariant natural transformation

$$\alpha_G \colon h^{lf}(X, G^{-\infty}(R)) \longrightarrow G^{-\infty}_{\Gamma,0}(X, R)$$

from the equivariant locally finite homology $h^{lf}(X, G^{-\infty}(R))$ to the equivariant bounded G-theory $G^{-\infty}_{\Gamma,0}(X, R)$, see Definition II.14, loc. cit. If Γ is torsion-free and acts freely cocompactly on X, one also has weak equivalences

$$h^{lf}(X, G^{-\infty}(R))^\Gamma \simeq B\Gamma_+ \wedge G^{-\infty}(R).$$

The fixed point spectra and the induced maps fit in the commutative diagram

$$\begin{array}{ccccc}
B\Gamma_+ \wedge G^{-\infty}(R) & \xrightarrow{\alpha_K^\Gamma} & K^{-\infty}_{\Gamma,0}(X, R)^\Gamma & \xrightarrow{\simeq} & K^{-\infty}(R[\Gamma]) \\
{\scriptstyle \alpha_G^\Gamma}\big\downarrow & & \big\downarrow & & \big\downarrow{\scriptstyle C} \\
G^{-\infty}_{\Gamma,0}(X, R)^\Gamma & \xleftarrow[s_{\Gamma *}]{\simeq} & K^{-\infty}\mathbf{B}_\times(R[\Gamma]) & \xrightarrow{=} & G^{-\infty}(R[\Gamma])
\end{array} \qquad (\dagger\dagger)$$

If we restrict our attention to the equivariant objects in $K^{-\infty}_{\Gamma,0}(X, R)^\Gamma$ that have the parametrization function map the generating set B to a single point then the map

$$i \colon K^{-\infty}_{\Gamma,0}(X, R)^\Gamma \longrightarrow G^{-\infty}_{\Gamma,0}(X, R)^\Gamma$$

is well-defined and fits as the diagonal in the left square.

Remark 5.25. When R is a regular Noetherian ring, the assembly map in boundedly controlled G-theory

$$\alpha_G \colon h^{lf}(X, G^{-\infty}(R)) \longrightarrow G^{-\infty}_{\Gamma,0}(X, R)$$

can be identified up to homotopy with the composition

$$h^{lf}(X, K^{-\infty}(R)) \xrightarrow{\alpha_K} K^{-\infty}_{\Gamma,0}(X, R) \xrightarrow{i} G^{-\infty}_{\Gamma,0}(X, R).$$

Now there is a homotopy commutative square

$$\begin{array}{ccc} B\Gamma_+ \wedge K^{-\infty}(R) & \xrightarrow{A_G} & G^{-\infty}(R[\Gamma]) \\ \simeq\downarrow & & \downarrow\simeq \\ h^{lf}(X, G^{-\infty}(R))^\Gamma & \xrightarrow{\alpha_G^\Gamma} & G^{-\infty}_{\Gamma,0}(X, R)^\Gamma \end{array}$$

Controlled excision is the main technical tool from controlled K-theory used to prove integral Novikov conjectures. It is used to see that in specific cases the equivariant assembly map in bounded K-theory is a homotopy equivalence. In particular, the argument in [**6**] applies to groups of finite asymptotic dimension which have a finite classifying space.

Theorem 5.26. *The fixed point map of spectra*

$$\alpha_G^\Gamma \colon h^{lf}(X, G^{-\infty}(R))^\Gamma \longrightarrow G^{-\infty}_{\Gamma,0}(X, R)^\Gamma$$

is a split injection for any geometrically finite group Γ of finite asymptotic dimension and a Noetherian ring R.

Proof. The main step in the proofs of the Novikov conjecture in algebraic K-theory [**4, 6**] to which we referred above is the application of homotopy fixed points to reduce the study of the map A_K to the nonequivariant study of the equivariant map α_K. This is shown to be a weak equivalence by using controlled excision to compute the target. With the excision results from section 3 and the equivariant properties established here, the proofs can be repeated verbatim obtaining splittings of the assembly maps α_G for the same collection of groups. □

Corollary 5.27 (Novikov conjecture in G-theory)**.** *The G-theoretic assembly map*

$$A_G \colon B\Gamma_+ \wedge G^{-\infty}(R) \longrightarrow G^{-\infty}(R[\Gamma])$$

is a split injection for any geometrically finite group Γ of finite asymptotic dimension and a regular Noetherian ring R.

Proof. If R is regular Noetherian, the two maps A_G and α_G^Γ can be identified as in Remark 5.25. □

References

[1] D.R. Anderson and H.J. Munkholm, *Boundedly controlled topology*, Lecture Notes in Mathematics **1323**, Springer-Verlag (1988).

[2] H. Bass, *Algebraic K-theory*, W. A. Benjamin, Inc. (1968).

[3] M. Cardenas and E.K. Pedersen, *On the Karoubi filtration of a category*, K-theory, **12** (1997), 165–191.

[4] G. Carlsson, *Bounded K-theory and the assembly map in algebraic K-theory*, in *Novikov conjectures, index theory and rigidity, Vol. 2* (S.C. Ferry, A. Ranicki, and J. Rosenberg, eds.), Cambridge U. Press (1995), 5–127.

[5] G. Carlsson and B. Goldfarb, *On homological coherence of discrete groups*, J. Algebra **276** (2004), 502–514.

[6] ———, *The integral K-theoretic Novikov conjecture for groups with finite asymptotic dimension*, Inventionnes Math. **157** (2004), 405–418.

[7] ———, *Algebraic K-theory of geometric groups*, in preparation.

[8] D.R. Farkas, *Group rings: an annotated questionnaire*, Comm. Algebra **8** (1980), 585–602.

[9] P. Gabriel and M. Zisman, *Calculus of fractions and homotopy theory*, Springer-Verlag (1967).

[10] A.I. Generalov, *Relative homological algebra. Cohomology of Categories, posets and coalgebras*, in Handbook of Algebra, Vol. 1 (M. Hazewinkel, ed.), 1996, Elsevier Science, 611–638.

[11] I. Hambleton and E.K. Pedersen, *Compactifying infinite group actions*, Contemp. Math. **258** (2000), 203–212.

[12] R. Hartshorne, *Algebraic geometry*, Springer-Verlag (1977).

[13] B. Keller, *Chain complexes and stable categories*, Manuscripta Math. **67** (1990), 379–417.

[14] ———, *Derived categories and their uses*, in Handbook of Algebra, Vol. 1 (M. Hazewinkel, ed.), 1996, Elsevier Science, 671–701.

[15] W. Lück, *Dimension theory of arbitrary modules over finite von Neumann algebras and L^2-Betti numbers II: Applications to Grothendieck groups, L^2-Euler characteristics and Burnside groups*, J. reine angew. Math. **496** (1998), 213–236.

[16] S. Mac Lane, *Categories for the working mathematician*, Springer-Verlag (1971).

[17] E.K. Pedersen and C. Weibel, *A nonconnective delooping of algebraic K-theory*, in *Algebraic and geometric topology* (A. Ranicki, N. Levitt, and F. Quinn, eds.), Lecture Notes in Mathematics **1126**, Springer-Verlag (1985), 166–181.

[18] ———, *K-theory homology of spaces*, in *Algebraic topology* (G. Carlsson, R.L. Cohen, H.R. Miller, and D.C. Ravenel, eds.), Lecture Notes in Mathematics **1370**, Springer-Verlag (1989), 346–361.

[19] ———, unpublished.

[20] D. Quillen, *Higher algebraic K-theory: I*, in *Algebraic K-theory I* (H. Bass, ed.), Lecture Notes in Mathematics **341**, Springer-Verlag (1973), 77–139.

[21] M. Schlichting, *Delooping the K-theory of exact categories and negative K-groups*, U. Paris VII thesis (2000).

[22] ———, *Delooping the K-theory of exact categories*, Topology **43** (2004), 1089–1103.

[23] R.E. Staffeldt, *On the fundamental theorems of algebraic K-theory*, K-theory **1** (1989), 511–532.

[24] R.W. Thomason, *The homotopy limit problem*, in *Proceedings of the Northwestern homotopy theory conference* (H.R. Miller and S.B. Priddy, eds.), Cont. Math. **19** (1983), 407–420.

[25] R.W. Thomason and Thomas Trobaugh, *Higher algebraic K-theory of schemes and of derived categories*, in *The Grothendieck Festschrift, Vol. III*, Progress in Mathematics **88**, Birkhäuser (1990), 247–435.

[26] F. Waldhausen, *Algebraic K-theory of spaces*, in *Algebraic and geometric topology* (A. Ranicki, N. Levitt, and F. Quinn, eds.), Lecture Notes in Mathematics **1126**, Springer-Verlag (1985), 318–419.

This article may be accessed via WWW at `http://tcms.org.ge/Journals/JHRS/`

Gunnar Carlsson
`gunnar@math.stanford.edu`

Department of Mathematics
Stanford University
Stanford
CA 94305

Boris Goldfarb
`goldfarb@math.albany.edu`

Department of Mathematics and Statistics
SUNY
Albany
NY 12222

Journal of Homotopy and Related Structures, vol. 6(1), 2011, pp.161–173

A-HOMOLOGY, A-HOMOTOPY AND SPECTRAL SEQUENCES

ENZO MIGUEL OTTINA

(*communicated by Shmuel Weinberger*)

Abstract

Given a CW-complex A we define an 'A-shaped' homology theory which behaves nicely towards A-homotopy groups allowing the generalization of many classical results. We also develop a relative version of the Federer spectral sequence for computing A-homotopy groups. As an application we derive a generalization of the Hopf-Whitney theorem.

1. Introduction

Given pointed topological spaces A and X, the A-homotopy groups of X are defined as $\pi_n^A(X) = [\Sigma^n A, X]$, that is, the homotopy classes of pointed maps from the reduced n-th suspension of A to X. These groups appear naturally in different situations, for example in Quillen's model categories [**9**], and generalize the homotopy groups with coefficients [**7, 8**].

The A-homotopy groups have also been studied indirectly by some authors as homotopy groups of function spaces. Among them, we mention M. Barratt, R. Brown and H. Federer. The first one proves in [**1**] that homotopy groups of function spaces can be described as a central extension of certain homology groups, and computes in [**2**] these homotopy groups in several cases using Whitney's tube systems [**11**]. In [**3**], Brown works in a simplicial setting obtaining results which are used to study homotopy types of function spaces. As an application, he obtains different proofs for some of Barratt's results. In [**5**], Federer introduces a spectral sequence which converges to the homotopy groups of function spaces. Clearly, the Federer spectral sequence may also be understood as a tool to compute A-homotopy groups when A is locally compact and Hausdorff.

In the first part of this article we delve deeply into this spectral sequence taking a different approach to that of Federer's: we focus our attention on A-homotopy groups of spaces rather than on homotopy groups of function spaces. We develop a relative version of the Federer spectral sequence and obtain as a first application a generalization of the Hopf-Whitney theorem (2.5).

The homotopy groups and the homology groups of a topological space are related, for example, by the Hurewicz theorem, or more generally, by the Whitehead exact

---

I would like to thank G. Minian for many valuable comments and suggestions on this article
Received May 10, 2010, revised March 20, 2011; published on May 15, 2011.
2000 Mathematics Subject Classification: 55N35, 55Q05, 55T05.
Key words and phrases: CW-complexes, Homology Theories, Homotopy Groups, Spectral Sequences.

sequence [**10**]. Therefore, it is natural to think that the A-homotopy groups should also have their homological counterpart. The main objective of the second part of this article is to define a suitable 'A-shaped' homology theory and give results which show the relationship between this homology theory and the A-homotopy groups. This is achieved in section 3 where, given a CW-complex A, we define the A-homology groups of a CW-complex X generalizing singular homology groups. We obtain many generalization of classical results, among the most important of which we mention a Hurewicz-type theorem relating the A-homotopy groups with the A-homology groups (3.9) and a homological version of the Whitehead theorem which states that, under certain hypotheses, a map between CW-complexes which induces isomorphisms in the A-homology groups is a homotopy equivalence (3.10).

Finally, we define a Hurewicz-type map between the A-homotopy groups and the A-homology groups and embed it in a long exact sequence generalizing the exact sequence constructed by Whitehead in [**10**].

Throughout this article, all spaces are supposed to be pointed and path-connected. Homology and cohomology will mean reduced homology and reduced cohomology respectively. Also, if X is a pointed topological space with base point x_0, we will simply write $\pi_n(X)$ instead of $\pi_n(X, x_0)$.

2. A relative version of the Federer spectral sequence

In this section we introduce a relative version of the Federer spectral sequence which will be used later. As one of its first applications, we will obtain a relative version of the Hopf-Whitney theorem.

Recall that the A-homotopy groups of a (pointed) topological space X are defined by $\pi_n^A(X) = [\Sigma^n A, X]$, that is the (pointed) homotopy classes of maps from $\Sigma^n A$ to X. Similarly, the relative A-homotopy groups of a (pointed) topological pair (Y, B) are defined by $\pi_n^A(Y, B) = [(C\Sigma^{n-1}A, \Sigma^{n-1}A), (Y, B)]$.

Now we state and prove the main result of this section.

Theorem 2.1. *Let (Y, B) be a topological pair such that $\pi_2(Y, B)$ is an abelian group and let A be a finite-dimensional (and path-connected) CW-complex. Then there exists a homological spectral sequence $\{E^a_{p,q}\}_{a\geqslant 1}$, with $E^2_{p,q}$ satisfying*

- $E^2_{p,q} \cong H^{-p}(A; \pi_q(Y, B))$ *for $p + q \geqslant 2$ and $p \leqslant -1$.*
- $E^2_{p,q}$ *is isomorphic to a subgroup of $H^{-p}(A; \pi_q(Y, B))$ if $p+q = 1$ and $p \leqslant -1$.*
- $E^2_{p,q} = 0$ *if $p + q \leqslant 0$ or $p \geqslant 0$.*

which converges to $\pi^A_{p+q}(Y, B)$ for $p + q \geqslant 2$.

We will call $\{E^a_{p,q}\}_{a\geqslant 1}$ the *relative Federer spectral sequence associated to A and* (Y, B).

Proof. We may suppose that A has only one 0-cell. For $r \leqslant -1$, let $A^r = *$, and for $r \in \mathbb{N}$, let J_r be an index set for the r-cells of A. For $\alpha \in J_r$ let g^r_α be the attaching map of the cell e^r_α.

For $r \in \mathbb{N}$ let $Z_r = \bigvee_{J_r} S^r \cong A^r/A^{r-1}$. The long exact sequences associated to the cofiber sequences $A^{r-1} \to A^r \overset{\overline{q}_r}{\to} Z_r$, $r \in \mathbb{N}$, may be extended as follows

$$\cdots \to \pi_2^{A^{r-1}}(Y,B) \overset{\partial_r}{\to} \pi_1^{Z_r}(Y,B) \overset{\eta}{\to} \frac{\pi_1^{Z_r}(Y,B)}{\operatorname{Im}\partial_r} \overset{0}{\to} \frac{\pi_1^{Z_{r-1}}(Y,B)}{\operatorname{Im}\partial_{r-1}} \overset{\text{id}}{\to} \frac{\pi_1^{Z_{r-1}}(Y,B)}{\operatorname{Im}\partial_{r-1}} \to 0$$

where η is the quotient map.

These extended exact sequences yield an exact couple (A_0, E_0, i, j, k) where the bigraded groups $A_0 = \bigoplus_{p,q\in\mathbb{Z}} A^1_{p,q}$ and $E_0 = \bigoplus_{p,q\in\mathbb{Z}} E^1_{p,q}$ are defined by

$$A^1_{p,q} = \begin{cases} \pi_{p+q+1}^{A^{-p-1}}(Y,B) & \text{if } p+q \geqslant 1 \\ \pi_1^{Z_{-p-1}}(Y,B)/\operatorname{Im}\partial_{-p-1} & \text{if } p+q=0 \\ 0 & \text{if } p+q \leqslant -1 \end{cases}$$

and

$$E^1_{p,q} = \begin{cases} \pi_{p+q}^{Z_{-p}}(Y,B) & \text{if } p+q \geqslant 1 \\ \pi_1^{Z_{-p-1}}(Y,B)/\operatorname{Im}\partial_{-p-1} & \text{if } p+q=0 \\ 0 & \text{if } p+q \leqslant -1 \end{cases}$$

Note that all these groups are abelian, except perhaps for $\pi_2^{A^r}(Y,B)$, $r \in \mathbb{N}$. Hence, E_0 is an abelian group. Therefore, the exact couple (A_0, E_0, i, j, k) induces a spectral sequence $(E^a_{p,q})_{p,q}$, $a \geqslant 1$, which converges to $\pi_n^A(Y,B)$ for $n \geqslant 2$ since A is finite-dimensional.

Note also that

$$E^1_{p,q} = \pi_{p+q}^{Z_{-p}}(Y,B) = \prod_{J_{-p}} \pi_q(Y,B) \cong C^{-p}(A;\pi_q(Y,B))$$

for $p+q \geqslant 1$ and $p \leqslant -1$, where $C^*(A;\pi_q(Y,B))$ denotes the cellular cohomology complex of A with coefficients in $\pi_q(Y,B)$.

The isomorphism $\gamma : E^1_{p,q} = \pi_{p+q}^{Z_{-p}}(Y,B) \to C^{-p}(A;\pi_q(Y,B))$ is given by

$$\gamma([f])(e_\alpha^{-p}) = [f \circ \mathrm{C}\Sigma^{p+q-1} i_\alpha]$$

where $i_\alpha : S^{-p} \to Z_{-p}$ denotes the inclusion in the α-th copy of S^{-p}. Note also that $E^2_{p,q} = 0$ if $p+q \leqslant 0$ or $p \geqslant 0$.

We wish to prove now that $E^2_{p,q} \cong H^{-p}(A;\pi_q(Y,B))$ for $p+q \geqslant 2$ and $p \leqslant -1$.

We consider the morphism

$$\delta : E^1_{p,q} \cong C^{-p}(A;\pi_q(Y,B)) \to E^1_{p-1,q} \cong C^{-p+1}(A;\pi_q(Y,B))$$

coming from the spectral sequence. We will prove that $\delta = d^*$ for $n = p+q \geqslant 2$ and $p \leqslant -1$, where d^* is the cellular boundary map. This is equivalent to saying that the following diagram commutes

$$\begin{array}{ccccc} \pi_n^{Z_{p'}}(Y,B) & \xrightarrow{(\overline{q}_{p'})^*} & \pi_n^{A^{p'}}(Y,B) & \xrightarrow{(\underset{J_{p'+1}}{+} g_\beta^{p'+1})^*} & \pi_{n-1}^{Z_{p'+1}}(Y,B) \\ \gamma\downarrow\cong & & & & \gamma\downarrow\cong \\ C^{p'}(A;\pi_{n+p'}(Y,B)) & & \xrightarrow{\quad d^* \quad} & & C^{p'+1}(A;\pi_{n+p'}(Y,B)) \end{array}$$

where $p' = -p$.

If $[h] \in \pi_n^{Z_{p'}}(Y,B)$ and $e_\alpha^{p'+1}$ is a $(p'+1)$-cell of A, then

$$\begin{aligned} \left(\gamma(\underset{\beta\in J_{p'+1}}{+} g_\beta^{p'+1})^* q^*(h)\right)(e_\alpha^{p'+1}) &= \gamma(h\mathrm{C}\Sigma^{n-1}q(\underset{\beta\in J_{p'+1}}{+} \mathrm{C}\Sigma^{n-1}g_\beta^{p'+1}))(e_\alpha^{p'+1}) = \\ &= [h\mathrm{C}\Sigma^{n-1}q\mathrm{C}\Sigma^{n-1}g_\alpha^{p'+1}]. \end{aligned}$$

On the other hand,

$$\begin{aligned} d^*(\gamma([h]))(e_\alpha^{p'+1}) &= (\gamma([h]))(d(e_\alpha^{p'+1})) = \sum_{\beta\in J_{p'}} \deg(q_\beta g_\alpha^{p'+1})(\gamma([h]))(e_\beta^{p'}) = \\ &= \sum_{\beta\in J_{p'}} \deg(q_\beta g_\alpha^{p'+1})[h\mathrm{C}\Sigma^{n-1}i_\beta] \end{aligned}$$

where $q_\beta : A^{p'} \to S^{p'}$ is the map that collapses $A^{p'} - e_\beta^{p'}$ to a point.

Let $r = n-1+p'$. Since the morphism $\bigoplus_{\beta\in J_{p'}} (\Sigma^{n-1}q_\beta)_*$ is the inverse of the isomorphism $\bigoplus_{\beta\in J_{p'}} (\Sigma^{n-1}i_\beta)_* : \bigoplus_{\beta\in J_{p'}} \pi_r(S^r) \to \pi_r(\Sigma^{n-1}Z_{p'})$ we obtain that

$$\begin{aligned} [\Sigma^{n-1}\overline{q}_{p'}\Sigma^{n-1}g_\alpha^{p'+1}] &= \bigoplus_{\beta\in J_{p'}} (\Sigma^{n-1}i_\beta)_*(\bigoplus_{\beta\in J_{p'}} (\Sigma^{n-1}q_\beta)_*([\Sigma^{n-1}\overline{q}_{p'}\Sigma^{n-1}g_\alpha^{p'+1}])) = \\ &= \bigoplus_{\beta\in J_{p'}} (\Sigma^{n-1}i_\beta)_*(\{[\Sigma^{n-1}q_\beta\Sigma^{n-1}g_\alpha^{p'+1}]\}_{\beta\in J_{p'}}) = \\ &= \sum_{\beta\in J_{p'}} [\Sigma^{n-1}i_\beta\Sigma^{n-1}q_\beta\Sigma^{n-1}g_\alpha^{p'+1}]. \end{aligned}$$

Hence,

$$\begin{aligned} [h\mathrm{C}\Sigma^{n-1}\overline{q}_{p'}\mathrm{C}\Sigma^{n-1}g_\alpha^{p'+1}] &= h_*([\mathrm{C}\Sigma^{n-1}\overline{q}_{p'}g_\alpha^{p'+1}]) = h_*(\mathrm{C}\Sigma^{n-1}(\sum_{\beta\in J_{p'}} [i_\beta q_\beta g_\alpha^{p'+1}])) = \\ &= \sum_{\beta\in J_{p'}} [h\mathrm{C}\Sigma^{n-1}(i_\beta q_\beta g_\alpha^{p'+1})] = \\ &= \sum_{\beta\in J_{p'}} [h\mathrm{C}\Sigma^{n-1}i_\beta][\mathrm{C}\Sigma^{n-1}(q_\beta g_\alpha^{p'+1})] = \\ &= \sum_{\beta\in J_{p'}} \deg(q_\beta g_\alpha^{p'+1})[h\mathrm{C}\Sigma^{n-1}i_\beta]. \end{aligned}$$

It follows that $E^2_{p,q} \cong H^{-p}(A; \pi_q(Y))$ for $p + q \geqslant 2$ and $p \leqslant -1$.

The same argument works for the case $p + q = 1$, $p \leqslant -2$, and we obtain a commutative diagram

$$\begin{array}{ccccc}
\pi_1^{Z_{p'}}(Y,B) & \xrightarrow{(\overline{q}_{p'})^*} & \pi_1^{A^{p'}}(Y,B) & \xrightarrow{(\underset{J_{p'+1}}{+} g_\beta^{p'+1})^*} & \pi_1^{\underset{J_{p'+1}}{\bigvee} S^{p'}}(Y,B) \\
\gamma \downarrow \cong & & & & \downarrow \cong \\
C^{p'}(A; \pi_{p'+1}(Y,B)) & & \xrightarrow{\quad d^* \quad} & & C^{p'+1}(A; \pi_{p'+1}(Y,B))
\end{array}$$

Then $E^2_{p,q} = \ker d^1_{p,q}/\mathrm{Im}\, d^1_{p+1,q} = \mathrm{Im}\,\partial_{-p}/\mathrm{Im}\, d^1_{p+1,q}$. By exactness, $\mathrm{Im}\,\partial_{-p} = \ker q^*$. Thus, $\mathrm{Im}\,\partial_{-p} \subseteq \ker d^*$ since the previous diagram commutes. Moreover, if $p \leqslant -2$ by the previous case the map $d^1_{p+1,q}$ coincides, up to isomorphisms, with the map $d^* : C^{p'-1}(A; \pi_{p'+1}(Y,B)) \to C^{p'}(A; \pi_{p'+1}(Y,B))$, and in case $p = -1$, both maps are trivial. Therefore, $E^2_{p,q}$ is isomorphic to a subgroup of $H^{-p}(A; \pi_q(Y,B))$ if $p + q = 1$ and $p \leqslant -1$. □

Of course, applying this theorem to the topological pair (CY, Y) we obtain the following absolute version, which is similar to Federer's result.

Corollary 2.2. *Let Y be a topological space with abelian fundamental group and let A be a finite-dimensional (and path-connected) CW-complex. Then there exists a homological spectral sequence $\{E^a_{p,q}\}_{a\geqslant 1}$, with $E^2_{p,q}$ satisfying*

- $E^2_{p,q} \cong H^{-p}(A; \pi_q(Y))$ *for* $p + q \geqslant 1$ *and* $p \leqslant -1$.
- $E^2_{p,q}$ *is isomorphic to a subgroup of* $H^{-p}(A; \pi_q(Y))$ *if* $p + q = 0$ *and* $p \leqslant -1$.
- $E^2_{p,q} = 0$ *if* $p + q < 0$ *or* $p \geqslant 0$.

which converges to $\pi^A_{p+q}(Y)$ for $p + q \geqslant 1$.

We will call $\{E^a_{p,q}\}_{a\geqslant 1}$ the *Federer spectral sequence associated to A and Y.*

Note that the relative version of the Federer spectral sequence is natural in the following sense. If A is a finite-dimensional CW-complex, (Y, B) and (Y', B') are topological pairs such that the groups $\pi_2(Y, B)$ and $\pi_2(Y', B')$ are abelian and $f : (Y, B) \to (Y', B')$ is a continuous map, then f induces a morphism between the relative Federer spectral sequence associated to A and (Y, B) and the one associated to A and (Y', B'). Indeed, f induces morphisms between the extended long exact sequences of the proof above and hence a morphism between the exact couples involved, which gives rise to the morphism between the spectral sequences mentioned above.

Moreover, if A' is another finite-dimensional CW-complex and $g : A \to A'$ is a cellular map, then g also induces morphisms between the extended long exact sequences mentioned before and therefore, a morphism between the relative Federer spectral sequence associated to A and (Y, B) and the one associated to A' and (Y, B). If g is not cellular we may replace it by a homotopic cellular map to obtain the induced morphism. Of course, by the description of the second page of our spectral sequence, the map g itself will also induce the same morphism from page two onwards.

Clearly, the same holds for the absolute version.

Remark 2.3.
(1) Looking at the extended exact sequences of the proof of 2.1 we obtain that the relative Federer spectral sequence converges to the trivial group in degree 1. Thus, the groups $E^2_{p,q}$, with $p+q=1$, become all trivial in E^∞.
(2) As we have mentioned above, the spectral sequence given by Federer in [**5**] is similar to our absolute version. But since we work with pointed topological spaces our version enables us to compute homotopy groups of function spaces only when the base point is the constant map. However, the hypothesis we require on the space Y ($\pi_1(Y)$ is abelian) are weaker than Federer's ($\pi_1(Y)$ acts trivially on $\pi_n(Y)$ for all $n \in \mathbb{N}$). Moreover, our approach in terms of A-homotopy groups admits the relative version given before.

Just as a simple example of application of the Federer spectral sequence consider the following, which is a reformulation of a well-known result for homotopy groups with coefficients.

Example 2.4. If A is a Moore space of type (G, m) (with G finitely generated) and X is a path-connected topological space with abelian fundamental group, in the Federer spectral sequence we get

$$E^2_{-p,q} = \begin{cases} \hom(G, \pi_q(X)) & \text{if } p = m \\ \mathrm{Ext}(G, \pi_q(X)) & \text{if } p = m+1 \\ 0 & \text{otherwise} \end{cases} \qquad \text{for } -p+q \geqslant 1.$$

Hence, from the corresponding filtrations, we deduce that, for $n \geqslant 1$, there are short exact sequences of groups

$$0 \longrightarrow \mathrm{Ext}(G, \pi_{n+m+1}(X)) \longrightarrow \pi^A_n(X) \longrightarrow \hom(G, \pi_{n+m}(X)) \longrightarrow 0$$

As a corollary, if G is a finite group of exponent r then $\alpha^{2r} = 0$ for every $\alpha \in \pi^A_n(X)$. For example, if X is a path-connected topological space with abelian fundamental group, then every element in $\pi^{\mathbb{P}^2}_n(X)$ $(n \geqslant 1)$ has order 1, 2 or 4.

We will now apply 2.2 to obtain an extension to the Hopf-Whitney theorem.

Theorem 2.5. *Let K be a path-connected CW-complex of dimension $n \geqslant 2$ and let Y be $(n-1)$-connected. Then there exists a bijection $[K, Y] \leftrightarrow H^n(K; \pi_n(Y))$.*

In addition, if K is the suspension of a path-connected CW-complex (or if Y is a loop space), then the groups $[K, Y]$ and $H^n(K; \pi_n(Y))$ are isomorphic.

Moreover, this isomorphism is natural in K and in Y.

Proof. The first part is the Hopf-Whitney theorem (cf. [**6**]). The second part can be proved easily by means of the Federer spectral sequence. Concretely, suppose that $K = \Sigma K'$ with K' path-connected. Let $\{E^a_{p,q}\}$ denote the Federer spectral sequence associated to K' and Y. Then $E^2_{p,q} = 0$ for $q \leqslant n-1$ since Y is $(n-1)$-connected, and $E^2_{p,q} = 0$ for $p \leqslant -n$ since $\dim K' = n-1$. Hence, $E^2_{-(n-1),n} \cong H^{n-1}(K'; \pi_n(Y))$ survives to E^∞. As it is the only nonzero entry in the diagonal $p+q=1$ of E^2 it follows that

$$[K,Y] \cong \pi_1^{K'}(Y) \cong E^2_{-(n-1),n} \cong H^{n-1}(K'; \pi_n(Y)) \cong H^n(K; \pi_n(Y)).$$

Finally, naturality follows from naturality of the Federer spectral sequence. □

In a similar way, from theorem 2.1 we obtain the following relative version of the Hopf-Whitney theorem, which not only is interesting for its own sake but also will be important for our purposes.

Theorem 2.6. *Let K be the suspension of a path-connected CW-complex of dimension $n-1 \geqslant 1$ and let (Y,B) be an n-connected topological pair. Then there exists an isomorphism of groups*

$$[(CK,K);(Y,B)] \leftrightarrow H^n(K; \pi_{n+1}(Y,B)).$$

which is natural in K and in (Y,B).

Proof. Suppose that $K = \Sigma K'$ with K' path-connected. Let $\{E^a_{p,q}\}$ denote the relative Federer spectral sequence associated to K' and (Y,B). Then $E^2_{p,q} = 0$ for $q \leqslant n$ since (Y,B) is n-connected, and $E^2_{p,q} = 0$ for $p \leqslant -n$ since $\dim K' = n-1$. Hence, $E^2_{-(n-1),n+1} = H^{n-1}(K'; \pi_{n+1}(Y,B))$ survives to E^∞. As it is the only nonzero entry in the diagonal $p+q=2$ of E^2 it follows that

$$\begin{aligned}[(CK,K);(Y,B)] &= \pi_2^{K'}(Y,B) \cong E^2_{-(n-1),n+1} \cong H^{n-1}(K'; \pi_{n+1}(Y,B)) \cong \\ &\cong H^n(K; \pi_{n+1}(Y,B)).\end{aligned}$$

and naturality follows again from naturality of the Federer spectral sequence. □

We will give now another application of 2.2. We will denote by $\mathcal{T_P}$ the class of torsion abelian groups whose elements have orders which are divisible only by primes in a set $\mathcal{P}$ of prime numbers.

Proposition 2.7. *Let A be a finite-dimensional CW-complex such that $H_n(A)$ is finitely generated for all $n \in \mathbb{N}$ and let X be a path-connected topological space such that $\pi_1(X)$ is abelian. If $H_n(A) \in \mathcal{T_P}$ for all $n \in \mathbb{N}$ then $\pi_n^A(X) \in \mathcal{T_P}$ for all $n \in \mathbb{N}$.*

Proof. By 2.2, it suffices to prove that $H^{-p}(A; \pi_q(X)) \in \mathcal{T_P}$ for all $p,q \in \mathbb{Z}$ such that $p+q \geqslant 0$ and $p \leqslant -1$.

By the universal coefficient theorem

$$H^{-p}(A; \pi_q(X)) \cong \hom(H_{-p}(A), \pi_q(X)) \oplus \operatorname{Ext}(H_{-p-1}(A), \pi_q(X)).$$

Since A is $\mathcal{T_P}$-acyclic and $H_n(A)$ is finitely generated for all $n \in \mathbb{N}$ it follows that $\hom(H_{-p}(A), \pi_q(X)) \in \mathcal{T_P}$ and $\operatorname{Ext}(H_{-p-1}(A), \pi_q(X)) \in \mathcal{T_P}$ for all $p \leqslant -1$ and $q \geqslant 0$. Thus, $H^{-p}(A; \pi_q(X)) \in \mathcal{T_P}$ for all $p \leqslant -1$ and $q \geqslant 0$. □

3. A-homology

In this section we define an '*A-shaped*' reduced homology theory, which we call A-homology and which coincides with the singular homology theory in case $A = S^0$. Our definition enables us to obtain generalizations of several classical results. For example, we prove a Hurewicz-type theorem (3.9) relating the A-homotopy groups with the A-homology groups.

We also give a homological version of the Whitehead theorem which states that, under certain hypotheses, a map between CW-complexes which induces isomorphisms in the A-homology groups is a homotopy equivalence (3.10).

Finally, we define a Hurewicz-type map between the A-homotopy groups and the A-homology groups and embed it in a long exact sequence 3.11 generalizing the Whitehead exact sequence [**10**].

We begin with a simple remark which will be used later.

Remark 3.1. Let $p : (E, e_0) \to (B, b_0)$ be a quasifibration, let $F = p^{-1}(b_0)$ and let A be a CW-complex. Since p induces isomorphisms $p_* : \pi_i(E, F, e_0) \to \pi_i(B, b_0)$ for all $i \in \mathbb{N}$ and $\pi_i(E, F, e_0) \cong \pi_{i-1}(P(E, e_0, F), c_{e_0})$ and $\pi_i(B, b_0) \cong \pi_{i-1}(\Omega B, c_{b_0})$ it follows that the induced map $\hat{p} : (P(E, e_0, F), c_{e_0}) \to (\Omega B, c_{b_0})$ is a weak equivalence. Thus, $\hat{p}$ induces isomorphisms $\hat{p}_* : \pi_i^A(P(E, e_0, F), c_{e_0}) \to \pi_i^A(\Omega B, c_{b_0})$.

Since $\pi_i^A(E, F, e_0) \cong \pi_{i-1}^A(P(E, e_0, F), c_{e_0})$ and $\pi_i^A(B, b_0) \cong \pi_{i-1}^A(\Omega B, c_{b_0})$ we obtain that $p_* : \pi_i^A(E, F, e_0) \to \pi_i^A(B, b_0)$ is an isomorphism for all $i \in \mathbb{N}$.

Our definition of A-homology groups is inspired by the Dold-Thom theorem.

Definition 3.2. Let A be a CW-complex and let X be a topological space. For $n \in \mathbb{N}_0$ we define the *n-th A-homology group of X* as

$$H_n^A(X) = \pi_n^A(SP(X))$$

where $SP(X)$ denotes the infinite symmetric product of X.

Theorem 3.3. *The functor $H_*^A(\_)$ defines a reduced homology theory on the category of (path-connected) CW-complexes.*

Proof. It is clear that $H_*^A(\_)$ is a homotopy functor. If (X, B, x_0) is a pointed CW-pair, then by the Dold-Thom theorem, the quotient map $q : X \to X/B$ induces a quasifibration $\hat{q} : SP(X) \to SP(X/B)$ whose fiber is homotopy equivalent to $SP(B)$. Since A is a CW-complex there is a long exact sequence

$$\cdots \longrightarrow \pi_n^A(SP(B)) \longrightarrow \pi_n^A(SP(X)) \longrightarrow \pi_n^A(SP(X/B)) \longrightarrow \pi_{n-1}^A(SP(B)) \longrightarrow \cdots$$

It remains to show that there exists a natural isomorphism $H_n^A(X) \cong H_{n+1}^A(\Sigma X)$ and that $H_n^A(X)$ are abelian groups for $n = 0, 1$. The natural isomorphism follows from the long exact sequence of above applied to the CW-pair (CX, X). Note that $H_n^A(CX) = 0$ since CX is contractible and H_n^A is a homotopy functor. The second part follows immediately, since $H_0^A(X) \cong H_1^A(\Sigma X) \cong H_2^A(\Sigma^2 X)$. The group structure on $H_0^A(X)$ is induced from the one on $H_1^A(X)$ by the corresponding natural isomorphism. □

The proof above encourages us to define the relative A-homology groups of a CW-pair (X, B) by $H_n^A(X, B) = \pi_n^A(SP(X/B))$ for $n \geqslant 1$. As shown before, there exist long exact sequences of A-homology groups associated to a CW-pair (X, B).

Federer's spectral sequence can be applied as a first method of computation of A-homology groups. Indeed, given a finite CW-complex A and a CW-complex X, the associated Federer spectral sequence $\{E^a_{p,q}\}$ converges to the A-homotopy groups of $SP(X)$ (note that $\pi_1(SP(X))$ is abelian). In this case we obtain that $E^2_{p,q} = H^{-p}(A, \pi_q(SP(X))) = H^{-p}(A, H_q(X))$ if $p+q \geqslant 1$ and $p \leqslant -1$. Moreover, we will show later a explicit formula to compute A-homology groups of CW-complexes.

We exhibit now some examples.

Example 3.4. If A is a finite-dimensional CW-complex and X is a Moore space of type (G, n) then $SP(X)$ is an Eilenberg-Mac Lane space of the same type. Hence, by the Federer spectral sequence

$$H_r^A(X) = \pi_r^A(SP(X)) = H^{n-r}(A, \pi_n(SP(X))) = H^{n-r}(A, G) \qquad \text{for } r \geqslant 1.$$

In particular, $H_r^A(S^n) = H^{n-r}(A, \mathbb{Z})$.

We also deduce that if X is a Moore space of type (G, n) and A is $(n-1)$-connected, then $H_r^A(X) = 0$ for all $r \geqslant 1$.

Example 3.5. Let A be a Moore space of type (G, m) (with G finitely generated) and let X be a path-connected CW-complex. As in example 2.4, for $n \geqslant 1$, there are short exact sequences of abelian groups

$$0 \longrightarrow \mathrm{Ext}(G, H_{n+m+1}(X)) \longrightarrow H_n^A(X) \longrightarrow \hom(G, H_{n+m}(X)) \longrightarrow 0$$

As a consequence, if G is a finite group of exponent r, then $\alpha^{2r} = 0$ for every $\alpha \in H_n^A(X)$.

It is well known that if a CW-complex does not have cells of a certain dimension j, then its j-th homology group vanishes. As one should expect, a similar result holds for the A-homology groups. Concretely, if A is an l-connected CW-complex of dimension k and X is a CW-complex, applying the Federer spectral sequence to the space $SP(X)$ one can obtain that:

1. If $\dim(X) = m$, then $H_r^A(X) = 0$ for $r \geqslant m - l$.
2. If X does not have cells of dimension less than m', then $H_r^A(X) = 0$ for $r \leqslant m' + l - k$.

Following the idea of example 3.4 we will show now a explicit formula to compute A-homology groups.

Proposition 3.6. *Let A be a finite-dimensional CW-complex and let X be a connected CW-complex. Then for every $n \in \mathbb{N}_0$, $H_n^A(X) = \bigoplus_{j \in \mathbb{N}} H^{j-n}(A, H_j(X))$.*

Proof. Since $SP(X)$ has the weak homotopy type of $\prod_{n \in \mathbb{N}} K(H_n(X), n)$ and A is a

CW-complex we obtain that

$$\begin{aligned} H_n^A(X) &= \pi_n^A(SP(X)) \cong \pi_n^A(\prod_{j\in\mathbb{N}} K(H_j(X),j)) \cong \prod_{j\in\mathbb{N}} \pi_n^A(K(H_j(X),j)) \cong \\ &\cong \prod_{j\in\mathbb{N}} H^{j-n}(A,H_j(X)) = \bigoplus_{j\in\mathbb{N}} H^{j-n}(A,H_j(X)) \end{aligned}$$

where the last isomorphism follows from the Federer spectral sequence. □

Now we show that, in case A is compact, H_*^A satisfies the wedge axiom. This can be proved in two different ways: using the definition of A-homotopy groups or using the above formula. We choose the first one.

Proposition 3.7. *Let A be a finite CW-complex, and let $\{X_i\}_{i\in I}$ be a collection of CW-complexes. Then*

$$H_n^A\left(\bigvee_{i\in I} X_i\right) = \bigoplus_{i\in I} H_n^A(X_i).$$

Proof. The space $SP(\bigvee_{i\in I} X_i)$ is homeomorphic to $\prod_{i\in I}^{w} SP(X_i)$ with the weak product topology, i.e. $\prod_{i\in I}^{w} SP(X_i)$ is the colimit of the products of finitely many factors. Since A is compact, $\pi_n^A(\prod_{i\in I}^{w} SP(X_i)) \cong \bigoplus_{i\in I} \pi_n^A(SP(X_i))$ and the result follows. □

We will prove now some of the main results of this article. We begin with a Hurewicz-type theorem relating the A-homology groups with the A-homotopy groups.

Theorem 3.8. *Let A be the suspension of a path-connected CW-complex of dimension $k-1 \geqslant 1$ and let (X,B) be an n-connected CW-pair with $n \geqslant k$. Suppose, in addition, that B is simply-connected and non-empty. Then $H_r^A(X,B) = 0$ for $r \leqslant n-k$ and $\pi_{n-k+1}^A(X,B) \cong H_{n-k+1}^A(X,B)$.*

Proof. By the Hurewicz theorem, $H_r(X,B) = 0$ for $r \leqslant n$ and $H_{n+1}(X,B) \cong \pi_{n+1}(X,B)$. Since (X,B) is a CW-pair, by the Dold-Thom theorem we obtain that $\pi_r(SP(X/B)) \cong H_r(X/B) \cong H_r(X,B) = 0$ for $r \leqslant n$.

Since A is a CW-complex of dimension $k \leqslant n$, then, $H_r^A(X,B) = \pi_r^A(SP(X/B)) = 0$ for $r \leqslant n-k$. Also,

$$\begin{aligned} \pi_{n-k+1}^A(X,B) &= [(C\Sigma^{n-k}A,\Sigma^{n-k}A);(X,B)] \cong H^n(\Sigma^{n-k}A,\pi_{n+1}(X,B)) \cong \\ &\cong H^n(\Sigma^{n-k}A,H_{n+1}(X,B)) \cong H^{n+1}(\Sigma^{n-k+1}A,\pi_{n+1}(SP(X/B))) \cong \\ &\cong [\Sigma^{n-k+1}A,SP(X/B)] = \pi_{n-k+1}^A(SP(X/B)) = H_{n-k+1}^A(X,B) \end{aligned}$$

where the first and fourth isomorphisms hold by 2.6 and 2.5 respectively. □

Moreover, by naturality of 2.5 and 2.6 it follows that the isomorphism above is the morphism induced in π_n^A by the map which is the composition of the quotient map $(X,B) \to (X/B,*)$ with the inclusion map $(X/B,*) \to (SP(X/B),*)$.

Clearly, from this relative A-Hurewicz theorem we can deduce the following absolute version.

Theorem 3.9. *Let A be the suspension of a path-connected CW-complex with $\dim A = k \geqslant 2$ and let X be an n-connected CW-complex with $n \geqslant k$. Then $H_r^A(X) = 0$ for $r \leqslant n-k$ and $\pi_{n-k+1}^A(X) \cong H_{n-k+1}^A(X)$.*

Moreover, the morphism $i_* : \pi_{n-k+1}^A(X) \to \pi_{n-k+1}^A(SP(X)) = H_{n-k+1}^A(X)$ induced by the inclusion map $i : X \to SP(X)$ is an isomorphism.

Thus, the morphism $i_* : \pi_n^A(X) \to \pi_n^A(SP(X))$ can be thought as a Hurewicz-type map and will be called *A-Hurewicz homomorphism*. Not only is it natural, but also it can be embedded in a long exact sequence, as we will show later (3.11).

We give now a homological version of the Whitehead theorem, which states that, under certain hypotheses, a continuous map between CW-complexes inducing isomorphisms in A-homotopy groups is a homotopy equivalence.

Theorem 3.10. *Let A' be a path-connected and locally compact CW-complex of dimension $k-1 \geqslant 0$ such that $H_{k-1}(A') \neq 0$ and let $A = \Sigma A'$. Let $f : X \to Y$ be a continuous map between simply-connected CW-complexes which induces isomorphisms $f_* : H_r^A(X) \to H_r^A(Y)$ for all $r \in \mathbb{N}$ and $f_* : \pi_i(X) \to \pi_i(Y)$ for all $i \leqslant k+1$. Then f is a homotopy equivalence.*

Proof. Replacing Y by the mapping cylinder of f, we may suppose that f is an inclusion map and hence (Y, X) is $(k+1)$-connected. We will prove by induction that (Y, X) is n-connected for all $n \in \mathbb{N}$.

Suppose that (Y, X) is n-connected for some $n \geqslant k+1$. Then

$$\pi_{n-k}^A(Y, X) = [(C\Sigma^{n-k-1}A, \Sigma^{n-k-1}A), (Y, X)] = 0$$

because $\dim(C\Sigma^{n-k-1}A) = n$. Moreover, by 3.8 we obtain that

$$\pi_{n-k+1}^A(Y, X) \cong H_{n-k+1}^A(Y, X) = 0.$$

Now, by 2.6,

$$\begin{aligned} 0 &= \pi_{n-k+1}^A(Y, X) = H^n(\Sigma^{n-k}A, \pi_{n+1}(Y, X)) = H^k(A, \pi_{n+1}(Y, X)) = \\ &= \hom(H_k(A), \pi_{n+1}(Y, X)) \oplus \operatorname{Ext}(H_{k-1}(A), \pi_{n+1}(Y, X)). \end{aligned}$$

Then $\hom(H_k(A), \pi_{n+1}(Y, X)) = 0$. By the hypotheses on A' it follows that $H_{k-1}(A') = H_k(A)$ has $\mathbb{Z}$ as a direct summand. Hence, $\pi_{n+1}(Y, X) = 0$ and thus (Y, X) is $(n+1)$-connected.

In consequence, (Y, X) is n-connected for all $n \in \mathbb{N}$. Then the inclusion map $f : X \to Y$ is a weak equivalence and since X and Y are CW-complexes it follows that f is a homotopy equivalence. □

To finish, we will make use of a modern construction of the exact sequence Whitehead introduced in [**10**] to embed the A-Hurewicz homomorphism defined above in a long exact sequence. As a corollary we will obtain another proof of theorem 3.9 together with an extension of it. A different way to obtain the Whitehead exact sequence is given in [**4**].

Let X be a CW-complex and let ΓX be the homotopy fiber of the inclusion $i : X \to SP(X)$. Hence, there is a long exact sequence

$$\cdots \longrightarrow \pi_n^A(\Gamma X) \longrightarrow \pi_n^A(X) \xrightarrow{i_*} \pi_n^A(SP(X)) \longrightarrow \pi_{n-1}^A(\Gamma X) \longrightarrow \cdots$$

and by definition $\pi_n^A(SP(X)) = H_n^A(X)$ and i_* is the A-Hurewicz homomorphism.
Thus, we have proved the following.

Proposition 3.11. *Let A and X be CW-complexes. Then, there is a long exact sequence*

$$\cdots \longrightarrow \pi_n^A(\Gamma X) \longrightarrow \pi_n^A(X) \xrightarrow{i_*} H_n^A(X) \longrightarrow \pi_{n-1}^A(\Gamma X) \longrightarrow \cdots$$

Using this long exact sequence we will give another proof of theorem 3.9. Recall that in [**10**], given a CW-complex Z, Whitehead defines the group $\Gamma_n(Z)$ as the kernel of the canonical morphism $\pi_n(Z^n) \to \pi_n(Z^n, Z^{n-1})$ which by exactness coincides with the image of the morphism $j_* : \pi_n(Z^{n-1}) \to \pi_n(Z^n)$, where $j : Z^{n-1} \to Z^n$ is the inclusion map. It is known that $\Gamma_n(Z) \cong \pi_n(\Gamma Z)$.

Now let A be a CW-complex of dimension $k \geqslant 2$ and let X be an n-connected topological space with $n \geqslant k$. Replacing X by a homotopy equivalent CW-complex Y with $Y^n = *$, it follows that $\Gamma_r(X) = 0$ for $r \leqslant n+1$. Hence, ΓX is $(n+1)$-connected.

Therefore, $\pi_r^A(\Gamma X) = 0$ for $r \leqslant n-k+1$. Thus, from the exact sequence above we obtain that $i_* : \pi_{n-k+1}^A(X) \to H_{n-k+1}^A(X)$ is an isomorphism. Moreover, $i_* : \pi_{n-k+2}^A(X) \to H_{n-k+2}^A(X)$ is an epimorphism.

Summing up, we have proved the following.

Theorem 3.12. *Let A be a path-connected CW-complex with $\dim A = k \geqslant 2$ and let X be an n-connected CW-complex with $n \geqslant k$. Let $i : X \to SP(X)$ be the inclusion map. Then $H_r^A(X) = 0$ for $r \leqslant n-k$, $i_* : \pi_{n-k+1}^A(X) \to H_{n-k+1}^A(X)$ is an isomorphism and $i_* : \pi_{n-k+2}^A(X) \to H_{n-k+2}^A(X)$ is an epimorphism.*

References

[1] M. Barratt, *Track groups I*, Proc. London Math. Soc. (3) 5 (1955), 71–106.

[2] M. Barratt, *Track groups II*, Proc. London Math. Soc. (3) 5 (1955), 285–329.

[3] R. Brown, *On Künneth suspensions*, Proc. Cambridge Philos. Soc. 60 (1964) 713–720.

[4] R. Brown and P. Higgins, *Colimit theorems for relative homotopy groups*, J. Pure Appl. Algebra 22 (1981) 11-41.

[5] H. Federer, *A study of function spaces by spectral sequences*, Trans. Amer. Math. Soc. 82 (1956), 340–361.

[6] R. Mosher and M. Tangora, *Cohomology operations and applications in homotopy theory*, Harper & Row, Publishers. 1968.

[7] J. A. Neisendorfer, *Primary homotopy theory*, Memoirs A.M.S. 232. Amer. Math. Soc. 1980.

[8] F. P. Peterson, *Generalized cohomotopy groups*, Amer. Jour. Math. 78 (1956), 259–281.

[9] D. Quillen, *Homotopical Algebra*, Lecture Notes in Mathematics. Vol. 43. Springer. 1967.

[10] J. H. C. Whitehead, *A certain exact sequence*, Annals of Mathematics 52 (1950) 51–110.

[11] H. Whitney, *Relations between the second and third homotopy groups of a simply-connected space*, Annals of Mathematics 50 (1949), 180–202.

This article may be accessed via WWW at `http://tcms.org.ge/Journals/JHRS/`

Enzo Miguel Ottina
`emottina@uncu.edu.ar`

Instituto de Ciencias Básicas
Universidad Nacional de Cuyo
Mendoza, Argentina.

Journal of Homotopy and Related Structures, vol. 6(1), 2011, pp.175–176

ERRATUM TO "OPERADS IN ITERATED MONOIDAL CATEGORIES," J. OF HOMOTOPY AND RELATED STRUCTURES, VOL.2 (1), 2007, PP.1 - 43

STEFAN FORCEY, JACOB SIEHLER AND E. SETH SOWERS

(*communicated by James Stasheff*)

Abstract

Theorem 5.4 of the paper [**1**] in question is incorrect as stated. In fact we state here instead that the characterization of minimal 2-fold operads in the natural numbers is an open question.

1. Description of error.

In Example 5.3 of the paper [**1**] it is pointed out that a nontrivial 2-fold operad in $\mathbb{N}$ is a nonzero sequence $\{\mathcal{C}(j)\}_{j\geqslant 0}$ of natural numbers which has the property that for any $j_1 \dots j_k$, $\max(\mathcal{C}(k), \sum \mathcal{C}(j_i)) \leqslant \mathcal{C}(\sum j_i)$ and for which $\mathcal{C}(1) = 0$.

Recall that minimal examples are formed by choosing a starting term or terms and then determining each later n^{th} term. These are minimal in the sense that the principle which determines each of the later terms in succession is that of choosing the minimal next term out of all possible such terms. For a starting finite sequence $0, a_2, \dots, a_l$ which obeys the axioms of a 2-fold operad so far, the operad $\mathcal{C}_{0,a_2,\dots,a_l}$ is found by defining terms $\mathcal{C}_{a_1,\dots,a_l}(n)$ for $n > l$ to be the maximum of all the values of $\max(\mathcal{C}(k), \sum_{i=1}^{k} \mathcal{C}(j_i))$ where the sum of the j_i is n.

In addition to the examples in the text, here is another.

$$C = \mathcal{C}_{0,3,6,8} = (\emptyset, 0, 3, 6, 8, 9, 12, 14, 16, 18, 20 \dots).$$

It is clear from the above example that Theorem 5.4 of [**1**] is incorrect as stated.

1.1 Theorem (Incorrect). *If "arbitrary" starting terms $0, a_2, \dots, a_k \in \mathbb{N}$ are given (themselves of course obeying the axioms of a 2-fold operad), then the nth term of the 2-fold operad $\mathcal{C}_{0,a_2,\dots,a_k}$ in $\mathbb{N}$ obeys*

$$a_n = a_q + pa_k \text{ where } n = pk + q, \text{ for } p \in \mathbb{N}, 0 \leqslant q < k.$$

The incorrect theorem would predict that $C(5) = 8, C(6) = 11, C(7) = 14, C(8) = 16, C(9) = 16$ and $C(10) = 19$.

Received April 23, 2011, revised April 25, 2011; published on May 15, 2011.

Key words and phrases: enriched categories, n-categories, iterated monoidal categories, operads.

The failure of the theorem's proof is in faulty induction in the step purporting to prove the inequality $a_n \leqslant a_q + pa_k$. However the other inequality, originally proven first, still holds:

1.2 Theorem. *If "arbitrary" starting terms $0, a_2, \dots, a_k \in \mathbb{N}$ are given (themselves of course obeying the axioms of a 2-fold operad), then the nth term of the 2-fold operad $C_{0,a_2,\dots,a_k}$ in $\mathbb{N}$ obeys*

$$a_n \geqslant a_q + pa_k \text{ where } n = pk + q, \text{ for } p \in \mathbb{N}, 0 \leqslant q < k.$$

Proof. We need to show that

$$a_n = \max_{j_1+\cdots+j_l=n} \{\max(a_l, \sum_{i=1}^{l} a_{j_i})\} = a_q + pa_k$$

where $n = pk + q$, for $p \in \mathbb{N}$, $0 \leqslant q < k$. First we note that $a_q + pa_k$ appears as a term in the overall max, so that $a_n \geqslant a_q + pa_k$. □

Other less strict versions of the theorem may hold as well. For now we leave as an open question the characterization of the minimal 2-fold operads in the natural numbers.

References

[1] S. Forcey, J. Siehler and E. S. Sowers, *Operads in iterated monoidal categories*, Journal of Homotopy and Related Structures, vol.2 (1), 2007, pp.1 - 43

This article may be accessed via WWW at `http://tcms.org.ge/Journals/JHRS/`

Stefan Forcey, Jacob Siehler and E. Seth Sowers
`sf34@uakron.edu`

Department of Mathematics
University of Akron
Akron OH, 44325-4002,
USA

Journal of Homotopy and Related Structures, vol. 6(1), 2011, pp.177–182

HOMOTOPY DG ALGEBRAS INDUCE HOMOTOPY BV ALGEBRAS

JOHN TERILLA, THOMAS TRADLER AND SCOTT O. WILSON

(*communicated by James Stasheff*)

Abstract

Let TA denote the space underlying the tensor algebra of a vector space A. In this short note, we show that if A is a differential graded algebra, then TA is a differential Batalin-Vilkovisky algebra. Moreover, if A is an A_∞ algebra, then TA is a commutative BV_∞ algebra.

1. Main Statement

Let (A, d_A) be a complex over a commutative ring R. Our convention is that d_A is of degree $+1$. The space $TA = \bigoplus_{n\geqslant 0} A^{\otimes n}$ is graded by declaring monomials of homogeneous elements $a_1 \otimes \cdots \otimes a_n \in A^{\otimes n}$ to be of degree $|a_1| + \cdots + |a_n| + n$.

There is a shuffle product $\bullet : TA \otimes TA \to TA$ generated by

$$(a_1 \otimes \cdots \otimes a_n) \bullet (a_{n+1} \otimes \cdots \otimes a_{n+m}) := \sum_{\sigma \in S(n,m)} (-1)^\kappa \cdot a_{\sigma^{-1}(1)} \otimes \cdots \otimes a_{\sigma^{-1}(n+m)},$$

where $S(n, m)$ is the set of all (n, m)-shuffles, *i.e.* $S(n, m)$ is the set of all permutations $\sigma \in \Sigma_{n+m}$ with $\sigma(1) < \cdots < \sigma(n)$ and $\sigma(n + 1) < \cdots < \sigma(n + m)$, (*cf.* [**6**]). Here $(-1)^\kappa$ is the Koszul sign, which introduces a factor of $(|a_i|+1)(|a_j|+1)$ whenever the elements a_i and a_j move past one another in a shuffle. Note that for degree zero elements of A, this Koszul sign is just $sgn(\sigma)$, the sign of the permutation σ. The shuffle product makes TA into a graded commutative associative algebra. Recall that TA is also a coalgebra under the usual tensor coproduct.

There is a differential $d : TA \to TA$ (of degree $+1$) given by extending the differential $d_A : A \to A$ as a coderivation of the tensor coproduct, see *e.g.* [**7**]:

$$d(a_1 \otimes \cdots \otimes a_n) = \sum_{i=0}^{n} (-1)^{|a_1|+\cdots+|a_{i-1}|+i-1} a_1 \otimes \cdots \otimes d_A(a_i) \otimes \cdots \otimes a_n$$

The second author was partially supported by the Max-Planck Institute in Bonn, Germany. The third author was supported in part by a grant from The City University of New York PSC-CUNY Research Award Program. We would like to thank Gabriel Drummond-Cole and Bruno Vallette for useful discussions about BV_∞ algebras.
Received December 21, 2010, revised March 02, 2011; published on July 12, 2011.
2000 Mathematics Subject Classification: 16E45, 17B60, 18G55.
Key words and phrases: Batalin-Vilkovisky BV algebra, homotopy BV algebra, tensor algebra, shuffle product, A-infinity algebra.

and together with the shuffle product, the triple $(TA, d, \bullet)$ is a differential graded commutative associative algebra.

If $\mu_A : A \otimes A \to A$ is an associative product, then there is another differential $\Delta = \tilde{\mu}_A : TA \to TA$, of degree -1, given by extending the multiplication as a coderivation,

$$\Delta(a_1 \otimes \cdots \otimes a_n) = \sum_{i=1}^{n-1} (-1)^{|a_1|+\cdots+|a_i|+i-1} a_1 \otimes \cdots \otimes \mu_A(a_i, a_{i+1}) \otimes \cdots \otimes a_n.$$

In Section 2 we show:

Theorem 1. *If (A, d_A, μ_A) is a differential graded algebra, then $(TA, d, \Delta, \bullet)$ defines a dBV algebra. The construction is functorial: If $f : A \to B$ is a morphism of differential associative algebras, then the induced map from TA to TB is a morphism of dBV algebras.*

Recall that a dBV algebra $(X, d, \Delta, \bullet)$ is a differential graded commutative associative algebra $(X, d, \bullet)$, with d of degree $+1$, and differential Δ of degree -1 such that d graded commutes with Δ (so that $d\Delta + \Delta d = 0$), and finally the deviation $\{,\}$ of Δ from being a derivation of $\bullet$,

$$\{x, y\} = (-1)^{|x|}\Delta(x \bullet y) - (-1)^{|x|}\Delta(x) \bullet y - x \bullet \Delta(y)$$

satisfies,

$$\{x, y\} = -(-1)^{(|x|+1)(|y|+1)}\{y, x\} \qquad \text{(Anti-symmetry)},$$
$$\{x \bullet y, z\} = x \bullet \{y, z\} + (-1)^{|y|(|z|+1)}\{x, z\} \bullet y \text{(Leibniz relation)}.$$

The Leibniz relation can be read as saying that bracketing with a fixed element (on the right) is a graded derivation of the product $\bullet$. These relations imply that bracketing with a fixed element on the left is also a graded derivation

$$\{x, y \bullet z\} = \{x, y\} \bullet z + (-1)^{(|x|+1)y} y \bullet \{x, z\}$$

and also imply that bracketing with a fixed element is a graded derivation of the bracket,

$$\{x, \{y, z\}\} = \{\{x, y\}, z\} + (-1)^{(|x|+1)(|y|+1)}\{y, \{x, z\}\} \qquad \text{(Jacobi identity)}.$$

A morphism of dBV algebras X and Y is a map $f : X \to Y$ that preserves the structures d, Δ, and $\bullet$.

Remark 1. In the special case where μ_A is graded commutative, Δ becomes a derivation of $\bullet$ and, thus, the bracket $\{,\}$ is zero. This is well known in the literature, see for example [**5**]. We were surprised we could not find in the literature the fact that TA becomes a dBV algebra when μ_A is not necessarily commutative. There is, however, a similar "Lie" version which is well known: the symmetric algebra of the underlying vector space of a Lie algebra is a BV algebra (see [**8**]).

Theorem 1 generalizes naturally. If $(A, \mu_1, \mu_2, \mu_3, \ldots)$ is an A_∞ algebra, then for each $k = 1, 2, \ldots$, the linear map $\mu_k : A^{\otimes k} \to A$ can be extended to a coderivation of degree $3 - 2k$ of the tensor coproduct $\Delta_{3-2k} : TA \to TA$. In Section 3 we show:

Theorem 2. *If $(A, \mu_1, \mu_2, \mu_3, \ldots)$ is an A_∞ algebra, then $(TA, \bullet, \Delta_1, \Delta_{-1}, \Delta_{-3}, \ldots)$ defines a commutative BV_∞ algebra.*

Remark 2. A commutative BV_∞ algebra, as defined by Kravchenko [**4**], is a generalization of a dBV algebra, and a special case of a BV_∞ algebra, as shown in [**3**]. (See also [**1**].) The precise definition is given in Section 3, where we show the requisite property that Δ_{3-2k} has operator-order k with respect to the shuffle product.

From a logical point of view, it is probably better to prove Theorem 2 first, from which Theorem 1 follows, see Remark 3 below. However, we prefer to give a direct proof of Theorem 1 using the traditional definition of a dBV algebra, making this an easy to read self-contained section. This also has the advantage of giving an explicit formula for the bracket $\{,\}$, and gives us the opportunity to illustrate explicitly how the signs are checked in this context.

2. Proof of the Theorem 1

The identities $d^2 = 0$, $\Delta^2 = 0$, $\bullet$ being associative and graded commutative, and d being a derivation of $\bullet$ are all straightforward. The (graded) anti-symmetry of the bracket follows formally from the (graded) symmetry of $\bullet$. The functoriality statement is immediate. It remains to show that the bracket $\{,\}$ satisfies the Leibniz relation.

We abbreviate $a_{i_1} \otimes \cdots \otimes a_{i_k}$ by $a_{i_1,\ldots,i_k}$, and $\sigma^{-1}(i)$ by σ_i^{-1} for a permutation $\sigma \in \Sigma_k$. First, we may calculate the bracket as

$$\{a_{1,\ldots,n}, a_{n+1,\ldots,n+m}\} = \sum_{\sigma\in S(n,m)} \pm\Delta(a_{\sigma_1^{-1},\ldots,\sigma_{n+m}^{-1}})$$
$$- (\pm\Delta(a_{1,\ldots,n}) \bullet a_{n+1,\ldots,n+m}) - (\pm a_{1,\ldots,n} \bullet \Delta(a_{n+1,\ldots,n+m}))$$

We claim that every term in the last two expressions cancels with precisely one term in $\sum_{\sigma\in S(n,m)} \pm\Delta(a_{\sigma_1^{-1},\ldots,\sigma_{n+m}^{-1}})$ so that $\{a_{1,\ldots,n}, a_{n+1,\ldots,n+m}\}$ equals

$$\sum_{\sigma\in S(n,m)} \sum_{j\in C_\sigma^{\{1,\ldots,n\},\{n+1,\ldots,n+m\}}} \pm a_{\sigma_1^{-1},\ldots,\sigma_{j-1}^{-1}} \otimes \mu_A(a_{\sigma_j^{-1}}, a_{\sigma_{j+1}^{-1}}) \otimes a_{\sigma_{j+1}^{-1},\ldots,\sigma_{n+m}^{-1}},$$

where the set $C_\sigma^{I,J}$ is defined, for a permutation $\sigma \in \Sigma_k$ and disjoint set of indices $I \cup J \subseteq \{1, \ldots, k\}$ with $I \cap J = \varnothing$, by

$$C_\sigma^{I,J} = \{j \quad : \quad \sigma_j^{-1} \in I \text{ and } \sigma_{j+1}^{-1} \in J, \text{ or } \sigma_j^{-1} \in J \text{ and } \sigma_{j+1}^{-1} \in I\}.$$

In other words, μ_A is applied in the above sum whenever exactly one of the two elements $a_{\sigma_j^{-1}}$ and $a_{\sigma_{j+1}^{-1}}$ is taken from $a_1, \ldots, a_n$, and the other element is taken from $a_{n+1}, \ldots, a_{n+m}$. Since the correct terms appear exactly once, the only difficulty is to check the cancellation by signs, which we leave to the end of this section.

Assuming this, if we abbreviate the expression $a_{i_1,\ldots,i_{j-1}} \otimes \mu_A(a_{i_j}, a_{i_{j+1}}) \otimes a_{i_{j+1},\ldots,i_k}$ by $a^{(j,j+1)}_{i_1,\ldots,i_k}$, then we can write,

$$\{a_{1,\ldots,n}, a_{n+1,\ldots,n+m}\} = \sum_{\sigma\in S(n,m)} \sum_{j\in C_\sigma^{\{1,\ldots,n\},\{n+1,\ldots,n+m\}}} \pm a^{(j,j+1)}_{\sigma_1^{-1},\ldots,\sigma_{n+m}^{-1}}$$

With this, we can check that $\{a_{1,\ldots,n} \bullet a_{n+1,\ldots,n+m}, a_{n+m+1,\ldots,n+m+p}\}$ equals

$$
\begin{aligned}
&= \sum_{\sigma\in S(n,m)} \pm\{a_{\sigma_1^{-1},\ldots,\sigma_{n+m}^{-1}}, a_{n+m+1,\ldots,n+m+p}\} \\
&= \sum_{\rho\in S(n,m,p)} \sum_{j\in C_\rho^{\{1,\ldots,n+m\},\{n+m+1,\ldots,n+m+p\}}} \pm a^{(j,j+1)}_{\rho_1^{-1},\ldots,\rho_{n+m+p}^{-1}} \\
&= \sum_{\rho\in S(n,m,p)} \sum_{j\in C_\rho^{\{1,\ldots,n\},\{n+m+1,\ldots,n+m+p\}}} \pm a^{(j,j+1)}_{\rho_1^{-1},\ldots,\rho_{n+m+p}^{-1}} \\
&\quad + \sum_{\rho\in S(n,m,p)} \sum_{j\in C_\rho^{\{n+1,\ldots,n+m\},\{n+m+1,\ldots,n+m+p\}}} \pm a^{(j,j+1)}_{\rho_1^{-1},\ldots,\rho_{n+m+p}^{-1}} \\
&= a_{1,\ldots,n} \bullet \{a_{n+1,\ldots,n+m}, a_{n+m+1,\ldots,n+m+p}\} \\
&\qquad\qquad \pm\{a_{1,\ldots,n}, a_{n+m+1,\ldots,n+m+p}\} \bullet a_{n+1,\ldots,n+m},
\end{aligned}
$$

where $S(n,m,p) \subseteq \Sigma_{n+m+p}$ consists of those permutations $\rho \in \Sigma_{n+m+p}$ that satisfy $\rho(1) < \cdots < \rho(n)$, $\rho(n+1) < \cdots < \rho(n+m)$, and $\rho(n+m+1) < \cdots < \rho(n+m+p)$. By a careful consideration of the signs similar to the check below, it follows that the Leibniz identity holds.

Now, we check the sign mentioned above. If we shuffle $a_{n+1,\ldots,n+j}$ past a_i, for $1 \leqslant i \leqslant n$ and $1 \leqslant j \leqslant m$, and then apply Δ, we obtain the term

$$a_{1,\ldots,i-1} \otimes a_{n+1,\ldots,n+j} \otimes \mu_A(a_i, a_{i+1}) \otimes a_{i+2,\ldots n} \otimes a_{n+j+1,\ldots,n+m}$$

with sign

$$(-1)^{(|a_i|+\cdots+|a_n|+n-i+1)(|a_{n+1}|+\cdots+|a_{n+j}|+j)+(|a_1|+\cdots+|a_{i-1}|+|a_{n+1}|+\cdots+|a_{n+j}|+|a_i|+(i-1+j))}$$

while in the other order, Δ then shuffle, we obtain the same term with sign

$$(-1)^{(|a_1|+\cdots+|a_i|+i+1)+(|\mu(a_i,a_{i+1})|+|a_{i+2}|+\cdots+|a_n|+n-i)(|a_{n+1}|+\cdots+|a_{n+j}|+j)}$$

and these agree. This special case implies the general case, for any shuffle, since a more general shuffle introduces the same additional sign in both cases.

Similarly, shuffling $a_{i+1,\ldots,n}$ past a_{n+j+1} for $1 \leqslant i < n$ and $1 \leqslant j < m$, and then applying Δ, we obtain the term

$$a_{1,\ldots,i} \otimes a_{n+1,\ldots,n+j-1} \otimes \mu_A(a_{n+j}, a_{n+j+1}) \otimes a_{i+1,\ldots,n} \otimes a_{n+j+2,\ldots,n+m}$$

with sign

$$(-1)^{(|a_{i+1}|+\cdots+|a_n|+n-i)(|a_{n+1}|+\cdots+|a_{n+j+1}|+j+1)+(|a_1|+\cdots+|a_i|+|a_{n+1}|+\cdots+|a_{n+j}|+i+j+1)}$$

while in the other order we obtain the same term with sign

$$(-1)^{(|a_{n+1}|+\cdots+|a_{n+j-1}|+j-1)+(|a_{i+1}|+\cdots+|a_n|+n-i)(|a_{n+1}|+\cdots+|a_{n+j-1}|+|\mu(a_{n+j},a_{n+j+1})|+j)}$$

These differ by $(-1)^{|a_1|+\cdots+|a_n|+n}$, as expected. Again, this special case implies the general case, as before. This completes the proof of Theorem 1.

3. Proof of Theorem 2

Let $(X, \bullet)$ be a graded commutative associative algebra. An operator $\Delta : X \to X$ has operator-order n if and only if

$$\sum (-1)^{n+1-r+\kappa} \Delta\left(x_{i_1} \bullet \cdots \bullet x_{i_r}\right) \bullet x_{i_{r+1}} \bullet \cdots \bullet x_{i_{n+1}} = 0$$

where the sum is taken over nonempty subsets $\{i_1, \ldots, i_r : i_1 < \ldots < i_r\} \subseteq \{1, \ldots, n+1\}$ and $\{1, \ldots, n+1\} \setminus \{i_1, \ldots, i_r\}$ has been ordered $i_{r+1} < \cdots < i_{n+1}$, and κ comes from the usual Koszul sign rule.

If Δ has operator-order one, then it is a derivation of $\bullet$. If Δ has operator-order two, then its deviation from being a derivation of $\bullet$, is a derivation of $\bullet$. This means that if we define $\{\,,\,\}$ to be the deviation of Δ from being a derivation of $\bullet$, then $\{\,,\,\}$ and $\bullet$ satisfy the Leibniz relation.

Remark 3. Using this fact, one can prove Theorem 1 without reference to the bracket—here is an outline: any map $\mu_A : A \otimes A \to A$ becomes an order 2 operator $\Delta : TA \to TA$ with respect to the shuffle product when it is lifted as a coderivation of the tensor coproduct (as we will show in the lemma below). It is straightforward to check that μ_A being associative implies that $\Delta^2 = 0$, since Δ^2 is the lift of the associator of μ_A to a coderivation. So, if (A, d_A, μ_A) is a differential graded algebra, with μ_A of degree zero, Δ has degree -1, and since d_A is a derivation of μ_A, then $d : TA \to TA$ and $\Delta : TA \to TA$ commute. That proves that $(TA, d, \Delta, \bullet)$ is a dBV algebra.

To generalize: a Kravchenko commutative BV_∞ algebra consists of a graded commutative differential graded algebra $(X, d, \bullet)$ and a collection $\{\Delta_k : X \to X\}_{k=1,-1,-3,-5,\ldots}$ of operators satisfying

- $\Delta_1 = d$,
- each Δ_{3-2k} has degree $3 - 2k$ and operator-order k,
- for each n, $\sum_{j+k=n} \Delta_j \Delta_k = 0$.

We use the degree convention in [**4**] but note that in [**3**] the opposite convention is used (there, d has degree -1 and the higher Δ operators have positive degree). As a special case, a dBV algebra is a Kravchenko commutative BV_∞ algebra with $\Delta_{-3} = \Delta_{-5} = \cdots = 0$.

To prove Theorem 2, assume that $(A, \mu_1, \mu_2, \mu_3 \ldots)$ is an A_∞ algebra. By definition of an A_∞ algebra, each μ_k lifts to a degree $3 - 2k$ coderivation $\Delta_{3-2k} : TA \to TA$, with $\Delta_1 = d$ and relations $\sum_{j+k=n} \Delta_j \Delta_k = 0$. Thus it only remains to prove, that each Δ_{3-2k} has order k with respect to the shuffle product $\bullet$. This follows from the following general lemma.

Lemma. *Let $f : A^{\otimes n} \to A$ be any linear map and let $F : TA \to TA$ be the lift of f to a coderivation. Then F has order n with respect to the shuffle product.*

Proof. Let $X^1, \ldots, X^{n+1}$ be monomials in TA. So, $X^i = a^i_1 \otimes \cdots \otimes a^i_{s_i}$ with each $a^i_\ell \in A$. Then $(-1)^{n+1-r+\kappa} F(X^{i_1} \bullet \cdots \bullet X^{i_r}) \bullet X^{i_{r+1}} \cdots \bullet X^{i_{n+1}}$ consists of a sum of terms of the form

$$\pm \ldots \otimes f(a^{i'_1}_{\ell_1} \otimes \cdots \otimes a^{i'_n}_{\ell_n}) \otimes \ldots \text{ (the rest of the } a^i_\ell\text{'s are outside of } f), \tag{1}$$

where f is applied to $a^{i'_1}_{\ell_1} \otimes \cdots \otimes a^{i'_n}_{\ell_n}$, and the remaining tensor products are applied outside of f. The list $\{i'_1, \ldots, i'_n\}$ may contain repetition, and we may order the list from smallest to largest without repetition as $\{i_1, \ldots, i_k\}$. Every term of the form (1) which contains only the indices $\{i_1, \ldots, i_k\}$ inside f, appears for each index set $J = \{j_1, \ldots, j_q\}$ with $\{i_1, \ldots, i_k\} \subseteq J \subseteq \{1, \ldots, n+1\}$ exactly once in the sum of $(-1)^{n+1-q+\kappa} F(X^{j_1} \bullet \cdots \bullet X^{j_q}) \bullet X^{j_{q+1}} \cdots \bullet X^{j_{n+1}}$. Now, for a fixed expression in Equation (1) induced by different index sets J, the only difference in the sign of (1) is a factor of $(-1)^q$, where $q = |J|$, and all other signs coincide for varying J. We thus need to show that summing $(-1)^{|J|}$ over all J with $\{i_1, \ldots, i_k\} \subseteq J \subseteq \{1, \ldots, n+1\}$ vanishes. Since there are exactly $n+1-k$ choose

$q-k$ such subsets J with q elements, we obtain that

$$\sum_J (-1)^{|J|} = \sum_{q=k}^{n+1} \binom{n+1-k}{q-k} \cdot (-1)^q = (-1)^k \cdot \sum_{q'=0}^{n+1-k} \binom{n+1-k}{q'} \cdot (-1)^{q'}$$
$$= (-1)^k \cdot (-1+1)^{n+1-k} = 0,$$

where we used the binomial theorem in the second to last equality. This completes the proof of the lemma. □

References

[1] G. Drummond-Cole, *Homotopy Batalin-Vilkovisky algebras, trivializing circle actions, and moduli space*, The City University of New York 2010.

[2] G. Drummond-Cole and B. Vallette, *Minimal homotopy BV*, preprint.

[3] I. Galvez-Carrillo, A. Tonks and B. Vallette, *Homotopy Batalin-Vilkovisky algebras*, arXiv 0907.2246, http://adsabs.harvard.edu/abs/2009arXiv0907.2246G

[4] O. Kravchencko, *Deformations of Batalin-Vilkovisky algebras*, Banach Center Publ. v. 51, 131–139, Polish Acad. Sci. 2000.

[5] J.-L. Loday, *Cyclic Homology*, Grundlehren der mathematischen Wissenschaften 301, Springer Verlag 1992.

[6] S. MacLane, *Homology*, Classics in Mathematics (Reprint of 1975 edition), Springer Verlag 1995.

[7] J. Stasheff, *The intrinsic bracket on the deformation complex of an associative algebra*, J. Pure Appl. Algebra 89 (1993), 231–235.

[8] D. Tamarkin and B. Tsygan, *Noncommutative differential calculus, homotopy BV algebras and formality conjectures*, Methods Func. Anal. Topology 6 (2) (2000), 85–100.

This article may be accessed via WWW at `http://tcms.org.ge/Journals/JHRS/`

John Terilla
`jterilla@qc.cuny.edu`

Queens College, City University of New York
Flushing NY 11367

Thomas Tradler
`ttradler@citytech.cuny.edu`

NYC College of Technology, City University of New York
Brooklyn NY 11201

Scott O. Wilson
`scott.wilson@qc.cuny.edu`

Queens College, City University of New York
Flushing NY 11367

Journal of Homotopy and Related Structures, vol. 6(2), 2011, pp.183–209

EQUIVARIANT DOLD-THOM TOPOLOGICAL GROUPS

MARCELO A. AGUILAR AND CARLOS PRIETO

(*communicated by Paul Goerss*)

Abstract

Let M be a covariant coefficient system for a finite group G. In this paper we analyze several topological abelian groups, some of them new, whose homotopy groups are isomorphic to the Bredon-Illman G-equivariant homology theory with coefficients in M. We call these groups equivariant Dold-Thom topological groups and we show that they are unique up to homotopy. We use one of the new groups to prove that the Bredon-Illman homology satisfies the infinite-wedge axiom and to make some calculations of the 0th equivariant homology.

1. Introduction

Given a finite group G, a covariant coefficient system M for G, and a pointed G-set S, we defined in [**2**] an abelian group $F^G(S, M)$. Using this construction, we associate to a pointed G-space X a simplicial abelian group $F^G(\mathcal{S}(X), M)$. Its geometric realization, denoted by $F^G(X, M)$, is a topological group whose homotopy groups are isomorphic to the reduced Bredon-Illman G-equivariant homology of X with coefficients in M. Given a G-equivariant ordinary covering map $p : E \longrightarrow X$, we also defined a continuous transfer $t_p^G : F^G(X^+, M) \longrightarrow F^G(E^+, M)$, that induces a transfer in equivariant homology, when M is a Mackey functor.

In [**3**] we showed that when the G-space X is a strong ρ-space (e.g. a G-simplicial complex or a finite dimensional countable locally finite G-CW-complex), there is a topology in the abelian groups $F^G(X^\delta, M)$, where X^δ stands for the underlying set of X. With this topology, $F^G(X, M)$ is a topological group, which we denote by $\mathbb{F}^G(X, M)$. This is a smaller group than the group $F^G(X, M)$ mentioned above. It has the property that its homotopy groups are isomorphic to the reduced Bredon-Illman G-equivariant homology of X with coefficients in M, provided that M is a homological Mackey functor. Furthermore, we proved in [**4**] that these topological groups admit a continuous transfer for G-equivariant ramified covering maps, whose total space and base space are strong ρ-spaces.

In this paper we present a different topology for the abelian group $F^G(X^\delta, M)$ and we denote the resulting topological group by $\mathcal{F}^G(X, M)$. We prove that for any

The authors were partially supported by PAPIIT grant IN101909.
Received February 07, 2011, revised April 20, 2011; published on July 17, 2011.
2000 Mathematics Subject Classification: 55N91, 55P91, 14F43.
Key words and phrases: Equivariant homology, homotopy groups, coefficient systems, Mackey functors, topological abelian groups.

pointed G-space X of the homotopy type of a G-CW-complex, and for any coefficient system, the homotopy groups of $\mathfrak{F}^G(X,M)$ are isomorphic to the reduced Bredon-Illman G-equivariant homology of X with coefficients in M. The assumptions on X and M are much weaker than those needed to define $\mathbb{F}^G(X,M)$. However, in such a generality, it does not seem possible to construct a continuous transfer even for ordinary covering G-maps.

These new topological groups $\mathfrak{F}^G(X,M)$ can be used to prove the infinite wedge axiom for the Bredon-Illman homology. They also allow us to make some calculations of the 0th homology groups of a G-space X.

Those topological groups whose homotopy groups are isomorphic to the reduced Bredon-Illman G-equivariant homology of X with coefficients in M will be called *Dold-Thom topological groups*. Hence $F^G(X,M)$, $\mathbb{F}^G(X,M)$, and $\mathfrak{F}^G(X,M)$ are Dold-Thom topological groups. Furthermore, all these groups are algebraically subgroups of other topological groups, so they also have another natural topology, namely the subspace topology. These topological groups will be denoted by $\overline{F}^G(X,M)$, $\overline{\mathbb{F}}^G(X,M)$, and $\overline{\mathfrak{F}}^G(X,M)$. The first two were studied in [**3**]. In this paper we prove that these groups are isomorphic to the former, if the coefficient system M takes values on k-modules, where k is a field of characteristic 0 or a prime p that does not divide the order of G. We also show that the Dold-Thom topological groups are unique up to homotopy. Other examples of Dold-Thom topological groups were constructed by Dold and Thom [**7**], McCord [**15**], Lima-Filho [**12**], dos Santos [**10**], and Nie [**17**].

The paper is organized as follows. In Section 2 we give the basic definitions that are needed. Then in Section 3 we define the new topological groups $\mathfrak{F}^G(X,M)$ and $\overline{\mathfrak{F}}^G(X,M)$, and we prove that the former is a Dold-Thom topological group. Then in Section 4 we compare the new topological groups with the previously defined ones. In Section 5 we analyze the case of coefficients in k-$\mathcal{M}$od, with the field k as explained above. In Section 6 we prove the wedge axiom and we compute the 0th equivariant homology in some cases. Finally in Section 7, we study the Dold-Thom topological groups and we show that two Dold-Thom topological groups, which are locally connected and have the homotopy type of a CW-complex, are homotopy equivalent.

2. Preliminaries

We shall work in the category of k-spaces, which will be denoted by k-$\mathcal{T}$op. We understand by a k-*space* a topological space X with the property that a set $W \subset X$ is closed if and only if $f^{-1}W \subset Z$ is closed for any continuous map $f : Z \longrightarrow X$, where Z is any compact Hausdorff space (see [**14, 18**]). Given any space X, one can clearly associate to it a k-space $k(X)$ using the condition above, which is weakly homotopy equivalent to X. The product of two spaces X and Y in this category is $X \times Y = k(X \times_{\text{top}} Y)$, where $X \times_{\text{top}} Y$ is the usual topological product. Two important properties of this category are that *if* $p : X \twoheadrightarrow X'$ *is an identification and* X *is a* k-*space, then* X' *is a* k-*space*, and *if* $p : X \twoheadrightarrow X'$ *and* $q : Y \twoheadrightarrow Y'$ *are identifications, then* $p \times q : X \times Y \twoheadrightarrow X' \times Y'$ *is an identification too.* If X

is a k-space, we shall say that $A \subset X$ has the *relative* k-*topology* (in k-$\mathcal{T}$op) if $A = k(A_{\rm rel})$, where $A_{\rm rel}$ denotes A with the (usual) relative topology in $\mathcal{T}$op. This topology is characterized by the following property: *Let Y be a* k-*space. Then a map $f : Y \longrightarrow A$ is continuous if and only if the composite $i \circ f : Y \longrightarrow X$ is continuous, where $i : A \hookrightarrow X$ is the inclusion* (see [**18**]).

In what follows, we shall denote by G-$\mathcal{T}$op$_*$ the category of topological pointed G-spaces such that G acts trivially on the base point, or correspondingly G-k-$\mathcal{T}$op$_*$. $\mathcal{T}$opab will denote the category of topological abelian groups in the category of k-spaces. Recall that a *covariant coefficient system* is a covariant functor $M : \mathcal{O}(G) \longrightarrow R\text{-}\mathfrak{Mod}$, where $\mathcal{O}(G)$ is the category of G-orbits G/H, $H \subset G$, and G-functions $\alpha : G/H \longrightarrow G/K$, and R is a commutative ring. A particular role will be played by the G-function $R_{g^{-1}} : G/H \longrightarrow G/gHg^{-1}$, given by *right translation* by $g^{-1} \in G$, namely

$$R_{g^{-1}}(aH) = aHg^{-1} = ag^{-1}(gHg^{-1})\,.$$

We shall often denote aH by $[a]_H$. Observe that if X is a G-set and $x \in X$, then the canonical bijection $G/G_x \longrightarrow G/G_{gx}$ is precisely $R_{g^{-1}}$. Here G_x denotes the *isotropy subgroup* of x, namely, the maximal subgroup of G that leaves x fixed.

Let S be a pointed G-set (where the base point x_0 remains fixed under the action of G) and M a covariant coefficient system. In [**2**] we defined an abelian group $F(S, M)$ as follows. Consider the following union

$$\widehat{M} = \bigcup_{H \subset G} M(G/H)\,.$$

Then

$$F(S, M) = \{u : S \longrightarrow \widehat{M} \mid u(x) \in M(G/G_x)\,,\ u(x_0) = 0\,,$$
$$\text{and } u(x) = 0 \text{ for almost every } x \in X\}\,.$$

Indeed, this group $F(S, M)$ is an R-module, whose structure is given by

$$(r \cdot u)(x) = ru(x) \in M(G/G_x)\,.$$

It has as canonical generators the functions

$$lx : S \longrightarrow \widehat{M} \quad \text{given by} \quad lx(x') = \begin{cases} l & \text{if } x' = x\,, \\ 0 & \text{if } x' \neq x\,, \end{cases}$$

where $x \in S$, $x \neq x_0$, and $l \in M(G/G_x)$. The group $F(S, M)$ is a functor of S as follows. Let $f : S \longrightarrow T$ be a pointed G-function. Then we define $f_* : F(S, M) \longrightarrow F(T, M)$ on generators by

$$f_*(lx) = M_*(\widehat{f_x})(l)f(x)\,,$$

that is, the homomorphism whose value on y is $M_*(\widehat{f_x})(l)$ if $y = f(x)$ and 0 otherwise, where $\widehat{f_x} : G/G_x \twoheadrightarrow G/G_{f(x)}$ is the canonical surjection.

There is an action of G on $F(S, M)$ given on generators by

$$g \cdot (lx) = M_*(R_{g^{-1}})(l)(gx)\,.$$

Then one can consider the submodule $F(S,M)^G$ of fixed points under the G-action. Let $f : S \longrightarrow T$ be as above. Since f_* is clearly a G-homomorphism, it restricts to a homomorphism between the submodules of fixed points, which we denote by

$$\overline{f}_*^G : \overline{F}^G(S,M) \longrightarrow \overline{F}^G(T,M)\,.$$

This makes $\overline{F}^G(-,M)$ into a functor $G\text{-}\mathcal{S}\mathrm{et}_* \longrightarrow R\text{-}\mathfrak{Mod}$

On the other hand, there is a surjective homomorphism $\beta_S : F(S,M) \twoheadrightarrow F(S,M)^G$ given by

$$\beta_S(lx) = \gamma_x^G(l) = \sum_{[g]\in G/G_x} M_*(R_{g^{-1}})(l)(gx)\,,$$

that is, essentially taking the sum over the orbit of x. One can now use this to give a different functorial structure on $F(S,M)^G$, defining for a G-function $f : S \longrightarrow T$ and a generator $\gamma_x^G(l)$,

$$f_*^G(\gamma_x^G(l)) = \gamma_{f(x)}^G M_*(\widehat{f}_x)(l)\,.$$

This makes $F^G(-,M)$ into a functor $G\text{-}\mathcal{S}\mathrm{et}_* \longrightarrow R\text{-}\mathfrak{Mod}$.

These three functors are related by the commutativity of the following diagram:

$$\begin{array}{ccccccc} S & & \overline{F}^G(S,M) & \overset{\iota_S}{\hookrightarrow} & F(S,M) & \overset{\beta_S}{\twoheadrightarrow} & F^G(S,M) \\ \downarrow{\scriptstyle f} & & \overline{f}_*^G\downarrow & & \downarrow{\scriptstyle f_*} & & \downarrow{\scriptstyle f_*^G} \\ T & & \overline{F}^G(T,M) & \underset{\iota_T}{\hookrightarrow} & F(S,M) & \underset{\beta_T}{\twoheadrightarrow} & F^G(S,M)\,. \end{array} \tag{2.1}$$

We shall use these groups to define different topological groups in the next section.

3. Topological groups and coefficient systems

First recall the following construction. Let X be a topological space and L an R-module. Then we have the R-module $F(X,L) = \{u : X \longrightarrow L \mid u(x_0) = 0, \text{ and } u(x) = 0 \text{ for almost every } x \in X\}$. Following [**1**] we have that this R-module can be topologized as follows (see also [**15**]). There is a surjective function

$$\mu : \coprod_k (L\times X)^k \twoheadrightarrow F(X,L)$$

given by $(l_1,x_1;\ldots;l_k,x_k) \mapsto l_1x_1 + \cdots + l_kx_k$. Then $F(X,L)$ has the identification topology.

Given a pointed G-space X, we denote by X^H the pointed subspace of elements of X that remain fixed under the group elements of H.

Definition 3.1. Let M be a covariant coefficient system and let X be any pointed G-space. For each subgroup $H \subset G$ consider the topological group $F(X^H, M(G/H))$ as defined above. Define $p_H : F(X^H, M(G/H)) \longrightarrow F(X,M)$ by $p_H(lx) = M_*(q_{H,x})(l)$, where $x \in X^H$, $l \in M(G/H)$, and $q_{H,x} : G/H \twoheadrightarrow G/G_x$, is the canonical projection. Now take the homomorphism

$$p_X : \prod_{H \subset G} F(X^H, M(G/H)) \twoheadrightarrow F(X, M)$$

given by $p_X((l_H x_H)_{H\subset G}) = \sum_{H\subset G} p_H(l_H x_H)$, where the product has the product topology of k-spaces. Given any generator $lx \in F(X,M)$, lx can be seen also as an element in $F(X^{G_x}, M(G/G_x))$. Therefore p_X is surjective. Give the R-module $F(X,M)$ the identification topology induced by p_X. We obtain a k-space, which we denote by $\mathfrak{F}(X,M)$. By giving $\overline{F}^G(X,M)$ the relative k-topology we obtain another k-space, which we denote by $\overline{\mathfrak{F}}^G(X,M)$. Moreover, by taking on $F^G(X,M)$ the identification topology given by the epimorphism β_X, we obtain another k-space, which we denote by $\mathfrak{F}^G(X,M)$.

Proposition 3.2. *The groups $\mathfrak{F}(X,M)$, $\overline{\mathfrak{F}}^G(X,M)$, and $\mathfrak{F}^G(X,M)$ are topological groups in the category of* k*-spaces.*

Proof. Consider the following diagram

$$\begin{array}{ccc}
\prod F(X^H, M(G/H)) \times \prod F(X^H, M(G/H)) & \xrightarrow{\prod +_H} & \prod F(X^H, M(G/H)) \\
\downarrow{\scriptstyle p_X \times p_X} & & \downarrow{\scriptstyle p_X} \\
\mathfrak{F}(X,M) \times \mathfrak{F}(X,M) & \xrightarrow[+]{} & \mathfrak{F}(X,M) \\
\downarrow{\scriptstyle \beta_X \times \beta_X} & & \downarrow{\scriptstyle \beta_X} \\
\mathfrak{F}^G(X,M) \times \mathfrak{F}^G(X,M) & \xrightarrow[+]{} & \mathfrak{F}^G(X,M) .
\end{array}$$

The function on the top is given by the product of the sum on each topological group $F(X^H, M(G/H))$ and therefore it is continuous. Since p_X and β_X are homomorphisms, both squares commute. Furthermore, since p_X and β_X are identifications, so are $p_X \times p_X$ and $\beta_X \times \beta_X$. Therefore $\mathfrak{F}(X,M)$ and $\mathfrak{F}^G(X,M)$ are topological groups. Finally, since $\overline{\mathfrak{F}}^G(X,M)$ has the relative k-topology, it is also a topological group. □

Proposition 3.3. *Let $f : X \longrightarrow Y$ be a pointed G-map. Then the homomorphism $f_* : \mathfrak{F}(X,M) \longrightarrow \mathfrak{F}(Y,M)$ is continuous. Thus the homomorphisms $\overline{f}_*^G : \overline{\mathfrak{F}}^G(X,M) \longrightarrow \overline{\mathfrak{F}}^G(Y,M)$ and $f_*^G : \mathfrak{F}^G(X,M) \longrightarrow \mathfrak{F}^G(Y,M)$ are also continuous.*

Proof. The following diagram commutes:

$$\begin{array}{ccc}
\prod_{H\subset G} F(X^H, M(G/H)) & \xrightarrow{\prod f_*^H} & \prod_{H\subset G} F(Y^H, M(G/H)) \\
\downarrow{\scriptstyle p_X} & & \downarrow{\scriptstyle p_Y} \\
\mathfrak{F}(X,M) & \xrightarrow[f_*]{} & \mathfrak{F}(Y,M) .
\end{array}$$

Indeed, if $(l_H x_H)_{H\subset G} \in \prod_{H\subset G} F(X^H, M(G/H))$ is a generator, then

$$p_Y(\prod f_*^H)((l_H x_H)_{H\subset G}) = p_Y((l_H f(x_H))_{H\subset G}) = \sum_{H\subset G} M_*(q_{f(x_H)})(l_H) f(x_H)\,,$$

and

$$f_* p_X((l_H x_H)_{H\subset G}) = f_*(\sum_{H\subset G} M_*(q_{x_H})(l_H) x_H) = \sum_{H\subset G} M_*(\widehat{f}_{x_H}) M_*(q_{x_H})(l_H) f(x_H)\,.$$

Here we denote q_{H,x_H} simply by q_{x_H}. Both composites are equal, since $\widehat{f}_{x_H} \circ q_{x_H} = q_{f(x_H)}$.

Since p_X is an identification and $\prod f_*^H$ is continuous, f_* is also continuous. □

The next result shows the homotopy invariance of the functors $\mathcal{F}$.

Proposition 3.4. *Let $f_0, f_1 : X \longrightarrow Y$ be G-homotopic pointed G-maps. Then $f_{0*}, f_{1*} : \mathcal{F}(X, M) \longrightarrow \mathcal{F}(Y, M)$ are G-homotopic homomorphisms. Moreover, also $\overline{f}_{0*}^G, \overline{f}_{1*}^G : \overline{\mathcal{F}}^G(X, M) \longrightarrow \overline{\mathcal{F}}^G(Y, M)$ and $f_{0*}^G, f_{1*}^G : \mathcal{F}^G(X, M) \longrightarrow \mathcal{F}^G(Y, M)$ are homotopic homomorphisms.*

Proof. For each $H \subset G$, take the restriction $f_\nu^H : X^H \longrightarrow Y^H$, and let $\mathcal{H} : X \times I \longrightarrow Y$ be a G-homotopy such that $\mathcal{H}(x, \nu) = f_\nu(x)$, $\nu = 0, 1$, and denote by $\mathcal{H}^H : X^H \times I \longrightarrow Y^H$ the restriction of $\mathcal{H}$, which is a homotopy between f_0^H and f_1^H. By [**15**], there is a homotopy $\widetilde{\mathcal{H}}^H : F(X^H, M(G/H)) \times I \longrightarrow F(Y^H, M(G/H))$ between $(f_0^H)_*$ and $(f_1^H)_*$, which is given by $\widetilde{\mathcal{H}}^H(u, t) = (\mathcal{H}_t^H)_*(u)$, where $\mathcal{H}_t^H(x) = \mathcal{H}^H(x, t)$, $x \in X^H$.

Define a homotopy $R : \prod F(X^H, M(G/H)) \times I \longrightarrow \prod F(Y^H, M(G/H))$ by $R((u_H)_{H\subset G}, t) = (\widetilde{\mathcal{H}}^H(u_H, t))_{H\subset G}$. Define $\mathcal{K} : \mathcal{F}(X, M) \times I \longrightarrow \mathcal{F}(Y, M)$ by $\mathcal{K}(u, t) = \mathcal{H}_{t*}(u)$ and consider the diagram

$$\begin{array}{ccc} \prod F(X^H, M(G/H)) \times I & \xrightarrow{R} & \prod F(Y^H, M(G/H)) \\ \Big\downarrow{\scriptstyle p_X \times \mathrm{id}} & & \Big\downarrow{\scriptstyle p_Y} \\ \mathcal{F}(X, M) \times I & \xrightarrow[\mathcal{K}]{} & \mathcal{F}(Y, M)\,. \end{array}$$

Since both R and $\mathcal{K}$ are homomorphisms for each fixed t, we may check the commutativity on a generator $((l_H x_H)_{H\subset G}, t) \in \prod F(X^H, M(G/H)) \times I$ as follows:

$$\begin{aligned} p_Y R((l_H x_H)_{H\subset G}, t) &= p_Y((l_H \mathcal{H}^H(x_H, t))_{H\subset G}) \\ &= \sum_{H\subset G} M_*(q_{\mathcal{H}^H(x_H,t)})(l_H) \mathcal{H}^H(x_H, t)\,, \\ \mathcal{K}(p_X \times \mathrm{id})((l_H x_H), t) &= \mathcal{K}\left(\sum_{H\subset G} M_*(q_{x_H})(l_H) x_H\right) \\ &= \sum_{H\subset G} M_*(\widehat{\mathcal{H}}_{t x_H}) M_*(q_{x_H})(l_H) \mathcal{H}(x_H, t)\,. \end{aligned}$$

Both expressions are equal, since $\mathcal{H}^H$ is the restriction of $\mathcal{H}$, and one clearly has that $q_{\mathcal{H}^H(x_H,t)} : G/H \longrightarrow G/G_{\mathcal{H}(x_H,t)}$ is equal to the composite $G/H \overset{q_{x_H}}{\twoheadrightarrow} G/G_{x_H} \overset{\widehat{\mathcal{H}}_{t x_H}}{\twoheadrightarrow} G/G_{\mathcal{H}(x_H,t)}$. Since $p_X \times \mathrm{id}$ is an identification, $\mathcal{K}$ is continuous an so it is a homotopy as desired.

The homotopy $\mathcal{K}$ is G-equivariant. Indeed, since $\mathcal{H}_{t*}$ is G-equivariant, one has $\mathcal{K}(g \cdot u, t) = \mathcal{H}_{t*}(g \cdot u) = g\mathcal{H}_{t*}(u) = g\mathcal{K}(u,t)$. Thus we can take the restriction of $\mathcal{K}$, $\overline{\mathcal{K}}^G : \overline{\mathcal{F}}^G(X,M) \times I \longrightarrow \overline{\mathcal{F}}^G(Y,M)$, which is a homotopy between $\overline{f}^G_{0*}$ and $\overline{f}^G_{1*}$, and by the naturality of β, we also have a commutative diagram

$$\begin{array}{ccc} \mathcal{F}(X,M) \times I & \xrightarrow{\mathcal{K}} & \mathcal{F}(Y,M) \\ \Big\downarrow {\scriptstyle \beta_X \times \mathrm{id}} & & \Big\downarrow {\scriptstyle \beta_Y} \\ \mathcal{F}^G(X,M) \times I & \xrightarrow[\mathcal{K}^G]{} & \mathcal{F}^G(Y,M), \end{array}$$

where $\mathcal{K}^G(v,t) = \mathcal{H}^G_{t*}(v)$. Thus $f^G_{0*}, f^G_{1*} : \mathcal{F}^G(X,M) \longrightarrow \mathcal{F}^G(Y,M)$ are also homotopic. □

We shall need the following G-equivariant version of the Whitehead Theorem.

Proposition 3.5. *Let Y and Z be G-spaces and $\varphi : Y \longrightarrow Z$ be a G-equivariant weak homotopy equivalence. Let $y_0 \in Y$ and $\varphi(y_0) \in Z$ be base points. Assume that (Y, y_0) and $(Z, \varphi(y_0))$ have the G-homotopy type of pointed G-CW-complexes. Then $\varphi : (Y, y_0) \longrightarrow (Z, \varphi(y_0))$ is a pointed G-homotopy equivalence.*

Proof. Consider the square

$$\begin{array}{ccc} (Y, y_0) & \xrightarrow{\varphi} & (Z, \varphi(y_0)) \\ {\scriptstyle \simeq_G} \Big\uparrow & & \Big\downarrow {\scriptstyle \simeq_G} \\ (C, c_0) & \dashrightarrow_{\psi} & (D, d_0), \end{array}$$

where (C, c_0) and (D, d_0) are pointed G-CW-complexes and ψ is defined so that the square commutes. Since the vertical arrows are G-homotopy equivalences, ψ is a G-equivariant weak homotopy equivalence.

Using [**6**, II(2.5)], one can show that ψ is a pointed G-homotopy equivalence. Therefore φ is also a pointed G-homotopy equivalence. □

As a consequence of the previous two results we have the following.

Corollary 3.6. *Let Y and Z be G-spaces and $\varphi : Y \longrightarrow Z$ be a G-equivariant weak homotopy equivalence. Let $y_0 \in Y$ and $\varphi(y_0) \in Z$ be base points. Assume that (Y, y_0) and $(Z, \varphi(y_0))$ have the G-homotopy type of pointed G-CW-complexes. Then φ induces a homotopy equivalence of topological groups $\varphi^G_* : \mathcal{F}^G(Y,M) \longrightarrow \mathcal{F}^G(Z,M)$.* □

Lemma 3.7. *Let (X, x_0) be a pointed G-space of the G-homotopy type of a pointed G-CW-complex, and let $\mathcal{S}(X)$ be the singular simplicial G-set of X. Let*

$\rho_X : |\mathcal{S}(X)| \longrightarrow X$ be given by $\rho([\sigma, s]) = \sigma(s)$, where $\sigma : \Delta^q \longrightarrow X$ and $s \in \Delta^q$. Then $\rho^G_{X*} : \mathfrak{F}^G(|\mathcal{S}(X)|, M) \longrightarrow \mathfrak{F}^G(X, M)$ is a homotopy equivalence of topological groups.

Proof. By [**5**, 1.9(e)], we have $|\mathcal{S}(X)|^H = |\mathcal{S}(X)^H|$, and clearly $|\mathcal{S}(X)^H| = |\mathcal{S}(X^H)|$. Hence $\rho^H_X : |\mathcal{S}(X)|^H \longrightarrow X^H$ coincides with $\rho_{X^H} : |\mathcal{S}(X^H)| \longrightarrow X^H$, which by Milnor's theorem (see [**13**]) is a weak homotopy equivalence. Therefore, ρ_X is a G-equivariant weak homotopy equivalence. Hence, by the previous corollary, the result follows. □

Definition 3.8. Assume now that Q is a simplicial pointed G-set and consider the simplicial group $F(Q, M)$ such that for any n, $F(Q, M)_n = F(Q_n, M)$. Given a morphism $\mu : \mathbf{m} \longrightarrow \mathbf{n}$ in Δ, we denote by $\mu^Q : Q_n \longrightarrow Q_m$ the induced pointed G-function. Then we define $\mu^{F(Q,M)} = \mu^Q_* : F(Q, M)_n \longrightarrow F(Q, M)_m$. Then one has two other simplicial groups $\overline{F}^G(Q, M)$ and $F^G(Q, M)$. $\overline{F}^G(Q, M)$ is the simplicial subgroup of $F(Q, M)$ induced by the restricted functor $\overline{F}^G(-, M)$ defined above. $F^G(Q, M)$ is the simplicial quotient group of $F(Q, M)$ induced by the quotient functor $F^G(-, M)$ defined in page 186.

Theorem 3.9. *There is a natural isomorphism of topological groups*

$$\Psi_Q : |F(Q, M)| \longrightarrow \mathfrak{F}(|Q|, M) \quad \text{given by} \quad \Psi_Q([lx, t]) = M_*(q_{x,t})(l)[x, t]\,,$$

where $q_{x,t} : G/G_x \twoheadrightarrow G/G_{[x,t]}$ is the canonical projection.

Proof. In [**2**, 2.6] we showed that Ψ_Q is a natural isomorphism of abelian groups. Thus we only need to prove that it is a homeomorphism. First recall [**5**, 1.9(e)] that $|Q^H| = |Q|^H$. Now we are going to see that the following diagram commutes.

$$\begin{array}{ccc}
|\prod_{H\subset G} F(Q^H, M(G/H))| & \xrightarrow{\cong} & \prod_{H\subset G} |F(Q^H, M(G/H))| \\
 & & \Big\downarrow {\scriptstyle \cong}\ {\scriptstyle \prod \psi_{Q^H}} \\
\Big\downarrow {\scriptstyle |p_Q|} & & \prod_{H\subset G} F(|Q^H|, M(G/H)) \\
 & & \Big\downarrow {\scriptstyle p_{|Q|}} \\
|F(Q, M)| & \xrightarrow[\Psi_Q]{} & \mathfrak{F}(|Q|, M)\,.
\end{array}$$

Take a generator $[(l_H x_H)_{H\subset G}, \tau] \in |\prod_{H\subset G} F(Q^H, M(G/H))|$. Then we can chase it down and we get $[\sum_{H\subset G} M_*(q_{x_H})(l_H)x_H, \tau]$, and then to the right to obtain $\sum_{H\subset G} M_*(q_{x_H,\tau})M_*(\widehat{p}_{x_H})(l_H)[x_H, \tau]$. If we first go right and down with $\prod \psi_{Q^H}$, we get $(l_H[x_H, \tau])$, and then down again with $p_{|Q|}$, we obtain $\sum_{H\subset G} M_*(r_{x_H})(l_H)[x_H, \tau]$, where

$$\begin{array}{ccc}
G/H & \xrightarrow{r_{x_H,\tau}} & G/G_{[x_H,\tau]} \\
 & {\scriptstyle p_{x_H}}\searrow \quad \nearrow{\scriptstyle q_{x_H,\tau}} & \\
 & G/G_{x_H} &
\end{array}$$

are the canonical projections. By [**13**], the isomorphism on the top is a homeomorphism, and by [**1**], each ψ_{Q^H} is also a homeomorphism. Since p_Q is a surjective simplicial map, its realization $|p_Q|$ is an identification. And since $p_{|Q|}$ is also an identification, then Ψ_Q is a homeomorphism. □

Corollary 3.10. *There are natural isomorphisms of topological groups*

$$\overline{\Psi}_Q^G : |\overline{F}^G(Q,M)| \longrightarrow \overline{\mathfrak{F}}^G(|Q|,M) \quad \textit{and} \quad \Psi_Q^G : |F^G(Q,M)| \longrightarrow \mathfrak{F}^G(|Q|,M)\,.$$

Proof. The following is clearly a commutative diagram:

$$\begin{array}{ccccc}
|\overline{F}^G(Q,M)| & \xhookrightarrow{|\iota_Q|} & |F(Q,M)| & \xrightarrow{|\beta_Q|} & |F^G(Q,M)| \\
\Big\downarrow \overline{\Psi}_Q^G & & \cong \Big\downarrow \Psi_Q & & \Big\downarrow \Psi_Q^G \\
\overline{\mathfrak{F}}^G(|Q|,M) & \xhookrightarrow{\iota_{|Q|}} & \mathfrak{F}(|Q|,M) & \xrightarrow{\beta_{|Q|}} & \mathfrak{F}^G(|Q|,M)
\end{array}$$

and the vertical arrows on the sides are isomorphisms of abelian groups (see [**2**, 2.6]). Since the one in the middle, by the previous theorem, is a homeomorphism, and the first horizontal arrows are embeddings, while the second ones are identifications, $\overline{\Psi}_Q^G$ and Ψ_Q^G are homeomorphisms too. □

The following is the main result of this section.

Theorem 3.11. *Let M be a covariant coefficient system for G and X a pointed G-space of the homotopy type of a G-CW-complex. Then the homotopy groups*

$$\pi_q(\mathfrak{F}^G(X,M))$$

are naturally isomorphic to the (reduced) Bredon-Illman G-equivariant homology groups $\widetilde{H}_q^G(X;M_*)$ with coefficients in M.

Proof. $F^G(\mathcal{S}(X),M)$ is a simplicial abelian group and hence, by [**13**], it is a chain complex with differential $\partial_q^G : F^G(\mathcal{S}_q(X),M) \longrightarrow F^G(\mathcal{S}_{q-1}(X),M)$ given by $\partial_q^G = \sum_{i=0}^q (-1)^i (d_i^{\mathcal{S}(X)})_*^G$. By [**2**, 4.5] this chain complex is isomorphic to Illman's chain complex $S^G(X,*;M)$ given in [**11**, p. 15], whose homology is by definition the Bredon-Illman G-equivariant reduced homology of X with coefficients in M, denoted by $\widetilde{H}_q^G(X;M)$.

We shall give an isomorphism

$$H_q(F^G(\mathcal{S}_*(X),M)) \longrightarrow \pi_q(\mathfrak{F}^G(X,M))\,.$$

To construct the isomorphism, we shall give several isomorphisms as depicted in the following diagram.

$$\begin{array}{ccccc}
H_q(F^G(\mathcal{S}(X),M)) & \xleftarrow[\cong]{i_*} & \pi_q(F^G(\mathcal{S}(X),M)) & \xrightarrow[\cong]{\Psi} & \pi_q(\mathcal{S}(|F^G(\mathcal{S}(X),M)|)) \\
\vdots & & & & \cong \Big\downarrow \Phi \\
\pi_q(\mathfrak{F}^G(X,M)) & \xleftarrow[(\rho_*^G)_*]{\cong} & \pi_q(\mathfrak{F}^G(|\mathcal{S}(X)|,M)) & \xleftarrow[\Psi_{\mathcal{S}(X)*}]{\cong} & \pi_q(|F^G(\mathcal{S}(X),M)|)
\end{array}$$

By [**13**, 22.1], i_* is an isomorphism. In particular, this shows that every cycle in $\widetilde{H}^G(X;M)$ is represented by a chain u, all of whose faces are zero. We call this a *special chain.*

The homomorphism Ψ, which is given by $\Psi(u)[\tau] = [u,\tau]$, where u is a special q-chain and $\tau \in \Delta^q$, is an isomorphism, as follows from [**13**, 16.6].

In order to define Φ, we must express $\Psi(u)$ as a map $\gamma : (\Delta[q], \dot{\Delta}[q]) \longrightarrow (\mathcal{S}|F^H(\mathcal{S}(X),M)|, *)$. By the Yoneda lemma, γ is the unique map such that $\gamma(\delta_q) = \Psi(u)$, where $\delta_q = \mathrm{id} : \overline{q} \longrightarrow \overline{q}$. The homomorphism Φ, defined by $\Phi[\gamma][f,\tau'] = \gamma(f)(\tau')$, for $f \in \Delta[q]_n$ and $\tau' \in \Delta^n$, is given by the adjunction between the realization functor and the singular complex functor (see [**13**, 16.1]).

By Theorem 3.9, $\Psi_{\mathcal{S}(X)}$ is an isomorphism of topological groups.

Finally, by Lemma 3.7, the homomorphism ρ_*^G is a homotopy equivalence. □

REMARK 3.12. Chasing along the diagram and using the canonical homeomorphism $|\Delta[q]| \longrightarrow \Delta^q$ given by $[f,\tau'] \mapsto f_\#(\tau')$, one obtains that the isomorphism $H_q(F^G(\mathcal{S}(X),M)) \longrightarrow \pi_q(\mathcal{F}^G(X,M))$ is given as follows. It maps a homology class $[u]$ represented by a special chain $u = \sum_\alpha \gamma^G_{\sigma_\alpha}(u(\sigma_\alpha))$ to the homotopy class $[\overline{u}]$ given by $\overline{u}(\tau) = \sum_\alpha \gamma^G_{\sigma_\alpha(\tau)}(M_*(p_\alpha)(u(\sigma_\alpha)))$, where $p_\alpha : G/G_{\sigma_\alpha} \longrightarrow G/G_{\sigma_\alpha(\tau)}$ is the quotient function and $\mathcal{S}(X)/G = \{[\sigma_\alpha]\}$.

4. Other topological abelian groups

A different way of topologizing the abelian groups $F(X^\delta,M)$ and $\overline{F}^G(X^\delta,M) = F^G(X^\delta,M)$, where X^δ denotes the underlying set of X, is as follows.

Definition 4.1. Let X be a pointed G-space (not necessarily a k-space) and M be a covariant coefficient system. Denote by $\mathcal{S}(X)$ the singular simplicial pointed G-set associated to X and consider the surjective map

$$\pi_X : |F(\mathcal{S}(X),M)| \twoheadrightarrow F(X^\delta,M) \quad \text{given by} \quad \pi_X([l\sigma,t]) = M_*(p_{\sigma,t})(l)\sigma(t)\,,$$

where $p_{\sigma,t} : G/G_\sigma \twoheadrightarrow G/G_{\sigma(t)}$ is the canonical projection.

Give $F(X^\delta,M)$ the identification topology defined by π_X and denote the resulting space by $\mathbb{F}(X,M)$. Moreover, denote by $\overline{\mathbb{F}}^G(X,M)$ the group $\overline{F}^G(X^\delta,M)$ with the relative k-topology induced by ι_X and denote by $\mathbb{F}^G(X,M)$ the group $F^G(X^\delta,M)$ with the quotient topology induced by β_X.

Consider the restriction of π_X

$$\widehat{\pi}^G_X : |\overline{F}^G(\mathcal{S}(X),M)| \twoheadrightarrow \overline{F}^G(X,M)$$

and denote by $\widehat{\mathbb{F}}^G(X,M)$ the resulting identification space.

We thus have a commutative diagram

$$\begin{array}{ccccccc}
|\overline{F}^G(\mathcal{S}(X),M)| & & \xhookrightarrow{\;|\iota_{\mathcal{S}(X)}|\;} & & |F(\mathcal{S}(X),M)| & \xrightarrow{|\beta_{\mathcal{S}(X)}|} & |F^G(\mathcal{S}(X),M)| \\
\downarrow{\scriptstyle \widehat{\pi}^G_X} & & & & \downarrow{\scriptstyle \pi_X} & & \downarrow{\scriptstyle \pi^G_X} \\
\widehat{\mathbb{F}}^G(X,M) & \xrightarrow[\mathrm{id}]{} & \overline{\mathbb{F}}^G(X,M) & \xhookrightarrow[\iota_X]{} & \mathbb{F}(X,M) & \xrightarrow[\beta_X]{} & \mathbb{F}^G(X,M)\,,
\end{array}$$

where the vertical maps are identifications. Thus $\mathrm{id} : \widehat{\mathbb{F}}^G(X,M) \longrightarrow \overline{\mathbb{F}}^G(X,M)$ is continuous.

Remark 4.2. The topological groups $\overline{F}^G(X,M)$ were defined in [**3**].

Recall [**4**] that a (pointed) G-space X is called a *strong ρ-space* if the map $\rho_X : |\mathcal{S}(X)| \longrightarrow X$, given by $\rho_X([\sigma,t]) = \sigma(t)$, is a G-retraction.

Proposition 4.3. *If X is a strong ρ-space, then* $\mathrm{id} : \mathbb{F}(X,M) \longrightarrow \mathfrak{F}(X,M)$ *is a homeomorphism, and hence the maps* $\mathrm{id} : \overline{\mathbb{F}}^G(X,M) \longrightarrow \overline{\mathfrak{F}}^G(X,M)$ *and* $\mathrm{id} : \mathbb{F}^G(X,M) \longrightarrow \mathfrak{F}^G(X,M)$ *are also homeomorphisms.*

Proof. Consider the diagram

$$\begin{array}{ccc} |F(\mathcal{S}(X),M)| & \xrightarrow[\cong]{\Psi_{\mathcal{S}(X)}} & \mathfrak{F}(|\mathcal{S}(X)|,M) \\ {\scriptstyle \pi_X}\downarrow & & \downarrow{\scriptstyle \rho_{X*}} \\ \mathbb{F}(X,M) & \xrightarrow[\mathrm{id}]{\cong} & \mathfrak{F}(X,M). \end{array}$$

One easily verifies that the diagram commutes. The map on the top is a homeomorphism by 3.9, the vertical map on the left is an identification by definition, and the vertical map on the right is an identification, since it is a retraction. Thus the identity on the bottom is a homeomorphism. □

Corollary 4.4. *If X is a strong ρ-space, then the topological groups $\mathbb{F}^G(X,M)$ and $\mathfrak{F}^G(X,M)$ are equal, as well as the topological groups $\overline{\mathbb{F}}^G(X,M)$ and $\overline{\mathfrak{F}}^G(X,M)$.* □

Example 4.5. If K is a simplicial G-set, then by [**4**], $|K|$ is a strong ρ-space. Hence the topological groups $\mathbb{F}(|K|,M)$ and $\mathfrak{F}(|K|,M)$ are equal, and thus also the topological groups $\mathbb{F}^G(|K|,M)$ and $\mathfrak{F}^G(|K|,M)$ are equal, as well as the topological groups $\overline{\mathbb{F}}^G(|K|,M)$ and $\overline{\mathfrak{F}}^G(|K|,M)$.

Proposition 4.6. *If X is a strong ρ-space, then $\mathbb{F}^G(X,M) = \mathfrak{F}^G(X,M)$.*

Proof. Since X is a strong ρ-space, the map $\rho_X : |\mathcal{S}(X)| \longrightarrow X$ is a G-retraction. Thus the epimorphism $\rho^G_X : \mathfrak{F}^G(|\mathcal{S}(X)|,M) \twoheadrightarrow \mathfrak{F}^G(X,M)$ is an identification. Consider the diagram

$$\begin{array}{ccc} |F^G(\mathcal{S}(X),M)| & \xrightarrow{\psi^G_M} & \mathfrak{F}^G(|\mathcal{S}(X)|,M) \\ {\scriptstyle \pi^G_X}\downarrow & & \downarrow{\scriptstyle \rho^G_*} \\ \mathbb{F}^G(X,M) & \xrightarrow[\mathrm{id}]{} & \mathfrak{F}^G(X,M). \end{array}$$

By definition of π^G_X it commutes and by 3.9, the arrow on the top is an isomorphism of topological groups. Since both vertical arrows are identifications, the result follows. □

There are other topological groups related to the ones previously defined, that were studied in [**3**].

Definition 4.7. Let X be a pointed G-space (not necessarily a k-space) and M be a covariant coefficient system for G. Define the topological group (in the category of k-spaces)

$$F(X, M) = |F(\mathcal{S}(X), M)|,$$

as well as the subgroup $\overline{F}^G(X, M) = |\overline{F}^G(\mathcal{S}(X), M)|$ and the quotient group $F^G(X, M) = |F^G(\mathcal{S}(X), M)|$. One clearly has continuous homomorphisms

$$\overline{F}^G(X, M) \overset{\iota_X}{\hookrightarrow} F(X, M) \overset{\beta_X}{\twoheadrightarrow} F^G(X, M)$$

given by $\iota_X = |\iota_{\mathcal{S}(X)}|$ and $\beta_X = |\beta_{\mathcal{S}(X)}|$. The first is an embedding and the second a quotient map of topological groups.

Remark 4.8. One has identifications of topological groups

$$\overline{F}^G(X, M) \twoheadrightarrow \widehat{\mathbb{F}}^G(X, M), \quad F(X, M) \twoheadrightarrow \mathbb{F}(X, M), \quad F^G(X, M) \twoheadrightarrow \mathbb{F}^G(X, M).$$

We shall see below under which conditions the topological groups $\widehat{\mathbb{F}}^G(X, M)$ and $\overline{\mathbb{F}}^G(X, M)$ are equal.

5. Coefficients in k-$\mathfrak{Mod}$

In this section we shall assume that $M : \mathcal{O}(G) \longrightarrow k\text{-}\mathfrak{Mod}$, where k is a field of characteristic 0 or a prime p that does not divide the order of G,

Recall that if S is a pointed G-set, then we have homomorphisms

$$\overline{F}^G(S, M) \overset{\iota_S}{\hookrightarrow} F(S, M) \overset{\beta_S}{\twoheadrightarrow} F^G(S, M).$$

We shall study the composite $\alpha_S = \beta_S \circ \iota_S : \overline{F}^G(S, M) \longrightarrow F^G(S, M)$.

Proposition 5.1. *Let S be a pointed G-set. Then α_S is a natural isomorphism.*

Proof. Take a generator

$$\gamma_x^G(l) = \sum_{[g] \in G/G_x} M_*(R_{g^{-1}})(l)(gx) \in \overline{F}^G(S, M).$$

Since $\gamma^G_{gx} M_*(R_{g^{-1}}) = \gamma^G_x$, we have

$$\begin{aligned}\alpha_S(\gamma^G_x(l)) &= \beta_S\left(\sum_{[g]\in G/G_x} M_*(R_{g^{-1}})(l)(gx)\right)\\ &= \sum_{[g]\in G/G_x} \beta_S\left(M_*(R_{g^{-1}})(l)(gx)\right)\\ &= \sum_{[g]\in G/G_x} \gamma^G_{gx} M_*(R_{g^{-1}})(l)\\ &= \sum_{[g]\in G/G_x} \gamma^G_x(l)\\ &= [G\!:\!G_x]\gamma^G_x(l)\\ &= \gamma^G_x([G\!:\!G_x]l)\,.\end{aligned}$$

Since $p \not| \, |G|$, the indexes $[G\!:\!G_x]$, seen as elements in k, are invertible elements, i.e., there exist the elements $[G\!:\!G_x]^{-1} \in k$. Since for every generator $\gamma^G_x(l)$, one has

$$\alpha_S(\gamma^G_x([G\!:\!G_x]^{-1}l) = \gamma^G_x(l)\,,$$

α_S is surjective. To see that it is injective, assume that $u \in \overline{F}^G(S, M)$ is such that $\alpha_S(u) = 0$. Hence

$$\alpha_S(u)(x) = [G\!:\!G_x]u(x) = 0\ \forall\, x\,.$$

Thus $[G\!:\!G_x]^{-1}[G\!:\!G_x]u(x) = u(x) = 0$ for all x and so $u = 0$.

The naturality of α_S follows from (2.1). □

Proposition 5.2. *Let X be a pointed G-space. Then*

(a) $\alpha_X = |\alpha_{\mathcal{S}(X)}| : \overline{F}^G(X, M) = |\overline{F}^G(\mathcal{S}(X), M)| \longrightarrow |F^G(\mathcal{S}(X), M)| = F^G(X, M)$ *and*

(b) $\alpha_X : \widehat{\mathbb{F}}^G(X, M) \longrightarrow \mathbb{F}^G(X, M)$

are isomorphisms of topological groups.

Proof. (a) follows immediately from Proposition 5.1. To see (b), consider the diagram

$$\begin{array}{ccc} |\overline{F}^G(\mathcal{S}(X), M)| & \xrightarrow[\cong]{|\alpha_{\mathcal{S}(X)}|} & |F^G(\mathcal{S}(X), M)| \\ {\scriptstyle \widehat{\pi}^G_X}\big\downarrow & & \big\downarrow{\scriptstyle \pi^G_X} \\ \widehat{\mathbb{F}}^G(X, M) & \xrightarrow[\alpha_X]{\cong} & \mathbb{F}^G(X, M)\,. \end{array}$$

To see that it is commutative, take a generator $[\gamma_\sigma^G(l),t] \in |\overline{F}^G(\mathcal{S}(X),M)|$. Then

$$\begin{aligned}\alpha_X\widehat{\pi}_X^G([\gamma_\sigma^G(l),t]) &= \alpha_X([G_{\sigma(t)}\!:\!G_\sigma]\gamma_{\sigma(t)}^G M_*(p_\sigma)(l))\\ &= [G_{\sigma(t)}\!:\!G_\sigma]\gamma_{\sigma(t)}^G M_*(p_\sigma)([G\!:\!G_{\sigma(t)}]l)\\ &= \gamma_{\sigma(t)}^G M_*(p_\sigma)([G\!:\!G_\sigma]l)\end{aligned}$$

$$\begin{aligned}\pi_X^G|\alpha_{\mathcal{S}(X)}|([\gamma_\sigma^G(l),t]) &= \pi_X^G([\gamma_\sigma^G([G\!:\!G_\sigma]l),t])\\ &= \gamma_{\sigma(t)}^G M_*(p_\sigma)([G\!:\!G_\sigma]l)\,.\end{aligned}$$

Thus we have the assertion.

Since both vertical arrows are identifications, $|\alpha_{\mathcal{S}(X)}|$ on the top is a homeomorphism and α_X on the bottom is bijective, then α_X is a homeomorphism too. □

Corollary 5.3. *The topological groups $\overline{\mathbb{F}}^G(X,M)$ and $\mathbb{F}^G(X,M)$ are naturally isomorphic.*

Proof. The isomorphism of topological groups α_X factors as the composite

$$\alpha_X : \widehat{\mathbb{F}}^G(X,M) \xrightarrow{\iota_X} \mathbb{F}(X,M) \xrightarrow{\beta_X} \mathbb{F}^G(X,M)\,.$$

Therefore $\alpha_X^{-1}\circ\beta_X : \mathbb{F}(X,M) \longrightarrow \widehat{\mathbb{F}}^G(X,M)$ is a left inverse for ι_X, so that ι_X is an embedding. Hence $\widehat{\mathbb{F}}^G(X,M) = \overline{\mathbb{F}}^G(X,M) \cong \mathbb{F}^G(X,M)$. □

6. Some applications of the homotopical Bredon-Illman homology

In this section we shall use the topological groups $\mathcal{F}^G(X,M)$ defined in Section 3, together with the G-equivariant weak homotopy equivalence axiom, that we proved in [**5**] using the groups $F^G(X,M)$, to show that the Bredon-Illman equivariant homology satisfies the (infinite) wedge axiom and to make some calculations. We start with a general result.

Lemma 6.1. *Let X have the homotopy type of a CW-complex. Then the connected components and the path-components of X coincide.*

Proof. Let $C : \mathcal{T}\mathrm{op} \longrightarrow \mathcal{S}\mathrm{et}$ be the functor which associates to a space X the set $C(X)$ of its connected components. This is clearly a homotopy functor.

Now let $\varphi : X \longrightarrow Y$ be a homotopy equivalence, where Y is a CW-complex. Consider the commutative diagram

$$\begin{array}{ccc}\pi_0(X) & \xrightarrow[\approx]{\varphi_*} & \pi_0(Y)\\ \big\downarrow & & \big\downarrow\\ C(X) & \xrightarrow[C(\varphi)]{\approx} & C(Y)\,.\end{array}$$

Since Y is locally path-connected, the arrow on the right is a bijection. Hence the arrow on the left is also a bijection. □

Definition 6.2. Let Λ be a set of indexes, and let $p(\Lambda)$ be the set of finite subsets of Λ. $p(\Lambda)$ is a directed set by inclusion. Let $\{F_\alpha \mid \alpha \in \Lambda\}$ be a family of pointed k-spaces. If $A \subset B$ are finite sets of indexes, then we have an inclusion $\prod_{\alpha \in A} F_\alpha \subset \prod_{\beta \in B} F_\beta$, defined by putting the base point $*_\beta$ in each factor F_β, whenever $\beta \notin A$. Define

$$\bigoplus_{\alpha \in \Lambda} F_\alpha = \operatorname*{colim}_{A \in p(\Lambda)} \prod_{\alpha \in A} F_\alpha \,.$$

Since the category k-$\mathcal{T}$op is closed under colimits, this new space lies in k-$\mathcal{T}$op. Notice that when $\Lambda = \mathbb{N}$ is the set of natural numbers, then the space defined above coincides with the *weak product* defined in [**19**].

REMARK 6.3. If the spaces F_α in the previous definition are topological abelian groups, then algebraically $\bigoplus_{\alpha \in \Lambda} F_\alpha$ is the direct sum of them, and the topology defined therein will give this sum a structure of a topological group (see next proposition). Thus, as topological groups,

$$\bigoplus_{\alpha \in \Lambda} F_\alpha = \operatorname*{colim}_{A \in p(\Lambda)} \bigoplus_{\alpha \in A} F_\alpha \,.$$

Proposition 6.4. *If for each $\alpha \in \Lambda$, F_α is a T_1 topological group, then $\bigoplus_{\alpha \in \Lambda} F_\alpha$ is a topological group.*

Proof. Since we are taking products in k-$\mathcal{T}$op, we consider a compact Hausdorff space K and any continuous map $f : K \longrightarrow \bigoplus_{\alpha \in \Lambda} F_\alpha \times \bigoplus_{\alpha \in \Lambda} F_\alpha$. We thus have to prove that the composite

$$K \xrightarrow{\ f\ } \bigoplus_{\alpha \in \Lambda} F_\alpha \times \bigoplus_{\alpha \in \Lambda} F_\alpha \xrightarrow{\ +\ } \bigoplus_{\alpha \in \Lambda} F_\alpha$$

is continuous.

To see this, notice that the family $\{\bigoplus_{\alpha \in A} F_\alpha \mid A \in p(\Lambda)\}$ has the following two properties:

(i) For any $A, B \in p(\Lambda)$ the index $C = A \cap B \in p(\Lambda)$ satisfies

$$\bigoplus_{\alpha \in A} F_\alpha \cap \bigoplus_{\beta \in B} F_\alpha = \bigoplus_{\gamma \in C} F_\alpha \,.$$

(ii) For each $A \in p(\Lambda)$, the set $\{B \in p(\Lambda) \mid \bigoplus_{\beta \in B} F_\beta \subset \bigoplus_{\alpha \in A} F_\alpha\}$ is finite.

Therefore, by [**9**, 15.10], there are indexes $A_1, \ldots, A_m$ and $B_1, \ldots, B_n$, such that $p_1 f(K) \subset \bigoplus_{\alpha \in A_1} F_\alpha \cup \cdots \cup \bigoplus_{\alpha \in A_m} F_\alpha$ and $p_2 f(K) \subset \bigoplus_{\beta \in B_1} F_\beta \cup \cdots \cup \bigoplus_{\beta \in B_n} F_\beta$, where p_1 and p_2 are the projections. Let $C = A_1 \cup \cdots \cup A_m \cup B_1 \cup \cdots \cup B_n$. Hence we have the following commutative diagram:

$$\begin{array}{ccccc}
 & & \bigoplus_{\gamma \in C} F_\gamma \times \bigoplus_{\gamma \in C} F_\gamma & \xrightarrow{\ +\ } & \bigoplus_{\gamma \in C} F_\gamma \\
 & \nearrow & & & \big\downarrow \\
K & \xrightarrow[f]{} & \bigoplus_{\alpha \in \Lambda} F_\alpha \times \bigoplus_{\alpha \in \Lambda} F_\alpha & \xrightarrow{\ +\ } & \bigoplus_{\alpha \in \Lambda} F_\alpha
\end{array}$$

where the sum on the top is continuous because a finite product of topological groups is a topological group, and the vertical inclusion on the right is continuous. Hence the composite on the bottom is continuous. □

Proposition 6.5. *Let X and Y be pointed spaces and L be an abelian group. Then there is an isomorphism of topological groups*

$$F(X \vee Y, L) \cong F(X, L) \times F(Y, L)\,.$$

Proof. Consider the sequences of spaces

$$X \overset{i}{\hookrightarrow} X \vee Y \overset{q}{\twoheadrightarrow} Y \qquad Y \overset{j}{\hookrightarrow} X \vee Y \overset{p}{\twoheadrightarrow} X$$

Then it follows that

$$F(X, L) \xrightarrow{i_*} F(X \vee Y, L) \xrightarrow{q_*} F(Y, L)$$

is a short exact sequence that splits. Namely, by functoriality one has that i_* is a split monomorphism, q_* is a split epimorphism and $q_* \circ i_* = 0$. Now, if $v \in F(X \vee Y, L)$ is such that $q_*(v) = 0$, then $q_*(v)(y) = v(x_0, y) = 0$. Thus $v = i_*(u)$, where $u \in F(X, L)$ is given by $u(x) = v(x, y_0)$. □

Coming back to Definition 6.2, we have the following generalization to the infinite case of the previous proposition.

Proposition 6.6. *Let X_α, $\alpha \in \Lambda$, be a family of pointed spaces. Then there is an isomorphism of topological groups $\bigoplus_{\alpha\in\Lambda} F(X_\alpha, L) \cong F(\bigvee_{\alpha\in\Lambda} X_\alpha, L)$.*

Proof. For each $A \in p(\Lambda)$, call $\psi_A : \bigoplus_{\alpha\in A} F(X_\alpha, L) \cong F(\bigvee_{\alpha\in A} X_\alpha, L) \longrightarrow F(\bigvee_{\alpha\in\Lambda} X_\alpha, L)$, where the isomorphism comes from Proposition 6.5 and the second map is induced by the canonical inclusion. If $A \subset B$, one easily verifies that $\psi_B = \psi_A \circ \psi_{A,B}$, where $\psi_{A,B} : \bigoplus_{\alpha\in A} F(X_\alpha, L) \longrightarrow \bigoplus_{\alpha\in B} F(X_\alpha, L)$ is induced by the inclusion $\bigvee_{\alpha\in A} X_\alpha \subset \bigvee_{\alpha\in B} X_\alpha$ modulo the isomorphism of 6.5.

By the universal property of the colimit, the maps (continuous homomorphisms) ψ_A induce a continuous homomorphism

$$\psi : \bigoplus_{\alpha\in\Lambda} F(X_\alpha, L) \longrightarrow F(\bigvee_{\alpha\in A} X_\alpha, L)\,.$$

In order to see that ψ is an isomorphism of topological groups, we now construct an inverse

$$\xi : F(\bigvee_{\alpha\in\Lambda} X_\alpha, L) \longrightarrow \bigoplus_{\alpha\in\Lambda} F(X_\alpha, L)$$

as follows. Let $u : \bigvee_{\alpha\in\Lambda} X_\alpha \longrightarrow L$ be an element in $F(\bigvee_{\alpha\in\Lambda} X_\alpha, L)$ and let $u_\alpha : X_\alpha \longrightarrow L$ be the restriction of u; that is, $u_\alpha = p_{\alpha *}(u)$, where $p_\alpha : \bigvee_{\alpha\in\Lambda} X_\alpha \longrightarrow X_\alpha$ is the canonical projection. Then $u_\alpha \neq 0$ only for finitely many values of α, i.e., only for $\alpha \in A$, and some $A \in p(\Lambda)$. Thus $(u_\alpha) \in \bigoplus_{\alpha\in\Lambda} F(X_\alpha, L)$ and one can define $\xi(u) = (u_\alpha)$. The homomorphism ξ is clearly an (algebraic) inverse of ψ. To see that ξ is continuous, define the function $\chi : L \times \bigvee_{\alpha\in\Lambda} X_\alpha \longrightarrow \bigoplus_{\alpha\in\Lambda} F(X_\alpha, L)$ by

$\chi(l, x_\alpha) = lx_\alpha$. Composing χ with the identification $\coprod_{\alpha\in\Lambda} L \times X_\alpha \twoheadrightarrow L \times \bigvee_{\alpha\in\Lambda} X_\alpha$ and then restricting to each $L \times X_\alpha$, we obtain the composite

$$L \times X_\alpha \xrightarrow{\mu_1} F(X_\alpha, L) \hookrightarrow \bigoplus_{\alpha\in\Lambda} F(X_\alpha, L)$$

which is continuous. Therefore χ is continuous. Now consider the commutative diagram

$$\begin{array}{ccc} (L \times \bigvee_{\alpha\in\Lambda} X_\alpha)^k & \xrightarrow{\chi^k} & (\bigoplus_{\alpha\in\Lambda} F(X_\alpha, L))^k \\ \downarrow & & \downarrow + \\ F(\bigvee_{\alpha\in\Lambda} X_\alpha, L) & \xrightarrow[\xi]{} & \bigoplus_{\alpha\in\Lambda} F(X_\alpha, L)\,. \end{array}$$

Since the vertical map on the left is an identification and the maps on the top and on the right are continuous (+ by 6.4), ξ is continuous. □

Now we can use the groups $\mathfrak{F}^G(X, M)$ to prove that the Bredon-Illman G-equivariant homology theory $\widetilde{H}^G_*(-; M)$ satisfies the wedge axiom. We need a lemma whose proof is not difficult.

Lemma 6.7. *Let $\{Y_\alpha; i_{\beta,\alpha} : Y_\alpha \hookrightarrow Y_\beta\}$ and $\{\overline{Y}_\alpha; \overline{i}_{\beta,\alpha} : \overline{Y}_\alpha \hookrightarrow \overline{Y}_\beta\}$ be diagrams in k-$\mathfrak{T}$op such that each $i_{\beta,\alpha}$ and $\overline{i}_{\beta,\alpha}$ are closed embeddings, and let $\{q_\alpha : Y_\alpha \twoheadrightarrow \overline{Y}_\alpha\}$ be a family of identifications such that $q_\beta \circ i_{\beta,\alpha} = \overline{i}_{\beta,\alpha} \circ q_\alpha$. Then the map $q : \operatorname{colim} Y_\alpha \twoheadrightarrow \operatorname{colim} \overline{Y}_\alpha$ induced in the colimits in k-$\mathfrak{T}$op is an identification.* □

The next lemma is a consequence of the previous one.

Lemma 6.8. *Let X_α be a pointed G-space for each $\alpha \in \Lambda$. Then the map*

$$\bigoplus_{\alpha\in\Lambda} \prod_{H\subset G} F(X^H_\alpha; M(G/H)) \xrightarrow{\oplus p_{X_\alpha}} \bigoplus_{\alpha\in\Lambda} \mathfrak{F}^G(X_\alpha, M)$$

is an identification. □

Now, as an application of our groups $\mathfrak{F}^G(X, M)$, we have the next results. First we have that the Bredon-Illman homology satisfies the wedge-axiom.

Proposition 6.9. *Let X_α, $\alpha \in \Lambda$, be an arbitrary family of pointed G-spaces. Then there is an isomorphism*

$$\mathfrak{F}^G(\bigvee_{\alpha\in\Lambda} X_\alpha, M) \cong \bigoplus_{\alpha\in\Lambda} \mathfrak{F}^G(X_\alpha, M)\,.$$

Proof. By Proposition 6.6, the inclusions $i_\alpha : X_\alpha \hookrightarrow \bigvee_\alpha X_\alpha$ induce an isomorphism φ of topological groups by

$$\begin{array}{ccc} \bigoplus_{\alpha\in\Lambda} \prod_{H\subset G} F(X^H_\alpha; M(G/H)) & & \\ \downarrow \cong & \overset{\varphi}{\searrow} & \\ \prod_{H\subset G} \bigoplus_{\alpha\in\Lambda} F(X^H_\alpha, M(G/H)) & \xrightarrow[\prod \psi_H]{\cong} & \prod_{H\subset G} F(\bigvee_{\alpha\in\Lambda} X^H_\alpha, M(G/H)) \end{array}$$

Similarly, we can define an isomorphism of abelian groups

$$\psi^G : \bigoplus_{\alpha\in\Lambda} \mathfrak{F}^G(X_\alpha; M) \longrightarrow \mathfrak{F}^G(\bigvee_{\alpha\in\Lambda} X_\alpha, M).$$

They fit into a commutative diagram

$$\begin{array}{ccc} \bigoplus_\alpha \prod_{H\subset G} F\left(X_\alpha^H; M(G/H)\right) & \xrightarrow{\varphi} & \prod_{H\subset G} F\left(\bigvee_\alpha X_\alpha^H, M(G/H)\right) \\ \Big\downarrow{\scriptstyle \oplus p_{X_\alpha}} & & \Big\downarrow{\scriptstyle p_{\vee X_\alpha}} \\ \bigoplus_\alpha \mathfrak{F}^G\left(X_\alpha, M\right) & \xrightarrow[\psi^G]{} & \mathfrak{F}^G\left(\bigvee_\alpha X_\alpha, M\right), \end{array}$$

where the vertical arrows are identifications (the left one by the previous lemma). Since the top arrow is a homeomorphism, then the bottom arrow is a homeomorphism too. □

Proposition 6.10. *Let X_α, $\alpha \in \Lambda$, be an arbitrary family of pointed G-spaces having the homotopy type of G-CW-complexes. Then there is an isomorphism*

$$\widetilde{H}_*^G(\bigvee_\alpha X_\alpha; M) \cong \bigoplus_\alpha \widetilde{H}_*^G(X_\alpha; M).$$

Proof. Under conditions (i) and (ii) given in the proof of 6.4, that are satisfied by the family $\{\bigoplus_{\alpha\in A} \mathfrak{F}^G(\bigvee_\alpha X_\alpha, M) \mid A \in p(\Lambda)\}$, it is proved in [**9**] that the homotopy groups commute with the colimit. Hence

$$\begin{aligned} \widetilde{H}_*^G(\bigvee_{\alpha\in\Lambda} X_\alpha; M) &\cong \pi_q(\mathfrak{F}^G(\bigvee_{\alpha\in\Lambda} X_\alpha; M)) && \text{(by 3.11)} \\ &\cong \pi_q(\bigoplus_{\alpha\in\Lambda} \mathfrak{F}^G(X_\alpha, M)) && \text{(by 6.9)} \\ &\cong \pi_q(\operatorname{colim}_A(\bigoplus_{\alpha\in A} \mathfrak{F}^G(X_\alpha, M))) && \text{(by 6.3)} \\ &\cong \operatorname{colim}_A(\pi_q(\bigoplus_{\alpha\in A} \mathfrak{F}^G(X_\alpha, M))) && \text{(by [9, 15.9])} \\ &\cong \operatorname{colim}_A(\bigoplus_{\alpha\in A} \pi_q(\mathfrak{F}^G(X_\alpha, M))) && \text{(since } A \text{ is finite)} \\ &\cong \bigoplus_{\alpha\in\Lambda} \pi_q(\mathfrak{F}^G(X_\alpha, M))) && \text{(by 6.3)} \\ &\cong \bigoplus_{\alpha\in\Lambda} \widetilde{H}_*^G(X_\alpha; M) && \text{(by 3.11).} \end{aligned}$$

□

In order to prove the wedge axiom in full generality, we need the following lemmas.

Lemma 6.11. *Let X and Y, and X' and Y' be well-pointed spaces and let $\varphi : X' \longrightarrow X$ and $\psi : Y' \longrightarrow Y$ be weak homotopy equivalences (mapping base points to base points). Then $\varphi \vee \psi : X' \vee Y' \longrightarrow X \vee Y$ is a weak homotopy equivalence.*

Proof. Consider the double attaching spaces $X \cup I \cup Y$ and $X' \cup I \cup Y'$, where the base point $x_0 \in X$ is identified with $0 \in I$, the base point $y_0 \in Y$ is identified with $1 \in I$, and similarly with the other union. Since the spaces are well pointed, the quotient maps $q : X \cup I \cup Y \twoheadrightarrow X \vee Y$ and $q' : X' \cup I \cup Y' \twoheadrightarrow X' \vee Y'$ that collapse I to the common base point are homotopy equivalences. The pairs $(X \cup [0,1), (0,1] \cup Y)$ and $(X' \cup [0,1), (0,1] \cup Y')$ are excisive in $X \cup I \cup Y$ and $X' \cup I \cup Y$, respectively. Hence, by [**9**, 16.24], the map $\varphi \cup \mathrm{id}_I \cup \psi : X \cup I \cup Y \longrightarrow X' \cup I \cup Y'$ is a weak homotopy equivalence. The following is clearly a commutative diagram

$$\begin{array}{ccc} X' \cup I \cup Y' & \xrightarrow{\varphi \cup \mathrm{id}_I \cup \psi} & X \cup I \cup Y \\ {\scriptstyle q'}\downarrow{\scriptstyle \simeq} & & {\scriptstyle \simeq}\downarrow{\scriptstyle q} \\ X' \vee Y' & \xrightarrow[\varphi \vee \psi]{} & X \vee Y \,. \end{array}$$

Therefore, $\varphi \vee \psi$ is a weak homotopy equivalence. □

As a consequence, we obtain the following result.

Lemma 6.12. *Let X_α and X'_α, $\alpha \in \Lambda$ be an arbitrary family of well-pointed G-spaces, and let $\varphi_\alpha : X'_\alpha \longrightarrow X_\alpha$ be an equivariant weak homotopy equivalences (mapping base points to base points). Then $\vee_{\alpha\in\Lambda}\varphi_\alpha : \bigvee_{\alpha\in\Lambda} X'_\alpha \longrightarrow \bigvee_{\alpha\in\Lambda} X_\alpha$ is an equivariant weak homotopy equivalence.*

Proof. First notice that since the base points are fixed under the G-action, $(\bigvee_{\alpha\in\Lambda} X_\alpha)^H = \bigvee_{\alpha\in\Lambda} X_\alpha^H$, and similarly $(\bigvee_{\alpha\in\Lambda} X'_\alpha)^H = \bigvee_{\alpha\in\Lambda} X'^H_\alpha$.

Let $p(\Lambda)$ be the set of finite subsets of Λ. Then, by 6.11, for every $A \in p(\Lambda)$, $\vee_{\alpha\in A}\varphi_\alpha^H : \bigvee_{\alpha\in A} X'^H_\alpha \longrightarrow \bigvee_{\alpha\in A} X^H_\alpha$ is a weak homotopy equivalence and induces isomorphisms between all the homotopy groups. Thus again, as in the proof of 6.10 and using [**9**, 15.9], $\vee_{\alpha\in\Lambda}\varphi_\alpha^H : \bigvee_{\alpha\in\Lambda} X'^H_\alpha \longrightarrow \bigvee_{\alpha\in\Lambda} X^H_\alpha$ induces isomorphisms in all homotopy groups and hence it is a weak homotopy equivalence, thus $\vee_{\alpha\in\Lambda}\varphi_\alpha : \bigvee_{\alpha\in\Lambda} X'_\alpha \longrightarrow \bigvee_{\alpha\in\Lambda} X_\alpha$ is an equivariant weak homotopy equivalence. □

Now we can prove the general wedge axiom for the Bredon-Illman G-equivariant homology.

Theorem 6.13. *Let X_α, $\alpha \in \Lambda$, be an arbitrary family of well-pointed G-spaces. Then there is an isomorphism $\bigoplus_{\alpha\in\Lambda} \widetilde{H}^G_q(X_\alpha; M) \cong \widetilde{H}^G_q(\bigvee_{\alpha\in\Lambda} X_\alpha; M)$ induced by the canonical inclusions $X_\alpha \hookrightarrow \bigvee X_\alpha$.*

Proof. For each $\alpha \in \Lambda$ there is an equivariant weak homotopy equivalence $\varphi_\alpha : \widetilde{X}_\alpha \longrightarrow X_\alpha$, where $\widetilde{X}_\alpha$ is a (pointed) G-CW-complex (for instance $\widetilde{X}_\alpha = |\mathcal{S}(X_\alpha)|$).

By 6.12, $\varphi = \vee_\alpha \varphi_\alpha : \bigvee_\alpha \widetilde{X}_\alpha \longrightarrow \bigvee_\alpha X_\alpha$ is an equivariant weak homotopy equivalence. Hence, by [**5**, 1.19], φ induces an isomorphism $\varphi_* : H^G_q(\bigvee_\alpha \widetilde{X}_\alpha; M) \longrightarrow H^G_q(\bigvee_\alpha X_\alpha)$. Recall that

$$H^G_q(\bigvee_\alpha \widetilde{X}_\alpha; M) = \widetilde{H}^G_q((\bigvee_\alpha \widetilde{X}_\alpha)^+; M) \quad \text{and} \quad H^G_q(\bigvee_\alpha X_\alpha) = \widetilde{H}^G_q((\bigvee_\alpha X_\alpha)^+; M)\,.$$

On the other hand, the cofiber sequences

$$\mathbb{S}^0 \hookrightarrow (\bigvee_\alpha \widetilde{X}_\alpha)^+ \twoheadrightarrow \bigvee_\alpha \widetilde{X}_\alpha \quad \text{and} \quad \mathbb{S}^0 \hookrightarrow (\bigvee_\alpha X_\alpha)^+ \twoheadrightarrow \bigvee_\alpha X_\alpha$$

induce short exact sequences that fit into the diagram

$$\begin{array}{ccccccccc} 0 & \longrightarrow & \widetilde{H}_0^G(\mathbb{S}^0; M) & \longrightarrow & \widetilde{H}_0^G((\bigvee_\alpha \widetilde{X}_\alpha)^+; M) & \longrightarrow & \widetilde{H}_0^G(\bigvee_\alpha \widetilde{X}_\alpha; M) & \longrightarrow & 0 \\ & & \| & & \cong \downarrow \varphi_*^+ & & \downarrow \varphi_* & & \\ 0 & \longrightarrow & \widetilde{H}_0^G(\mathbb{S}^0; M) & \longrightarrow & \widetilde{H}_0^G((\bigvee_\alpha X_\alpha)^+; M) & \longrightarrow & \widetilde{H}_0^G(\bigvee_\alpha X_\alpha; M) & \longrightarrow & 0 \end{array}$$

Thus, by the five-lemma, we have an isomorphism

$$\varphi_* : \widetilde{H}_0^G(\bigvee_\alpha \widetilde{X}_\alpha; M) \longrightarrow \widetilde{H}_0^G(\bigvee_\alpha X_\alpha; M).$$

On the other hand, since $\widetilde{H}_q^G(\mathbb{S}^0; M) = 0$ if $q > 0$, there are isomorphisms

$$\widetilde{H}_q^G((\bigvee_\alpha \widetilde{X}_\alpha)^+; M) \cong \widetilde{H}_q^G(\bigvee_\alpha X_\alpha; M)$$

$$\widetilde{H}_q^G((\bigvee_\alpha \widetilde{X}_\alpha)^+; M) \cong \widetilde{H}_q^G(\bigvee_\alpha X_\alpha; M)$$

and thus also an isomorphism

$$\varphi_* : \widetilde{H}_q^G(\bigvee_\alpha \widetilde{X}_\alpha; M) \longrightarrow \widetilde{H}_q^G(\bigvee_\alpha X_\alpha; M).$$

Similarly, for every $\alpha \in \Lambda$, there are isomorphisms

$$\varphi_{\alpha *} : \widetilde{H}_q^G(\widetilde{X}_\alpha; M) \longrightarrow \widetilde{H}_q^G(X_\alpha; M).$$

Since the G-spaces $\widetilde{X}_\alpha$ are G-CW-complexes, by 6.10 there is an isomorphism

$$\bigoplus_{\alpha \in \Lambda} \widetilde{H}_q^G(\widetilde{X}_\alpha; M) \cong \widetilde{H}_0^G(\bigvee_\alpha \widetilde{X}_\alpha; M)$$

for all $q \geqslant 0$, and by the isomorphisms above it follows that there is an isomorphism

$$\bigoplus_{\alpha \in \Lambda} \widetilde{H}_q^G(X_\alpha; M) \cong \widetilde{H}_q^G(\bigvee_\alpha X_\alpha; M)$$

as desired. □

Next we make some calculations using the homotopical approach to Bredon-Illman homology.

Proposition 6.14. *Let X be a pointed G-space of the homotopy type of a G-CW-complex. Assume that X^H is connected for each $H \subset G$. Then*

$$\widetilde{H}_0^G(X; M) = 0.$$

Proof. By Lemma 6.1, $C(X^H) = \pi_0(X^H)$ for each $H \subset G$. Hence

$$\pi_0(F(X^H, M(G/H))) \cong \widetilde{H}_0(X^H; M(G/H) = 0.$$

Thus the topological groups $F(X^H, M(G/H))$ are path-connected and so is their product. Since $\mathcal{F}^G(X, M)$ is a quotient of this product, its also path-connected. Hence $0 = \pi_0(\mathcal{F}^G(X, M)) \cong \widetilde{H}_0^G(X; M)$. □

Proposition 6.15. *Let X be a G-space of the homotopy type of a G-CW-complex. Assume that X^H is connected for each $H \subset G$. Consider the family $\mathcal{H}$ of all $H \subset G$ such that $X^H \neq \emptyset$. Then $H_0^G(X; M)$ is a quotient of $\bigoplus_{H\in\mathcal{H}} M(G/H)$. Furthermore, if X has a fixed point, then $H_0^G(X; M) \cong M(G/G)$.*

Proof. For each $H \in \mathcal{H}$, there is an isomorphism

$$\pi_0(F((X^H)^+, M(G/H))) \cong H_0(X^H; M(G/H)) \cong M(G/H).$$

Hence one has an epimorphism

$$\bigoplus_{H\in\mathcal{H}} M(G/H) \cong \pi_0(\prod_{H\in\mathcal{H}} F((X^H)^+, M(G/H))) \twoheadrightarrow \pi_0(\mathcal{F}^G(X^+, M)) \cong H_0^G(X; M).$$

Furthermore, if X has a fixed point x_0, then $X^+ = X \vee \mathbb{S}^0$, where we take x_0 as base point of X. By the exactness axiom, one always has $\widetilde{H}_*^G(X \vee Y; M) \cong \widetilde{H}_*^G(X; M) \oplus \widetilde{H}_*^G(Y; M)$. Then, in particular, $\widetilde{H}_0^G(X^+; M) \cong \widetilde{H}_0^G(X; M) \oplus \widetilde{H}_0^G(\mathbb{S}^0; M)$. Since by the previous proposition $\widetilde{H}_0^G(X; M) = 0$, one obtains

$$H_0^G(X; M) \cong \widetilde{H}_0^G(\mathbb{S}^0; M) \cong M(G/G).$$

□

7. Dold-Thom topological groups and the transfer

In this section we summarize the properties of the topological groups defined and explain which of them admit a transfer, either for ordinary or for ramified covering G-maps. We start with the following concept.

Definition 7.1. Let M be a covariant coefficient system for G. We shall call a functor $\mathcal{G}(-, M) : G\text{-}\mathcal{T}\mathrm{op} \longrightarrow \mathcal{T}\mathrm{opab}$ a *Dold-Thom topological group functor with coefficients in M for a subcategory $\mathcal{T} \subset G$-$\mathcal{T}$op* if there is a natural isomorphism

$$\varphi_X : \pi_*(\mathcal{G}(X, M)) \longrightarrow \widetilde{H}_*^G(X; M),$$

where the right-hand side denotes the reduced G-equivariant Bredon-Illman homology groups of X with coefficients in M, and X is an object of $\mathcal{T}$.

EXAMPLES 7.2. The following are examples of Dold-Thom topological group functors with coefficients in a coefficient system M:

1. The groups $F^G(X, M) = |F^G(\mathcal{S}(X), M)|$, for any G-space X, defined in [**2**].
2. The groups $\mathbb{F}^G(X, M)$ defined in Section 5 of [**3**], when M is a homological Mackey functor and X has the homotopy type of a G-CW-complex.

3. The groups $\mathcal{F}^G(X, M)$ as shown in 3.11.
4. The groups $|F^G(Q, M)|$ for $X = |Q|$ and Q a simplicial G-set, in particular, the simplicial G-set associated to a simplicial G-complex as shown in [**5**].
5. The groups $\mathrm{AG}(X)$ and $\mathrm{AG}(X; m)$ defined by Dold and Thom [**7**] in the nonequivariant case, where the coefficients are a cyclic group ($\mathbb{Z}$ and $\mathbb{Z}/m\mathbb{Z}$, respectively) and X is a (countable) simplicial complex. Here G does not stand for any group.
6. The groups $B(L, X)$ defined by McCord [**15**] in the nonequivariant case and coefficients in an abelian group L, where X is a weak Hausdorff k-space of the homotopy type of a CW-complex.
7. The groups $\mathrm{AG}(X)$ defined by Lima-Filho [**12**] for a G-CW-complex X and coefficients in $\mathbb{Z}$.
8. The groups $L \otimes X$ defined by dos Santos [**10**] for a pointed G-CW-complex X and coefficients in a G-module L.
9. The groups $\mathcal{G}X \otimes_{G\mathcal{F}} M$ defined in [**17**] for a pointed G-CW-complex X.

And if M takes values in k-$\mathcal{M}$od, where k is a field of characteristic 0 or a prime p that does not divide $|G|$:

10. The groups $\overline{F}^G(X, M)$, by 5.2(a) and 1.
11. The groups $\overline{\mathbb{F}}^G(X, M)$, if X has the homotopy type of a G-CW-complex, by 5.3 and 2.
12. The groups $\overline{\mathcal{F}}^G(X, M)$ for a strong ρ-space X, by 5.2(b), 5.3, 4.6, and 3.

Definition 7.3. Let M be a Mackey functor for the finite group G and $\mathcal{G}(-, M)$ be a Dold-Thom topological group functor with coefficients in (the covariant part of) M. Let $p : E \longrightarrow X$ be either an n-fold G-equivariant ordinary covering map (see [**2**]) or an n-fold G-equivariant ramified covering map (see [**4**]). By a *transfer* for p in $\mathcal{G}(-, M)$ we understand a continuous homomorphism

$$t_p^G : \mathcal{G}(X^+, M) \longrightarrow \mathcal{G}(E^+, M)\,,$$

which satisfies the following conditions:

(a) Pullback: If $f : Y \longrightarrow X$ is continuous and we take the pullback diagram

$$\begin{array}{ccc} E' & \xrightarrow{\ \widetilde{f}\ } & E \\ {\scriptstyle q}\big\downarrow & & \big\downarrow{\scriptstyle p} \\ Y & \xrightarrow[\ f\ ]{} & X\,, \end{array}$$

then $t_p^G \circ f_*^G = \widetilde{f}_*^G \circ t_q^G : \mathcal{G}(Y^+, M) \longrightarrow \mathcal{G}(E^+, M)$.

(b) Normalization: If $p = \mathrm{id}_X : X \longrightarrow X$, then $t_{\mathrm{id}_X}^G = \mathrm{id} : \mathcal{G}(X^+, M) \longrightarrow \mathcal{G}(X^+, M)$.

(c) Functoriality: If $p : E \longrightarrow X$ and $q : X \longrightarrow Y$ are G-equivariant ordinary, resp. ramified covering maps, then

$$t^G_{q \circ p} = t^G_p \circ t^G_q : \mathcal{G}(Y^+, M) \longrightarrow \mathcal{G}(E^+, M)\,.$$

(d) If M is homological, then the composite $p^G_* \circ t^G_p : \mathcal{G}(X^+, M) \longrightarrow \mathcal{G}(X^+, M)$ is multiplication by n.

EXAMPLES 7.4. The following Dold-Thom topological group functors have transfers for p:

1. If $p : E \longrightarrow X$ is an n-fold G-equivariant ordinary covering map and M is any Mackey functor, then there is a transfer $t^G_p : F^G(X^+, M) \longrightarrow F^G(E^+, M)$, as shown in [**2**].
2. If $p : E \longrightarrow X$ is an n-fold G-equivariant ramified covering map between strong ρ-spaces of the homotopy type of G-CW-complexes, and M is homological, then there is a transfer $t^G_p : \mathbb{F}^G(X^+, M) \longrightarrow \mathbb{F}^G(E^+, M)$, as shown in [**4**].
3. If $p : K \longrightarrow Q$ is a special G-equivariant simplicial ramified covering map and M is any Mackey functor, then there is a transfer $|t^G_p| : |F^G(Q^+, M)| \longrightarrow |F^G(K^+, M)|$, as shown in [**5**].
4. If $p : E \longrightarrow X$ is an n-fold G-equivariant ramified covering map between strong ρ-spaces of the homotopy type of G-CW-complexes, and M is homological, then there is a transfer $t^G_p : \mathfrak{F}^G(X^+, M) \longrightarrow \mathfrak{F}^G(E^+, M)$, by 4.3 and Example 2.

REMARK 7.5. Let $p : E \longrightarrow X$ be an n-fold G-equivariant covering map. The restrictions $p^H : E^H \longrightarrow X^H$ are not, in general, covering maps. Thus there are no transfers $F(X^{H+}, M(G/H)) \longrightarrow F(E^{H+}, M(G/H))$. Since the topology of the groups $\mathfrak{F}^G(X^+, M)$ and $\mathfrak{F}^G(E^+, M)$ is given in terms of that of the groups $F(X^{H+}, M(G/H))$ and $F(E^{H+}, M(G/H))$, it does not seem possible to prove the continuity of t^G_p. And even if the transfers t_{p^H} exist, they do not commute with the identifications. However, if as stated in Example 4 above, the spaces are ρ-spaces, then the groups $\mathfrak{F}^G(X^+, M)$ and $\mathfrak{F}^G(E^+, M)$ coincide with the groups $\mathbb{F}^G(X^+, M)$ and $\mathbb{F}^G(E^+, M)$, as shown in 4.4, and one can show that the transfer is continuous.

Now we study the homotopy type of the Dold-Thom topological groups. First we have the following general result.

Theorem 7.6. *Let A be a locally connected topological abelian group in the category k-$\mathfrak{T}$op that has the homotopy type of a CW-complex. Then there is a homotopy equivalence*

$$A \simeq \bigoplus_{q \geqslant 0} K(\pi_q(A), q)\,,$$

where $K(\pi_q(A), q)$ denotes the corresponding Eilenberg-Mac Lane space.

Proof. Since translation by a_0, $a_0 \in A$, is a homeomorphism in the k-topology, the connected component A_0 of $0 \in A$ is a closed subgroup of A.

By Lemma 6.1, the connected components of A coincide with the path-components. Thus A is the topological sum of all its path-components, and they are closed and open, because A is locally connected. Since all path-components of A are homeomorphic (via translation) to A_0, we have a homeomorphism

$$A \cong \pi_0(A) \times A_0 \,. \tag{7.7}$$

Since A has the homotopy type of a CW-complex, so does A_0. Consider

$$A_0 \overset{i}{\hookrightarrow} F(A_0, \mathbb{Z}) \overset{\nu}{\twoheadrightarrow} A_0 \,,$$

given by $i(a) = 1a$ and $\nu(u) = \sum_{a \in A_0} u(a)a$. Then i and ν are clearly continuous, and the composite $\nu \circ i$ is equal to the identity id_{A_0}. Applying the functor π_q we have the following

$$\begin{array}{ccccc} & & \overset{1}{\frown\!\!\!\longrightarrow} & & \\ \pi_q(A_0) & \xrightarrow[i_\#]{} & \pi_q(F(A_0,\mathbb{Z})) & \xrightarrow[\nu_\#]{} & \pi_q(A_0) \\ & \searrow^{h_q} & \big\downarrow \cong & \nearrow_{\lambda_q} & \\ & & \widetilde{H}_q(A_0;\mathbb{Z}) & & \end{array}$$

where λ_q is defined so that the triangle commutes, and thus we obtain a left inverse for the Hurewicz homomorphism.

Now consider a homotopy equivalence $A_0 \xrightarrow{\varphi} C_0$, where C_0 is a CW-complex, and the diagram

$$\begin{array}{ccccc} \pi_q(A_0) & \xrightarrow{h_q} & \widetilde{H}_q(A_0) & \xrightarrow{\lambda_q} & \pi_q(A_0) \\ {\scriptstyle \varphi_\#}\big\downarrow{\scriptstyle \cong} & & {\scriptstyle \varphi_\#}\big\downarrow{\scriptstyle \cong} & & {\scriptstyle \varphi_\#}\big\downarrow{\scriptstyle \cong} \\ \pi_q(C_0) & \xrightarrow[h_q]{} & \widetilde{H}_q(C_0) & \dashrightarrow_{\alpha_q} & \pi_q(C_0) \,, \end{array}$$

where α_q is so that the diagram commutes. Hence α_q is a left inverse to Hurewicz too. Since C_0 is a connected CW-complex, by Moore's theorem (see [**19**, IX(1.9)], we have $C_0 \simeq \bigoplus_{q \geqslant 1} K(\pi_q(C_0), q)$ (see 6.2). Hence, using (7.7), we get

$$A \cong \pi_0(A) \times A_0 \simeq \pi_0(A) \times \bigoplus_{q \geqslant 1} K(\pi_q(C_0), q) \approx \bigoplus_{q \geqslant 0} K(\pi_q(A_0), q) \,.$$

□

REMARK 7.8. This result was proved by John Moore in [**16**] for the case of connected simplicial abelian groups and furthermore by Dold and Thom [**8**] for the case of connected finite polyhedral abelian groups.

As a consequence of Theorem 7.6 we have the following.

Theorem 7.9. *Let $\mathcal{G}(-, M)$ is a Dold-Thom topological group functor. If X is a pointed G-space such that $\mathcal{G}(X, M)$ is locally connected and has the homotopy type* *of a CW-complex, then*

$$\mathcal{G}(X, M) \simeq \bigoplus_{q \geqslant 0} K(\widetilde{H}^G_q(X; M), q)\,.$$

Hence these groups are unique up to homotopy. □

Proposition 7.10. (a) *The topological groups $F^G(X, M)$ are CW-complexes for any pointed G-space X.*

(b) *If X is a pointed G-CW-complex, then $\mathcal{F}^G(X, M)$ is locally connected and has the homotopy type of a CW-complex.*

(c) *If X is a pointed G-simplicial complex or a finite-dimensional countable locally finite G-CW-complex, then $\mathbb{F}^G(X, M)$ is locally connected and has the homotopy type of a CW-complex.*

We conclude that in all these cases the topological groups have the homotopy type of $\bigoplus_{q \geqslant 0} K(\widetilde{H}^G_q(X; M), q)$.

Proof. (a) follows from the fact that $F^G(X, M)$ is the geometric realization $|F^G(\mathcal{S}(X), M)|$.

(b) First notice that the property of a space being locally connected is inherited by quotient spaces. Hence, if X is locally connected, so is also $\prod_{n \geqslant 1}(L \times X)^n$ for any abelian group L, and given the quotient map $\prod_{n \geqslant 1}(L \times X)^n \twoheadrightarrow F(X, L)$, the topological group $F(X, L)$ is locally connected.

Now, if X is a G-CW-complex, then X^H is locally connected for every $H \subset G$. Hence each topological group $F(X^H, M(G/H))$ is locally connected and since by definition there is a quotient map $\prod_{H \subset G} F(X^H, M(G/H)) \twoheadrightarrow \mathcal{F}^G(X, M)$, the topological group $\mathcal{F}^G(X, M)$ is also locally connected. Furthermore there is a G-homotopy equivalence $\rho_X : |\mathcal{S}(X)| \longrightarrow X$, which by the homotopy invariance 3.4 induces a homotopy equivalence $\rho^G_{X*} : \mathcal{F}^G(|\mathcal{S}(X)|, M) \longrightarrow \mathcal{F}^G(X, M)$. But by 3.10, there is an isomorphism of topological groups $\mathcal{F}^G(|\mathcal{S}(X)|, M) \cong |F^G(\mathcal{S}(X), M)|$, thus the first is a CW-complex, and hence $\mathcal{F}^G(X, M)$ has the homotopy type of a CW-complex.

(c) Actually we only need X to be a pointed G-CW-complex which is also a strong ρ-space, then by 4.4, $\mathbb{F}^G(X, M) = \mathcal{F}^G(X, M)$ and the result follows from (b).

In any case, the corresponding topological group satisfies the assumptions of Theorem 7.6 and we obtain the conclusion. □

References

[1] M. Aguilar and C. Prieto, *Topological abelian groups and equivariant homology,* Communications in Algebra **36** (2008), 434–454.

[2] M. Aguilar and C. Prieto, *Equivariant homotopical homology with coefficients in a Mackey functor,* Topology and its Applications **154** (2007), 2826–2848 (with a correction, ibid. **156** (2009), 1609–1613).

[3] M. Aguilar and C. Prieto, *Topological groups and Mackey functors,* Bol. Soc. Mat. Mex. **14** (2008), 233–262.

[4] M. Aguilar and C. Prieto, *The transfer for ramified covering G-maps,* Forum Math. **22** (2010) 1089–1115.

[5] M. Aguilar and C. Prieto, *Bredon homology and ramified covering G-maps,* Topology and its Applications **157** (2010) 401–416.

[6] T. tom Dieck, *Transformation Groups,* Walter de Gruyter, Berlin & New York, 1987.

[7] A. Dold and R. Thom, *Quasifaserungen und unendliche symmetrische Produkte,* Annals of Math. **67** (1958), 239–281.

[8] A. Dold and R. Thom, *Une généralisation de la notion d'espace fibré. Application aux produits symétriques infinis,* C. R. Acad. Sci. Paris **242** (1956), 1680–1682.

[9] B. Gray, *Homotopy Theory,* Academic Press New York San Francisco London 1975.

[10] P. F. dos Santos, *A note on the equivariant Dold-Thom theorem,* J. Pure Appl. Algebra **183** (2003), 299–312.

[11] S. Illman, *Equivariant singular homology and cohomology I,* Memoirs of the Amer. Math. Soc. Vol. 1, Issue 2, No. 156, 1975.

[12] P. Lima-Filho, *On the equivariant homotopy of free abelian groups on G-spaces and G-spectra, Math. Z.* **224** (1997), 567-601.

[13] J. P. May, *Simplicial Objects in Algebraic Topology,* The University of Chicago Press, Chicago London 1992.

[14] J. P. May, *A Concise Course in Algebraic Topology,* The University of Chicago Press, Chicago & London 1999.

[15] M. C. McCord, *Classifying spaces and infinite symmetric products,* Trans. Amer. Math. Soc. **146** (1969), 273–298.

[16] J. C. Moore, *Semi-simplicial complexes and Postnikov systems,* Symposium Internacional de Topología Algebraica, 232–247, UNAM y UNESCO, México 1958.

[17] Z. Nie, *A functor converting equivariant homology to homotopy,* Bull. London Math. Soc., **39**(3) (2007), 499–508.

[**18**] R. M. Vogt, *Convenient categories of topological spaces for homotopy theory*, Archiv der Mathematik **XXII** (1971), 545–555.

[**19**] G. W. Whitehead, *Elements of Homotopy Theory*, Springer-Verlag, New York Heidelberg Berlin 1978.

This article may be accessed via WWW at `http://tcms.org.ge/Journals/JHRS/`

Marcelo A. Aguilar
`marcelo@math.unam.mx`

Instituto de Matemáticas
Universidad Nacional Autónoma de México
04510 México, D.F., Mexico

Carlos Prieto
`cprieto@math.unam.mx`

Instituto de Matemáticas
Universidad Nacional Autónoma de México
04510 México, D.F., Mexico

Journal of Homotopy and Related Structures, vol. 6(2), 2010, pp.211–211

ERRATUM TO "POSSIBLE CONNECTIONS BETWEEN WHISKERED CATEGORIES AND GROUPOIDS, LEIBNIZ ALGEBRAS, AUTOMORPHISM STRUCTURES AND LOCAL-TO-GLOBAL QUESTIONS", VOL. 5(1) 2010, PP. 305–318.

RONALD BROWN

(*communicated by Hvedri Inassaridze*)

This Note is intended to correct a lack of attribution in the paper [**Bro10**]. That paper cited the paper [**Gil98**], but without the paper in hand, I failed to state that the definition of whiskered category is the same as the definition of semi-regular category in [**Gil98**]. Further, Proposition 5 of [**Bro10**] is the same as Proposition 1.2 of [**Gil98**].

'Resolutions of monoids' which is [**Bro10**, Section 6] is surely related to [**Gil98**, Section 3] on 'Monoid presentations of groups', but the precise relationship needs further work. Clearly we both started from ideas in the joint paper [**BG89a**].

References

[**Bro10**] Brown, R. 'Possible connections between whiskered categories and groupoids, many object lie algebras, automorphism structures and local-to-global questions'. *J. Homotopy and Related Structures* 5(1) (2010) 305-318.

[**BG89a**] Brown, R. and Gilbert, N. D. 'Algebraic models of 3-types and automorphism structures for crossed modules'. *Proc. London Math. Soc. (3)* **59** (1) (1989) 51–73.

[**Gil98**] Gilbert, N. D. 'Monoid presentations and associated groupoids'. *Internat. J. Algebra Comput.* **8** (2) (1998) 141–152.

This article may be accessed via WWW at `http://tcms.org.ge/Journals/JHRS/`

Ronald Brown
`r.brown@bangor.ac.uk`

School of Computer Science
Bangor University
Gwynedd LL57 2UT
UK

Received July 14, 2011; published on August 14, 2011.
2000 Mathematics Subject Classification: 18D05,18D10,18G50,55J99,57R30
Key words and phrases: whiskered category and groupoid, semi-regular category and groupoid, resolutions

Journal of Homotopy and Related Structures, vol. 6(2), 2011, pp.213–238

TWISTING COCHAINS AND HIGHER TORSION

KIYOSHI IGUSA

(*communicated by James Stasheff*)

Abstract

This paper gives a short summary of the central role played by Ed Brown's "twisting cochains" in higher Franz-Reidemeister (FR) torsion and higher analytic torsion. Briefly, any fiber bundle gives a twisting cochain which is unique up to fiberwise homotopy equivalence. However, when they are based, the difference between two of them is a higher algebraic K-theory class measured by higher FR torsion. Flat superconnections are also equivalent to twisting cochains.

This paper is dedicated to Edgar Brown.

Contents

Research for this paper was supported by NSF. I would like to thank Bernard Keller whose lectures and comments lead me the work of my colleague Ed Brown. I thank Ed Brown for many useful conversations about the topics in this paper. I also thank Jim Stasheff for his support and encouragement and for numerous helpful comments on several versions of this paper and I also want to thank Jonathan Block for explaining his sign conventions to me. This was very helpful for getting the correct signs in [**Igu09**] and I have changed the signs in this paper accordingly. Finally, I would like to thank the anonymous referee for numerous suggestions for improving this paper.
Received December 07, 2010, revised June 17, 2011; published on October 10, 2011.
2000 Mathematics Subject Classification: Primary 57R22, Secondary 19J10, 55R40.
Key words and phrases: higher Franz-Reidemeister torsion, higher analytic torsion, A-infinity functors, superconnections, algebraic K-theory, Volodin K-theory, Chen's iterated integrals.

Introduction

About 50 years ago Ed Brown [**Bro59**] constructed a small chain complex giving the homology of the total space E of a fiber bundle

$$F \to E \to B$$

whose base B and fiber F are finite cell complexes. It is given by the tensor product of chain complexes for F and B with the usual tensor product boundary map modified by a "twisting cochain." There are many ways to understand the meaning of the twisting cochain.

1. It is the difference between two A_∞ functors.
2. It is a combinatorial flat $\mathbb{Z}$-graded superconnection.
3. It is a family of chain complexes homotopy equivalent to F and parametrized by B.

If $F \to E \to B$ is a smooth bundle with compact manifold fiber and simply connected base then we get another twisting cochain given by fiberwise Morse theory. Comparison of these two twisting cochains gives an algebraic K-theory invariant of the bundle called the *higher Franz-Reidemeister (FR) torsion.* Higher FR-torsion distinguishes different smooth structures on the same topological manifold bundle. Therefore, this construction is a strictly differentiable phenomenon.

The purpose of this paper is to explain some of the basic properties of these constructions and unify them using a simplified version of Ed Brown's construction. A longer exposition can be found in [**Igu05**] which, in turn, gives a summary of the contents of [**Igu02**].

We summarize the contents of this paper. In Section 1 we review the definition of an A_∞ functor. A_∞ structures were first constructed by Stasheff [**Sta63**] and A_∞ categories first appeared in [**Fuk93**]. But here we take A_∞ functors only from ordinary categories to the category of $\mathbb{Z}$-graded projective modules over a ring R. The definition in this restricted case is given by a formula (Equation 2) due to Sugawara[**Sug60**]. We use an old construction of Eilenberg and MacLane[**EM53**] to make homology into an A_∞ functor when it is projective (Equation 3).

In Section 2 we define twisting cochains. We begin with the classical definition of Brown and we also review Brown's construction of the *twisted tensor product*

$$C_*(B) \otimes_\varphi C_*(F)$$

whose homology is equal to the homology of the total space of a fiber bundle $F \to E \to B$. Brown defined this to be the usual tensor product with boundary map ∂_φ twisted by φ (Equation 4). We give a variation of Brown's definitions which arises from certain A_∞ functors. When there is an underlying functor on the category $\mathcal{X}$ whose induced maps are all isomorphisms we get a coefficient sheaf $F(X, Y)$ over $\mathcal{X}$ and the higher homotopies in the A_∞ functor are cochains on $\mathcal{X}$ with values in F. Our twisting cochain is denoted ψ to distinguish it from the classical one of Brown. We use the twisted tensor product to construct a total complex (2.6) which is an actual functor approximating the A_∞ functor.

Section 3 describes Volodin K-theory[**Vol71**] and its relationship to twisting cochains. We construct the Volodin category $\mathcal{V}^b(R)$ and a generalization which we call the *Whitehead category* (3.4). This is a category of acyclic based free chain complexes over a given ring. This category carries a universal twisting cochain of a certain kind and this twisting cochain defines a cohomology class on the classifying space of the Whitehead category:

$$\tau_{2k} \in H^{4k}(\mathcal{W}h^h_\bullet(\mathbb{Q}, 1); \mathbb{R}).$$

Higher Franz-Reidemeister torsion is in Section 4. We show that, under the right conditions, a smooth fiber bundle with compact manifold fiber $M \to E \to B$ gives two canonical twisting cochains, the topologically defined twisting cochain of Ed Brown and a smoothly defined twisting cochain obtained by fiberwise Morse theory. The fiberwise mapping cone of the comparison map is fiberwise contractible. It gives a mapping of the base B into the Whitehead category provided that we have a basis for the topological twisting cochain. (The Morse theoretic twisting cochain has a basis coming from the critical points.) Such a basis can be chosen in the special case when $\pi_1 B$ acts trivially on the rational homology of M. This based free twisting cochain is classified by a map

$$B \to |\mathcal{W}h^h_\bullet(\mathbb{Q}, 1)|$$

and we can pull back the universal FR torsion class τ_{2k} to B to obtain the higher FR torsion invariant for the bundle. This invariant has been computed in many cases but here we give only one example: the case when the fiber is a closed oriented even dimensional manifold (Theorem 4.6).

The rest of the paper contains an elementary discussion of flat $\mathbb{Z}$-grade superconnections. The aim is to show that they are equivalent to twisting cochains. Section 5 derives a definition of an *infinitesimal twisting cochain*. This is basically a twisting cochain on very small simplices expressed in terms of differential forms. The prefix "super" refers to a $\mathbb{Z}/2\mathbb{Z}$ grading. However, a superconnection on a $\mathbb{Z}$-graded vector bundle will automatically obtain a $\mathbb{Z}$-grading. The Bismut-Lott definition of such a superconnection [**BL95**] also requires that this $\mathbb{Z}$-graded superconnection have total degree 1.

In Section 6, we view the endomorphism valued differential forms as operators on the vector bundle in the standard way to obtain flat superconnections. In the last section 8 we show that a flat superconnection is the differential in a cochain complex which is dual to Brown's twisted tensor product. Going backwards, the second to last section 7 explains how superconnections can be integrated over 1 and 2 simplices using Chen's iterated integrals to give the beginning of a simplicial twisting cochain. Complete details for integration of superconnections over arbitrary simplices can be found in [**Igu09**].

1. A_∞ functors

In this paper we consider the differential graded category $\mathcal{C}(R)$ of chain complexes of projective R-modules over an associative ring R and $\mathcal{G}r(R)$, the underlying graded category of $\mathbb{Z}$-graded projective R-modules. All R-modules will be right R-modules.

Both categories $\mathcal{C}(R), \mathcal{G}r(R)$ have graded hom sets given by

$$\mathrm{HOM}(C_*, D_*) = \bigoplus \mathrm{HOM}_n(C_*, D_*) = \bigoplus_n \prod_k \mathrm{Hom}_R(C_k, D_{n+k}).$$

But $\mathcal{C}(R)$ has a differential $m_1 : \mathrm{HOM}_n(C_*, D_*) \to \mathrm{HOM}_{n-1}(C_*, D_*)$ given by

$$m_1(f) = df - (-1)^n fd.$$

We will be considering functors from an ordinary category $\mathcal{X}$ into the differential graded category $\mathcal{C}(R)$, but we view these as functors $\mathcal{X} \to \mathcal{G}r(R)$ with additional structure. Usually we assume that the functor takes values in the full subcategory of $\mathcal{G}r(R)$ consisting either of nonnegatively graded R-modules with additional structure given by a degree -1 differential or nonpositively graded modules with degree 1 differential.

To fix a problem which arises in the notation we will use the nerve $\mathcal{N}_\bullet \mathcal{X}^{op}$ of the opposite category. Thus a *p-simplex* in $\mathcal{X}$ (an element of $\mathcal{N}_p \mathcal{X}^{op}$) will be a sequence of morphisms of the form:

$$X_0 \xleftarrow{f_1} X_1 \xleftarrow{f_2} \cdots \xleftarrow{f_p} X_p \tag{1}$$

To clarify the notation, the composition $X_p \to X_0$ is a morphism in $\mathcal{X}$ which is also a morphism $X_0 \to X_p$ in $\mathcal{X}^{op}$. Note that a 0-simplex consists of one object X_0 with no maps. The main purpose of this is to make the domain X_j of the front j-face of a $j+k$ simplex equal to the range of the back k-face.

The following definition of A_∞ functors, in particular Equation (2), in the restricted case when the domain is an ordinary category is due to Sugawara [**Sug60**].

Definition 1.1. An A_∞ *functor*

$$\Phi = (\Phi, \Phi_0, \Phi_1, \Phi_2, \cdots) : \mathcal{X} \to \mathcal{G}r(R)$$

on an ordinary category $\mathcal{X}$ is an operation which assigns to each object $X \in \mathcal{X}$ a $\mathbb{Z}$-graded projective R-module ΦX and to each sequence of composable morphisms (1) a morphism

$$\Phi_p(f_1, f_2, \cdots, f_p) : \Phi X_p \to \Phi X_0$$

of degree $p-1$ and satisfies the following cocycle condition for $p \geqslant 0$.

$$\sum_{i=0}^{p} (-1)^i \Phi_i(f_1, \cdots, f_i)\Phi_{p-i}(f_{i+1}, \cdots, f_p)$$

$$= \sum_{i=1}^{p-1} (-1)^i \Phi_{p-1}(f_1, \cdots, f_i f_{i+1}, \cdots, f_p) \tag{2}$$

For $p \geqslant 1$ this can be written as follows where $m_1(f) = \Phi_0 f - (-1)^{\deg f} f \Phi_0$ and $m_2(f, g) = f \circ g$.

$$m_1(\Phi_p) + \sum_{i=1}^{p-1} (-1)^i m_2(\Phi_i, \Phi_{p-i}) = \sum_{i=1}^{p-1} (-1)^i \Phi_{p-1}(1_{i-1}, m_2, 1_{p-i-1}).$$

For $p = 0, 1, 2, 3$ this equation has the following interpretation.

($p = 0$) $\Phi_0\Phi_0 = 0$, i.e., $(\Phi X, \Phi_0)$ is a chain complex.

($p = 1$) $\Phi_0\Phi_1(f) = \Phi_1(f)\Phi_0$, i.e.,

$$\Phi_1(f) : (\Phi X_1, \Phi_0) \to (\Phi X_0, \Phi_0)$$

is a chain map.

($p = 2$) $\Phi_0\Phi_2(f_1, f_2) + \Phi_2(f_1, f_2)\Phi_0 = \Phi_1(f_1)\Phi_1(f_2) - \Phi_1(f_1f_2)$, i.e.,

$$\Phi_2(f_1, f_2) : \Phi_1(f_1f_2) \simeq \Phi_1(f_1)\Phi_1(f_2)$$

is a chain homotopy.

($p = 3$) $\Phi_0\Phi_3(f_1, f_2, f_3) - \Phi_3(f_1, f_2, f_3)\Phi_0 =$

$$\Phi_2(f_1, f_2f_3) - \Phi_2(f_1, f_2)\Phi_1(f_3) + \Phi_1(f_1)\Phi_2(f_2, f_3) - \Phi_2(f_1f_2, f_3).$$

In other words, Φ_3 is a null homotopy of the coboundary of Φ_2.

Proposition 1.2. *Suppose that $\Phi_2 = 0$ and Φ_1 takes isomorphisms to isomorphisms. Then (Φ, Φ_0, Φ_1) is a functor from the category $\mathcal{X}$ to the category of projective R-complexes and degree 0 chain maps and (Φ, Φ_1) is a functor from $\mathcal{X}$ to the category of $\mathbb{Z}$-graded projective R-modules and degree 0 maps.*

Proof. If $\Phi_2 = 0$ then $\Phi_1(f)$ is a chain map $(\Phi X_1, \Phi_0) \to (\Phi X_0, \Phi_0)$ with the property that $\Phi_1(fg) = \Phi_1(f)\Phi_1(g)$. This implies $\Phi_1(id_X)$ is a projection operator. If Φ_1 takes isomorphisms to isomorphisms this must be the identity map on ΦX. □

A *natural transformation* of A_∞ functors is an A_∞ functor on the product category $\mathcal{X} \times I$ where I is the category with two objects 0, 1 and one nonidentity morphism $0 \to 1$. This is a family of chain maps $(\Phi X, \Phi_0) \to (\Phi' X, \Phi_0)$ which are natural only up to a system of higher homotopies. If these chain maps are homotopy equivalences we say that Φ, Φ' are A_∞ *homotopy equivalent* or *fiber homotopy equivalent*. (We view an A_∞ functor as a family of chain complexes over the nerve of $\mathcal{X}^{op}$.)

One way to construct an A_∞ functor on $\mathcal{X}$ is to start with an actual functor C from $\mathcal{X}$ to the category of projective R-complexes, then replace each $C(X)$ with a homotopy equivalent projective R-complex ΦX with differential Φ_0. Following Eilenberg and MacLane [**EM53**], the higher homotopies are given by

$$\Phi_p(f_1, \cdots, f_p) = q_0 C(f_1)\eta_1 C(f_2)\eta_2 \cdots \eta_{p-1} C(f_p) j_p \tag{3}$$

where $j_i : \Phi X_i \to C(X_i)$ and $q_i : C(X_i) \to \Phi X_i$ are homotopy inverse chain maps and $\eta_i : C(X_i) \to C(X_i)$ is a chain homotopy $id \simeq j_i \circ q_i$.

In the special case when the homology of $C(X)$ is projective (e.g., if R is a field) and $C(X)$ is either nonnegatively or nonpositively graded, the *homology complex*

$$\Phi X = H_*(C(X))$$

(with zero boundary map) gives an example of a homotopy equivalent chain complex. Using the construction above, we obtain the A_∞ *homology functor.*

Suppose that $p : E \to B$ is a fiber bundle where the base B is (the geometric realization of) a simplicial complex. Then for each simplex σ in B the inverse image $E|\sigma = p^{-1}(\sigma)$ is homeomorphic to $F \times \sigma$ and thus homotopy equivalent to the fiber F. Taking either a cellular chain complex or the total singular complex, we get a functor from the category $\operatorname{simp} B$ of simplices in B, with inclusions as morphisms, to the category of augmented chain complexes. When the homology of F is projective, the construction above gives an A_∞ homology functor on $\operatorname{simp} B$.

We will see that this A_∞ functor gives a twisting cochain on B which, by Ed Brown's twisted tensor product construction, gives a chain complex for the homology of E. But first we want to point out that the dual of an A_∞ functor is also an A_∞ functor.

Suppose that $R = K$ is a field. Then we have the degree-wise duality functor on $\mathcal{G}r(K)$ sending $V = \bigoplus V_n$ to V^* where

$$V_n^* = \operatorname{Hom}(V_{-n}, K).$$

Then morphisms of degree q are sent to morphisms of degree q and the order of composition is reversed. So we get a functor $\mathcal{G}r(K) \to \mathcal{G}r(K)^{op}$. Given an A_∞ functor $\Phi : \mathcal{X} \to \mathcal{G}r(K)$ we can compose with this duality functor:

$$\mathcal{X} \to \mathcal{G}r(K) \to \mathcal{G}r(K)^{op}$$

This is the same as a functor $\Phi^* : \mathcal{X}^{op} \to \mathcal{G}r(K)$.

Proposition 1.3. *The composition of an A_∞ functor $\Phi : \mathcal{X} \to \mathcal{G}r(K)$ with the degree-wise duality functor on $\mathcal{G}r(K)$ gives an A_∞ functor Φ^* on $\mathcal{X}^{op}$.*

Proof. This is very straightforward. The interesting point is that there is no change in signs. Apply duality to Equation (2) and reverse the order of the morphisms f_i^* (since they are begin composed in the opposite order in $\mathcal{X}^{op}$). We get:

$$\sum_{i=0}^{p}(-1)^i \Phi_{p-i}^*(f_p^*, \cdots, f_{i+1}^*)\Phi_i^*(f_i^*, \cdots, f_1^*) = \sum_{i=1}^{p-1}(-1)^i \Phi_{p-1}^*(f_p^*, \cdots, f_{i+1}^* f_i^*, \cdots, f_1^*)$$

Now multiply both sides by $(-1)^p$, replace i by $p-i$ and make the notation change: $g_i = f_{p-i+1}^*$ to see that Φ^* satisfies the definition of an A_∞ functor. □

2. Twisting cochains

First we review Ed Brown's original construction [**Bro59**]. Suppose that Λ is a commutative ring, K is a nonnegatively graded Λ-coalgebra and A is a graded Λ-algebra. Then Brown defined a twisting cochain to be a sum $\varphi = \sum_{p \geqslant 0} \varphi_p$ of Λ-linear maps $\varphi_p : K_p \to A_{p-1}$, so that

1. $\varphi_0 = 0$
2. $\partial \varphi_p = \varphi_{p-1}\partial - \sum_{1 \leqslant i \leqslant p-1} \varphi_i \cup' \varphi_{p-i}$ where

$$\varphi_i \cup' \varphi_{p-i} = \mu(\varphi_i \otimes \varphi_{p-i})\Delta$$

where μ is the multiplication in A and Δ is the comultiplication of K. When this expression is evaluated, a sign of $(-1)^i$ is produced by the Koszul sign rule. (See Definition 2.1 below.)

Brown also assumed that A has an augmentation, which we do not require.

Given a differential graded A-module M, Brown defined the *twisted tensor product*

$$K \otimes_\varphi M$$

to be the standard graded tensor product over Λ with differential ∂_φ given by

$$\partial_\varphi(x \otimes y) = \partial x \otimes y + (-1)^{\deg x} x \otimes \partial y - \sum_{(x)} (-1)^{\deg x_{(1)}} x_{(1)} \otimes \varphi(x_{(2)}) y \tag{4}$$

where we use Sweedler notation $\Delta x = \sum_{(x)} x_{(1)} \otimes x_{(2)}$ [**Swe69**].

Brown then showed that the total singular complex $S_*(E)$ (with coefficients in Λ) of the total space of a fiber bundle $F \to E \to B$ over a path connected space is homotopy equivalent to the twisted tensor product

$$S'_*(B) \otimes_\varphi C_*(F)$$

where $K = S'_*(B)$ is the subcomplex of the total singular complex of B consisting of singular simplices in B with all its vertices at the base point of B, $M = C_*(F)$ is any free Λ-complex homotopy equivalent to the total singular complex of F and A is the differential graded algebra $A = \mathrm{HOM}(M, M)$. (Brown took A to be the total singular complex of the loop space of B which acts on $S_*(F)$.)

The cochains $\varphi_p, p \geqslant 1$ are given as follows. Each 1-simplex σ of B is a loop in B which induces a holonomy $F \to F$. Then $\varphi_1 := \sigma_* - 1$ where $\sigma_* : C_*(F) \to C_*(F)$ is the induced chain map. Any 2-simplex $\sigma : \Delta^2 \to B$ gives a homotopy between one loop and the composition of the other two loops: $\partial_0\sigma\partial_2\sigma \simeq \partial_1\sigma$. We take $\varphi_2(\sigma)$ to be the corresponding chain homotopy $\partial_0\sigma_* \circ \partial_2\sigma_* \simeq \partial_1\sigma_*$. The construction of φ_p for $p \geqslant 3$ is similar and is a special case of the construction given below.

We now consider twisting cochains $\psi = \sum_{p \geqslant 0} \psi_p$ arising from A_∞ functors. In the case of a fiber bundle $p : E \to B$, the idea is that we take simplices in the base B with distinct vertices. For each vertex v, ψ_0 is the boundary map of the complex $C_*(F_v)$. Each 1-simplex gives, not an endomorphism of a single complex $M = C_*(F)$, but morphisms between complexes associated to vertices of the simplex. Thus $\mathrm{HOM}(M, M)$ is replaced by a category of graded R-modules. We also work over an associative ring R. This simply means that the tensor product $S_*(B) \otimes_\varphi M$ is replaced by a direct sum of (shifted) copies of (various) M, one for each free generator of $S_*(B)$. (See Definition 2.6 below).

Suppose that $(\Phi, \Phi_1) : \mathcal{X} \to \mathcal{G}r(R)$ is a functor so that $\Phi_1(f)$ is a degree 0 isomorphism for all $f : X \to Y$. In that case, the graded bifunctor

$$F(X, Y) = \mathrm{HOM}(\Phi X, \Phi Y) \tag{5}$$

gives a locally trivial coefficient system on the category $\mathcal{X}$.

Since we are using the nerve of the opposite category, a *p-cochain* on $\mathcal{X}$ with coefficients in a bifunctor F is a mapping ψ which assigns to each p-simplex

$$X_* = (X_0 \xleftarrow{f_1} X_1 \xleftarrow{f_2} \cdots \xleftarrow{f_p} X_p)$$

in $\mathcal{X}$ an element $\psi(X_*) \in F(X_p, X_0)$. The *coboundary* of ψ is the $p+1$ cochain given by

$$\delta\psi(X_0, \cdots, X_{p+1}) = (f_1)_*\psi(X_1, \cdots, X_{p+1}) + \sum_{i=1}^{p}(-1)^i\psi(X_0, \cdots, \widehat{X_i}, \cdots, X_{p+1}) + (-1)^p(f_{p+1})^*\psi(X_0, \cdots, X_p). \quad (6)$$

Definition 2.1. Given a functor $(\Phi, \Phi_1) : \mathcal{X} \to \mathcal{G}r(R)$ as above, a *twisting cochain* ψ on $\mathcal{X}$ with coefficients in (Φ, Φ_1) is a sum of cochains $\psi = \sum_{p\geqslant 0}\psi_p$ where ψ_p is a p-cochain on $\mathcal{X}$ with coefficients in the degree $p-1$ part F_{p-1} of the $\mathrm{HOM}(\Phi, \Phi)$ bifunctor F of (5) so that the following condition is satisfied.

$$\delta\psi = \psi \cup' \psi.$$

Here $\cup'$ is the cup product using the Koszul sign rule:

$$\psi_p \cup' \psi_q(X_0, \cdots, X_{p+q}) = (-1)^p\psi_p(X_0, \cdots, X_p)\psi_q(X_p, \cdots, X_{p+q})$$

since ψ_q has total odd degree.

To obtain a classical twisting cochain φ, we restrict to the case where $\mathcal{X}$ has a single object X, $\Phi_1(f)$ is the identity map on ΦX for all morphisms f and R is a commutative ring. We take K to be the free R-complex of the nerve of $\mathcal{X}^{op}$. This is the differential grade R-coalgebra $K = C_*(\mathcal{N}_\bullet\mathcal{X}^{op})$ which in degree p is freely generated by the set of p simplices $X_0 \leftarrow \cdots \leftarrow X_p$ in $\mathcal{X}$. We let M be the projective R-complex $M = (\Phi X, \psi_0)$. Then $A = \mathrm{HOM}(M, M)$ is a differential graded R-complex. The functor F is trivial and $\delta f = f\partial$. So, the equation $\delta\psi = \psi \cup' \psi$ becomes:

$$\psi_{p-1}\partial = \psi_0 \cup' \psi_p + \psi_p \cup' \psi_0 + \sum_{i=1}^{p-1}\psi_i \cup' \psi_{p-i}$$

Since $\psi_0 = \partial^M$, we have

$$\psi_0 \cup' \psi_p + \psi_p \cup' \psi_0 = \partial^M\psi_p + (-1)^p\psi_p\partial^M = \partial^A\psi_p$$

Therefore, $\varphi_0 = 0$ and $\varphi_p = \psi_p$ for all $p \geqslant 1$ gives a twisting cochain in the sense of Brown. The referee has pointed out that, if $\Phi_1(f)$ is not always the identity map on ΦX, we can still get a classical twisting cochain by letting $\varphi_0 = 0$, $\varphi_p = \psi_p$ for $p \geqslant 2$ and

$$\varphi_1 = \psi_1 + \Phi_1 - id_M.$$

By comparison of definitions we have the following.

Proposition 2.2. *ψ is a twisting cochain on $\mathcal{X}$ with coefficients in (Φ, Φ_1) if and only if*

$$(\Phi, \psi_0, \Phi_1 + \psi_1, \psi_2, \psi_3, \cdots)$$

is an A_∞ functor.

Consider again the A_∞ homology functor

$$\sigma \mapsto H_*(E_\sigma)$$

of a fiber bundle $E \to B$ over a triangulated space B. Suppose that the fiber F has projective homology. In this case ψ_0 and ψ_1 are both zero and the higher homotopies ψ_p for $p \geqslant 2$ are unique up to simplicial homotopy (over $B \times I$). In this case Ed Brown showed that in his twisted tensor product $S'_*(B) \otimes_\psi S_*(F)$, $S_*(F)$ can be replaced with $H_*(F)$ and ψ is given by the A_∞ homology functor constructed above.

An easy spectral sequence comparison argument shows that we may replace $S'_*(B)$ with any homotopy equivalent differential graded coalgebra. We take the cellular complex $C_*(B)$ given by a triangulation of B. Then the twisted tensor product $C_*(B) \otimes_\psi H_*(F)$ is the total complex of the usual bicomplex $C_p(B; H_q(F))$ with boundary map modified by the twisting cochain ψ as follows:

$$\partial_\psi(x \otimes y) = \partial x \otimes y - \sum_{p+q=\deg x} (-1)^p f_p(x) \otimes \psi_q(b_q(x))(y). \tag{7}$$

Here $x = (x_0 \supseteq x_1 \supseteq \cdots \supseteq x_n)$ is a simplex in the first barycentric subdivision of B, $f_p(x) = (x_0 \supseteq \cdots \supseteq x_p)$ is the front p-face of x and $b_q(x) = (x_p \supseteq \cdots \supseteq x_n)$ is the back q-face.

Theorem 2.3 (Brown[**Bro59**]). *Assuming that F has projective homology, the twisted tensor product gives the homology of the total space:*

$$H_*(C_*(B) \otimes_\psi H_*(F)) \cong H_*(E).$$

Remark 2.4. The mapping $f_p \otimes \psi_q b_q$ in (7) has bidegree $(-q, q-1)$. It gives the corresponding boundary map in the Serre spectral sequence for E which is given by filtering the twisted tensor product by reverse filtration of $H_*(F)$ (by subcomplexes $H_{*\geqslant n}(F)$) [**Igu02**].

Corollary 2.5. *When $E \to B$ is an oriented $n-1$ sphere bundle, the degree n part ψ_n of the twisting cochain ψ is a cocycle representing the Euler class of E:*

$$[\psi_n] = e^E \in H^n(B; R).$$

Proof. Since $\mathrm{HOM}(H_*(S^{n-1}), H_*(S^{n-1}))$ has elements only in degrees $0, n-1$, $\psi_k = 0$ for $k \neq n$. By definition of a twisting cochain we have

$$\delta\psi_n = (\psi \cup' \psi)_n = 0.$$

Therefore, ψ_n is an $n-1$ cocycle on B. Since it gives the differential in the Serre spectral sequence, it must represent the Euler class. □

In the present setting, Ed Brown's twisted tensor product is equivalent to the following construction.

Definition 2.6. The *total complex* $E(\psi; \Phi)$ of the twisting cochain ψ with coefficients in the functor (Φ, Φ_1) is given by

$$E(\psi; \Phi) = \bigoplus_{k \geqslant 0} \bigoplus_{(X_0 \leftarrow \cdots \leftarrow X_k)} (X_*) \otimes \Phi X_k$$

with boundary map ∂_ψ given by (7).

Remark 2.7. Note that every simplex $X_* = (X_0 \leftarrow \cdots \leftarrow X_k)$ gives a subcomplex of the total complex $E(\psi; \Phi)$ by:

$$E(X_*) = \bigoplus_{j \geqslant 0} \bigoplus_{a:[j] \to [k]} a^*(X_*) \otimes \Phi X_{a(j)}.$$

This is also the total complex of the A_∞ functor on $[k]$ (considered as a category with objects $0, 1, \cdots, k$ and morphisms $k \to k-1 \to \cdots \to 1 \to 0$) given by pulling back ψ along the functor $X_* : [k] \to \mathcal{X}$.

Note that this gives a functor from the category of simplices in $\mathcal{X}$ to the category of subcomplexes of the total complex with morphisms being inclusion maps. Thus, just as the A_∞ homology functor constructs an A_∞ functor out of an actual functor, the total complex construction gives an actual functor on simp$\mathcal{N}_\bullet\mathcal{X}^{op}$ from an A_∞ functor on $\mathcal{X}$.

Using the total complex, a twisting cochain on $\mathcal{X}$ can be viewed as a family of chain complexes parametrized by the nerve of $\mathcal{X}^{op}$. With some extra structure, this gives a map from the geometric realization of $\mathcal{X}$ to the Volodin K-theory space of R.

3. Volodin K-theory

Algebraic K-theory is related to twisting cochains in the following way. When two *based, upper triangular twisting cochains* are homotopy equivalent, there is an algebraic K-theory obstruction to deforming one into the other. Formally, we take the pointwise mapping cone. This gives a based free acyclic upper triangular twisting cochain on the category $\mathcal{X}$. This is equivalent to a mapping from the geometric realization $|\mathcal{X}|$ of $\mathcal{X}$ to a fancy version of the Volodin K-theory space of the ring R. To avoid confusion, we assume that R has the property that the rank of a free R-module is well defined, i.e., that $R^n \cong R^m$ implies $n = m$.

When the basis is only well-defined up to permutation and multiplication by elements of a subgroup G of the group of units of R, an acyclic twisting cochain on $\mathcal{X}$ defines a mapping from $|\mathcal{X}|$ into the fiber $\mathcal{W}h^h_\bullet(R, G)$ of the mapping

$$\Omega^\infty\Sigma^\infty(BG_+) \to BGL(R)^+ \times \mathbb{Z}.$$

The well-known basic case is the Whitehead group

$$Wh_1(G) = \pi_0\mathcal{W}h^h_\bullet(\mathbb{Z}[G], G)$$

which is the obstruction to G-collapse of a contractible f.g. based free R-complex. In this section we discuss the different versions of the Volodin construction, show how they are related to twisting cochains and identify the homotopy type of two of them.

The basic definition is sometimes called the "one index" case. It is a space of invertible matrices locally varying by upper triangular column operations. When this definition is expressed as a twisting cochain, the construction seems artificial, with only one term ψ_1 in the twisting cochain ψ. However, when the missing higher terms ψ_p are inserted we recover the general Volodin space.

Definition 3.1. For every $n \geqslant 2$ the *Volodin category* $\mathcal{V}^n(R)$ is the category whose objects are pairs (A, σ) consisting of an invertible $n \times n$ matrix $A \in GL(n, R)$ and a partial ordering σ of $\{1, 2, \cdots, n\}$. A morphism $(A, \sigma) \to (B, \tau)$ is an $n \times n$ matrix T with coefficients in R so that

1. $\sigma \subseteq \tau$, i.e., τ is a refinement of σ.
2. $AT = B$. (So the morphism $T = A^{-1}B$ is unique if it exists.)
3. $T = (t_{ij})$ is τ*-upper triangular* in the sense that
   (a) $t_{ii} = 1$ for $i = 1, \cdots, n$
   (b) $t_{ij} = 0$ unless $i \leqslant j$ in the partial ordering τ $(i \leqslant j \Leftrightarrow (i, j) \in \tau)$.

Note that composition is reverse matrix multiplication:

$$S \circ T = TS.$$

There is a *simplicial Volodin space* $V_\bullet^n(R)$ without explicit partial orderings. A p-simplex $g \in V_p^n(R)$ consists of a $p+1$ tuple of invertible $n \times n$ matrices

$$g = (g_0, g_1, \cdots, g_p)$$

so that for some partial ordering σ of $\{1, \cdots, n\}$ the matrices $g_i^{-1}g_j$ are all σ-upper triangular. There is a simplicial map

$$\mathcal{N}_\bullet \mathcal{V}^n(R) \to V_\bullet^n(R) \tag{8}$$

from the nerve of the Volodin category $\mathcal{V}^n(R)$ to the simplicial set $V_\bullet^n(R)$ given by forgetting the partial orderings. However, the collection of admissible partial orderings on any $g \in V_p^n(R)$ has a unique minimal element and therefore forms a contractible category. Consequently, (8) induces a homotopy equivalence

$$|\mathcal{V}^n(R)| \simeq |V_\bullet^n(R)|$$

If we stabilize matrices in the usual way by adding a 1 in the lower right corner we get the stable Volodin category

$$\mathcal{V}(R) = \varinjlim \mathcal{V}^n(R)$$

and the stable Volodin space $V_\bullet^\infty(R) = \varinjlim V_\bullet^n(R)$ which are related to Quillen K-theory by the following well-known theorem due to Vasserstein and Wagoner but best explained by Suslin [**Sus81**].

Theorem 3.2. $|\mathcal{V}(R)| \simeq |V_\bullet^\infty(R)| \simeq \Omega BGL(R)^+$ *where* $\Omega BGL(R)^+$ *is the loop space of the plus construction on the classifying space of* $GL(R) = GL(\infty, R)$.

The Volodin category $\mathcal{V}^n(R)$ has a canonical twisting cochain. It comes from the realization that an invertible matrix is the same as a based contractible chain complex with two terms.

Definition 3.3. The *canonical twisting cochain* on $\mathcal{V}^n(R)$ is given as follows.

1. Let $\Phi(A,\sigma) = C_*$ be the based free graded R-module with $C_0 = C_1 = R^n$ for every object (A,σ) of $\mathcal{V}^n(R)$.
2. $\Phi_1 = (id, id)$ is the identity chain map $C_* \to C_*$ for every morphism.
3. $\psi_0(A,\sigma) = A : R^n \to R^n$.
4. $\psi_1(T) = (0, T - I)$. So $I + \psi_1(T) = (I, T)$ gives a chain isomorphism:

$$\begin{array}{ccc} R^n & \xrightarrow{T} & R^n \\ {\scriptstyle B}\downarrow & & \downarrow{\scriptstyle A} \\ R^n & \xrightarrow{I} & R^n \end{array}$$

The higher homotopies $\psi_p, p \geqslant 2$, are all zero for $\mathcal{V}^n(R)$. However, there is a fancier version of the Volodin category with higher homotopies. We call it the "Whitehead category." This is very similar to the original definition of Volodin [**Vol71**]

Definition 3.4. If G is a subgroup of the group of units of a ring R then the *Whitehead category* $\mathcal{W}h_\bullet(R,G)$ is defined to be the simplicial category whose simplicial set of objects consists of pairs (P,ψ) where ψ is an upper-triangular twisting cochain on the category $[k]$ (as in Remark 2.7) with coefficients in the fixed graded based R-module:

$$R^P := \bigoplus R^{P_i}$$

where $P = \coprod P_i$ is a graded poset (a poset with a grading not necessarily related to the ordering). By *upper-triangular* we mean that $\psi(\sigma)(x)$ is a linear combination of $y < x \in P$ for all simplices σ in $[k]$. As in the Volodin category, Φ_1 is the identity mapping on R^P.

A *morphism* $(P,\psi) \to (Q,\varphi)$ in $\mathcal{W}h_k(R,G)$ consists of a graded order preserving monomorphism going the other way:

$$f : Q \to P$$

so that $S = P - f(Q)$ is a disjoint union of *expansion pairs* which are, by definition, pairs $x^+ > x^-$ otherwise unrelated to every other element of P so that $\psi_0(x^+) = x^- g$ for some $g \in G$, together with a function $\gamma : Q \to G$ so that φ differs from $\psi \circ f$ only by multiplication by γ, i.e., $f_*(\varphi_p(\sigma)(x)) = \psi_p(\sigma)(f(x))\gamma(x)$.

The following theorem, due to J. Klein and the author, is proved in [**Igu02**], Section 5.6.

Theorem 3.5 (Igusa-Klein). *There is a homotopy fiber sequence*

$$|\mathcal{W}h^h_\bullet(R,G)| \to \Omega^\infty\Sigma^\infty(BG_+) \to \mathbb{Z} \times BGL(R)^+$$

where $\mathcal{W}h^h_\bullet(R,G)$ is the simplicial full subcategory of $\mathcal{W}h_\bullet(R,G)$ consisting of (P,ψ) so that each chain complex (R^P,ψ_0) is contractible (i.e., has the homology of the empty set) and $BG_+ = BG \coprod pt$.

Remark 3.6. If we take G to be a finite group then $\Omega^\infty S^\infty(BG_+)$ is rationally trivial above degree 0 so $\mathcal{W}h^h_\bullet(R,G)$ has the rational homotopy type of the Volodin space:

$$|\mathcal{W}h^h_\bullet(R,G)| \simeq_{\mathbb{Q}} |\mathcal{V}(R)| \simeq \Omega BGL(R)^+.$$

In particular, if $R=\mathbb{Q}$, we get [**Bor74**]

$$|\mathcal{W}h^h_\bullet(\mathbb{Q},1)| \simeq_{\mathbb{Q}} \Omega BGL(\mathbb{Q})^+ \simeq_{\mathbb{Q}} BO.$$

Using the Borel regulator maps

$$K_{4k+1}\mathbb{Q} \to \mathbb{R}$$

given by continuous cohomology classes in $H^{2k}(BGL(\mathbb{C});\mathbb{R})$ we get the *universal real higher Franz-Reidemeister torsion* invariants

$$\tau_{2k} \in H^{4k}(\mathcal{W}h^h_\bullet(\mathbb{Q},1);\mathbb{R}).$$

These give characteristic classes for smooth bundles under certain conditions.

4. Higher FR torsion

We will discuss the circumstances under which we obtain well defined algebraic K-theory classes for a fiber bundle. If we have a smooth bundle $p: E \to B$ where E, B and the fiber $M_b = p^{-1}(b)$ are compact connected smooth manifolds and R is a commutative ring so that the fiber homology $H_*(M_b;R)$ is projective then we obtain two canonical twisting cochains on B.

The first is Brown's twisting cochain ψ with coefficients in the fiberwise homology bundle

$$\Phi(b) = H_*(M_b;R).$$

Recall that this requires the fiber homology to be projective.

The second is the fiberwise cellular chain complex $C_*(f_b)$ associated to a fiberwise *generalized Morse function* (GMF) $f: E \to \mathbb{R}$. These are defined to be smooth functions which, on each fiber M_b, have only Morse and birth-death singularities (cubic in one variable plus nondegenerate quadratic in the others). The fiberwise GMF is *not* well-defined up to homotopy. However, there is a canonical choice called a "framed function" ([**Igu87**], [**Igu02**], [**Igu05**]) which exists stably and is unique up to framed fiber homotopy. This gives the following.

Theorem 4.1. *Any compact smooth manifold bundle $E \to B$ gives a mapping*

$$C(f): B \to |\mathcal{W}h_\bullet(\mathbb{Z}[\pi_1 E], \pi_1 E)|$$

which is well-defined up to homotopy and fiber homotopy equivalent to the fiberwise total singular complex of E with coefficients in $\mathbb{Z}[\pi_1 E]$.

Remark 4.2. The *fiberwise total singular complex* of E is the functor which assigns to each simplex $\sigma: \Delta^k \to B$, the total singular complex of $E|\sigma$. The fiberwise framed function f is defined on a product space $E \times D^N$.

In order to compare the two constructions we need a representation

$$\rho : \pi_1 E \to U(R)$$

of $\pi_1 E$ into the group of units of R with respect to which the fiber homology $H_*(M_b; R)$ is projective over R. By the functorial properties of the Whitehead category we get a mapping

$$B \to |\mathcal{W}h_\bullet(R, G)|$$

where $G \subseteq U(R)$ is the image of ρ. By Theorem 4.1, this will be fiberwise homotopy equivalent to the A_∞ fiberwise homology functor $\Phi_1 + \psi$. The fiberwise mapping cone will be fiberwise contractible but it will not give a mapping to $\mathcal{W}h^h_\bullet(R, G)$ unless the fiberwise homology has a basis. This gives the following.

Corollary 4.3. *If $\pi_1 B$ acts trivially on the fiberwise homology $H_*(M_b; R)$ then a fiberwise mapping cone construction gives a mapping*

$$C(C(f)) : B \to |\mathcal{W}h^h_\bullet(R, G)|$$

which is well-defined up to homotopy.

Remark 4.4. A more precise statement is that we take the direct sum of the fiberwise mapping cone with a fixed contractible projective R-complex P_* with the property that $H_*(M_b; R) \oplus P_*$ is free in every degree.

When $R = \mathbb{Q}$, the construction of higher FR torsion extends to the case then $\pi_1 B$ acts *unipotently* on $H_*(M; \mathbb{Q})$ by which we mean that $H_*(M; \mathbb{Q})$ admits a filtration by $\pi_1 B$ submodules so that the action of $\pi_1 B$ on the successive subquotients is trivial.

Corollary 4.5. *Suppose that $E \to B$ is a compact smooth manifold bundle over a connected space B so that $\pi_1 B$ acts unipotently on the rational homology of the fiber M. Then we have a mapping*

$$B \to |\mathcal{W}h^h_\bullet(\mathbb{Q}, 1)|$$

which is well-defined up to homotopy and we can pull back universal higher torsion invariants to obtain well-defined cohomology classes

$$\tau_{2k}(E) \in H^{4k}(B; \mathbb{R})$$

which are trivial if the bundle is diffeomorphic to a product bundle.

It has been known for many years (by [**FH78**] using the stability theorem [**Igu88**]) that there are smooth bundles which are homeomorphic but not diffeomorphic to product bundles and that these exotic smooth structures are detected by algebraic K-theory. Therefore, when the higher FR torsion was successfully defined, it had already been known to be nonzero in these cases.

However, in these exotic examples the fiber M is either odd dimensional or even dimensional with boundary. We now have the complete calculation of the higher torsion in the case of closed oriented even dimensional fibers.

Theorem 4.6 (6.6 in [**Igu05**]). *Suppose that M^{2n} is a closed oriented even dimensional manifold and $M \to E \to B$ is a smooth bundle so that $\pi_1 B$ acts unipotently on the rational homology of M. Then the higher FR torsion invariants $\tau_{2k}(E)$ are well-defined and given by*

$$\tau_{2k}(E) = \frac{1}{2}(-1)^k \zeta(2k+1)\frac{1}{(2k)!}T_{2k}(E) \in H^{4k}(B;\mathbb{R})$$

where $\zeta(s) = \sum \frac{1}{n^s}$ is the Riemann zeta function and

$$T_{2k}(E) = tr_B^E\left(\frac{(2k)!}{2}ch_{2k}(T^vE \otimes \mathbb{C})\right) \in H^{4k}(B;\mathbb{Z})$$

with T^vE being the vertical tangent bundle of E, $ch_{2k}(T^vE \otimes \mathbb{C})$ stands for the degree $4k$ term in the Chern character of $T^vE \otimes \mathbb{C}$ and

$$tr_B^E : H^n(E;\mathbb{Z}) \to H^n(B;\mathbb{Z})$$

is the transfer (with $n = 4k$).

Remark 4.7. Note that $T_{2k}(E)$ is a tangential fiber homotopy invariant. This is in keeping with the belief that there are rationally no stable exotic smooth structures on bundles with closed oriented even dimensional fibers. (*Stable* means stable under product with large dimensional disks D^N. The exotic smooth structure on disk bundles and odd dimensional sphere bundles of [**FH78**] and the explicit examples given by Hatcher ([**Igu02**],[**Goe01**]) are stable.) For more details about this conjecture see [**GI10**]. In that paper we construct virtually all stable exotic smooth structures on bundles with closed odd dimensional fibers and explain why the even dimensional case is so different. See also [**Igu08**] and [**Goe08**] for an outline of those results.

In the special case when $n = 1$, M is an oriented surface and the bundle E is classified by a map of B into the classifying space BT_g of the Torelli group T_g where g is the genus of M. The tangential homotopy invariant T_{2k} is equal to the Miller-Morita-Mumford class in this case ([**Mum83**], [**Mor84**], [**Mil86**]). It is still unknown whether or not any of these classes (tautological classes in degree $4k$) is rationally nontrivial on the Torelli group.

There are several competing versions of higher FR torsion and the version described here is sometimes called *Igusa-Klein (IK) torsion* since the first computation was given in [**IK93**]. Dwyer, Weiss and Williams have defined three kinds of higher Reidemeister which are called *smooth, topological and homotopy DWW torsion* [**DWW03**]. Badzioch, Dorabiala, Klein and Williams have recently shown [**BDKW09**] that smooth DWW torsion is equivalent to IK torsion, making use of the axiomatic characterization of higher torsion given in [**Igu08b**].

Bismut and Lott [**BL95**] have defined higher analytic torsion invariants which have been computed in many cases ([**BL97**], [**BG01**], [**Bun00**], [**Ma97**], [**Goe01**]). In the case of closed oriented even dimensional fibers, the analytic torsion is always zero and Goette has now shown that the expression in Theorem 4.6 gives the difference between BL torsion and IK torsion in all cases. (See the survey article [**Goe08**].)

Higher analytic torsion is defined using flat $\mathbb{Z}$-graded superconnections. It was observed by Goette [**Goe01**] that these are infinitesimal twisting cochains or, as he puts it, that twisting cochains are combinatorial superconnections. We will explain this comment.

5. Flat superconnections

When we review the definition of a flat $\mathbb{Z}$-graded superconnection, we will see that it is the same as an infinitesimal twisting cochain. More precisely, the superconnection is the boundary map of the infinitesimal twisted tensor product. This gives one explanation of the supercommutator rules.

Instead of defining superconnections and showing their relationship to twisting cochains we will take the opposite approach. We ask the question: What is the natural definition of an "infinitesimal twisting cochain?" This question will lead us to the definition of a flat superconnection and we will see that the "superconnection complex" $(\Omega(B, V), D)$ is dual to a twisted tensor product.

Suppose that B is a smooth manifold and $C = \bigoplus_{n \geqslant 0} C_n$ is a nonnegatively graded complex vector bundle over B. Suppose we have a graded flat connection ∇ on C making each C_n into a locally constant coefficient sheaf for the twisting cochain that we want. The example that we keep in mind is when C is the fiberwise homology of a smooth manifold bundle $F \to E \xrightarrow{p} B$. By this we mean the graded vector bundle over B whose fiber over $b \in B$ is the homology of $p^{-1}(b)$. The dual bundle

$$C^* := \bigoplus_{n \geqslant 0} \operatorname{Hom}(C_n, \mathbb{C})$$

is the fiberwise cohomology bundle $H^*(F) \to C^* \to B$.

Now, imagine that B is subdivided into tiny simplices and we have a twisting cochain on B with coefficients in (C, ∇) which satisfies smoothness conditions to be added later. Then, at each vertex v we have a degree -1 endomorphism $\psi_0(v)$ of $C(v)$. This gives a degree 1 endomorphism $A_0 = \psi_0^*$ of the dual $C^*(v)$. Suppose we can extend this to a smooth family of such maps

$$A_0 \in \Omega^0(B, \operatorname{End}(C^*)) = \Omega^0(B, \operatorname{End}(C)^{op})$$

so that $A_0(x)$ has degree 1 and square zero ($A_0(x)^2 = 0$) at all $x \in B$.

Next, we take the edges of B. If an edge e goes from v_0 to v_1 the twisting cochain gives us a degree 0 map

$$C(v_0) \xleftarrow{\psi_1(e)} C(v_1)$$

so that $\psi_1(e)$ together with the map (parallel transport) given by the flat connection ∇ is a chain map. This chain map is the parallel transport of a non-flat connection ∇_1 which we now describe.

If we dualize $\psi_1(e)$ and take only the linear term (ignoring Δv^2 terms) we get a degree 0 map $A_1(\Delta v) : C^*(v_0) \to C^*(v_1)$ which is linear in Δv. To obtain the smooth version we need to take local coordinates for C so that parallel transport of

∇ is constant, i.e., so that, on C^*, $\nabla^* = d$. Then A_1 becomes a matrix 1-form on B (assuming the twisting cochain is smooth in a suitable sense)

$$A_1 \in \Omega^1(B, \mathrm{End}(C^*))$$

so that parallel transport by the new connection $\nabla_1 = d - A_1$ on C^* keeps A_0 invariant. (The change in sign comes from the fact that parallel transport by $d - A_1$ is given infinitesimally by $I + A_1(\Delta v)$ where $A_1(\Delta v)$ is evaluation of the matrix 1-form A_1 on the vector Δv.) This means that

$$[\nabla_1, A_0] = [d - A_1, A_0] = 0$$

$$dA_0 = [A_1, A_0] = A_1 A_0 + A_0 A_1.$$

We interpret this as an approximately commutative diagram:

$$\begin{array}{ccc} C^*(v_0) & \xrightarrow{I+A_1(\Delta v)} & C^*(v_1) \\ {\scriptstyle A_0}\uparrow & & \uparrow{\scriptstyle A_0+\Delta A_0} \\ C^*(v_0) & \xrightarrow{I+A_1(\Delta v)} & C^*(v_1) \end{array}$$

Higher order terms are needed to make the diagram actually commute. The linear terms give the following approximate equation:

$$\Delta A_0 \cong A_1(\Delta v) A_0 - A_0 A_1(\Delta v)$$

Since A_0 is odd, we get two changes of signs:

$$\Delta A_0 \cong -dA_0(\Delta v) \qquad A_1(\Delta v) A_0 = -(A_1 A_0)(\Delta v)$$

As $\Delta v \to 0$ we get the equation $dA_0 = A_1 A_0 + A_0 A_1$ as claimed.

At the next step, we take two small triangles in B forming a rectangle. The following diagram which commutes up to homotopy by $\psi_2^* = A_2$ indicates what is happening. Here $A_1 = A_1^x dx + A_1^y dy$ where A_1^x, A_1^y are (even) matrix 0-forms and $A_1^x \Delta x$ indicates multiplication by the scalar quantity Δx.

$$\begin{array}{ccc} C^*(v_1') & \xrightarrow{I+A_1^x\Delta x+\Delta A_1^x \Delta x} & C^*(v_2) \\ {\scriptstyle I+A_1^y\Delta y}\uparrow & & \uparrow{\scriptstyle I+A_1^y\Delta y+\Delta A_1^y \Delta y} \\ C^*(v_0) & \xrightarrow{I+A_1^x\Delta x} & C^*(v_1) \end{array}$$

This gives the following approximate equation where $\Delta x, \Delta y$ are scalar quantities and $\Delta v_x, \Delta v_y$ are the corresponding vector quantities giving our rectangle in B.

$$\begin{aligned} (A_0 A_2 + A_2 A_0)(\Delta v_x, \Delta v_y) &\cong A_1^y \Delta y + \Delta A_1^y \Delta y A_1^x \Delta x - A_1^x \Delta x + \Delta A_1^x \Delta x A_1^y \Delta y \\ &\cong A_1^y A_1^x \Delta x \Delta y + \frac{\partial A_1^y}{\partial x} \Delta x \Delta y - A_1^x A_1^y \Delta x \Delta y - \frac{\partial A_1^x}{\partial y} \Delta x \Delta y \end{aligned}$$

Since A_1^x, A_1^y are even, the right hand side can be written as

$$\left(dA_1 - A_1^2\right)(\Delta v_x, \Delta v_y)$$

In other words, we have

$$A_2 \in \Omega^2(B, \mathrm{End}(C^*))$$

satisfying the equation

$$dA_1 = A_0A_2 + A_1^2 + A_2A_0.$$

In general we will require that

$$dA_{n-1} = \sum_{p+q=n} A_pA_q.$$

(See [**Igu09**] for a full explanation.) This leads to the following definition.

Definition 5.1. An *infinitesimal twisting cochain* on B with coefficients in a graded vector bundle C^* with graded flat connection ∇ ($\nabla = \sum \nabla_k$ where $(-1)^k\nabla_k$ is a flat connection on C^k) is equal to a sequence of $\mathrm{End}(C^*)$-valued forms

$$A_p \in \Omega^p(B, \mathrm{End}_{1-p}(C^*)) = \Omega^0(B, \mathrm{End}_{1-p}(C^*)) \otimes_{\Omega^0(B)} \Omega^p(B)$$

of total degree 1 so that

$$\nabla A_{n-1} = \sum_{p+q=n} A_pA_q. \tag{9}$$

Next we pass to the algebra of operators on $\Omega(B, C^*)$ where we carefully distinguish between differential forms A and the operators $\widetilde{A}$ that they define to arrive at the Bismut-Lott definition of a flat $\mathbb{Z}$-grade superconnection.

6. Forms as operators

If $A \in \Omega(B, \mathrm{End}(C^*))$ is written as $A = \sum \varphi_i \otimes \alpha_i$ with fixed total degree $|A| = |\varphi_i| + |\alpha_i|$, let $\widetilde{A}$ be the linear operator on

$$\Omega(B, C^*) = \Omega^0(B, C^*) \otimes_{\Omega^0(B)} \Omega(B)$$

given by

$$\widetilde{A}(c \otimes \gamma) := \sum_i (-1)^{|c|\cdot|\alpha_i|}\varphi_i(c) \otimes \alpha_i \wedge \gamma \tag{10}$$

Proposition 6.1 (Prop. 1 in [**Qui85**]). *If $\omega \in \Omega^k(B)$ then*

$$\widetilde{A} \circ \omega = (-1)^{k|A|}\omega \circ \widetilde{A}.$$

Conversely, any linear operator on $\Omega(B, C^*)$ of fixed total degree having this property is equal to $\widetilde{A}$ for a unique $A \in \Omega(B, \mathrm{End}(C^*))$.

Proof. Since $\widetilde{A}$ acts only on the first tensor factor we get

$$\widetilde{A} \circ \omega = \widetilde{A\omega} = (-1)^{k|A|}\widetilde{\omega A} = (-1)^{k|A|}\omega \circ \widetilde{A}$$

as required. Conversely, any linear operator which is $\Omega(B)$-linear in this sense must be "local" and thus we may restrict to a coordinate chart U over which C^* has a basis of sections. This makes $\Omega(U, C^*|U)$ into a free module over $\Omega(U)$. Thus any

$\Omega(U)$-linear operator is uniquely given by $\widetilde{A}$ where $A \in \Omega(U, \mathrm{End}(C^*|U))$ is given by the value of the operator on the basis of sections of $C^*|U$. We can patch these together on intersections of coordinate charts since, by uniqueness, the differential forms defined using different coordinate charts will agree. □

Here is another straightforward calculation.

Proposition 6.2. $[d, \widetilde{A}] = d \circ \widetilde{A} - (-1)^{|A|}\widetilde{A} \circ d = \widetilde{dA}$.

If A' is another $\mathrm{End}(C^*)$ valued form on B then $\widetilde{AA'} = \widetilde{A} \circ \widetilde{A'}$. So

$$[d, \widetilde{A}_{n-1}] = \widetilde{dA}_{n-1} = \sum_{p+q=n} \widetilde{A}_p \circ \widetilde{A}_q$$

which, in coordinate free notation, is

$$[\nabla, \widetilde{A}] = \widetilde{A} \circ \widetilde{A}$$

Since $|A| = 1$ and $\nabla^2 = 0$, we get

$$(\nabla - \widetilde{A})^2 = (\nabla - \widetilde{A}) \circ (\nabla - \widetilde{A}) = 0.$$

This leads to the following definition due to Bismut and Lott [**BL95**]. (A similar definition appeared in [**Che75**].)

Definition 6.3. Let $V = \bigoplus_{n \geqslant 0} V^n$ be a graded complex vector bundle over a smooth manifold B. Then a *superconnection* on V is defined to be a linear operator D on $\Omega(B, V)$ of total degree 1 so that

$$D\alpha = d\alpha + (-1)^{|\alpha|}\alpha D$$

for all $\alpha \in \Omega(B)$. The superconnection D is called *flat* if

$$D^2 = 0.$$

If D is flat then $(\Omega(B, V), D)$ is a chain complex which we call the *superconnection complex*. We will see later that it is homotopy equivalent to the dual of a twisted tensor product. The superconnection complex is bigraded:

$$\Omega(B, V) = \bigoplus \Omega^p(B, V^q)$$

and the superconnection D has terms of degree $(k, 1-k)$ for $k \geqslant 0$. This gives a spectral sequence in the usual way with $E_1^{p,q} = \Omega^p(B, H^q(V, A_0))$ and

$$E_2^{p,q} = H^p(B; H^q(V, A_0)) \Rightarrow H^{p+q}(\Omega(B, V), D)$$

A flat superconnections on V corresponds to a contravariant A_∞ functor on B. To get a twisting cochain we need an ordinary graded flat connection ∇ on V. Then $D - \nabla$ gives a twisting cochain by reversing the above process.

The first step is to get out of the superalgebra framework by writing D as a sum

$$D = \nabla - \widetilde{A} = \nabla - \widetilde{A}_0 - \widetilde{A}_1 - \widetilde{A}_2 - \cdots$$

where $A_p \in \Omega^p(B, \mathrm{End}_{1-p}(V))$ corresponds to $\widetilde{A}_p$ by (10) and satisfies (9).

Next, we obtain a contravariant twisting cochain on the category of smooth simplices in B with coefficients in the category of cochain complexes by iterated integration of A_*. Then we dualize, relying on Proposition 1.3 to recover the original twisting cochain.

7. Chen's iterated integrals

This section gives a very short discussion and proof of the first two steps in the process of integrating a flat superconnection to obtain a twisting cochain. Details are fully explained in [**Igu09**] although the original idea is contained in Chen's work [**Che73**], [**Che75**], [**Che77**]. A more direct, less computational method of constructing the twisting cochain is explained in the next section.

Since ∇ is a flat connection, we can choose local coordinates so that V is a trivial bundle and $\nabla = d$. Starting with $p = 0$ we note that $A_0(x)$ is a degree 1 endomorphism of V_x with $A_0(x)^2 = 0$ making $C(x) = (V_x, A_0(x))$ into a cochain complex for all $x \in B$. Putting $n = 2$ in (9) we see that the curvature $(d - A_1)^2$ of the connection $d - A_1$ is null homotopic. Also we will see that parallel transport of this connection is a cochain map.

It is well-known that the parallel transport Φ_1 associated to the connection $d - A_1$ on V is given by an iterated integral of the matrix 1-form A_1. Given any piecewise smooth path $\gamma : [0, 1] \to B$, parallel transport is the family of degree zero homomorphisms $\Phi_1(t, s) : C(\gamma(s)) \to C(\gamma(t))$ so that $\Phi_1(s, s) = I = id_V$ and $d - A_1 = 0$, i.e.,

$$\frac{\partial}{\partial t}\Phi_1(t, s) = A_1/t\Phi_1(t, s)$$

$$\frac{\partial}{\partial s}\Phi_1(t, s) = -\Phi_1(t, s)A_1/s$$

where $A_1/t = A_1(\gamma(t))(\gamma') \in \mathrm{End}(C(\gamma(t)))$. The solution is given by Chen's iterated integral [**Che77**]:

$$\Phi_1(s_0, s_1) = I + \int_{s_0 \geqslant t \geqslant s_1} dt_1 A_1/t + \int_{s_0 \geqslant t_1 \geqslant t_2 \geqslant s_1} dt_1 dt_2 (A_1/t_1)(A_1/t_2)$$
$$+ \int_{s_0 \geqslant t_1 \geqslant t_2 \geqslant t_3 \geqslant s_1} dt_1 dt_2 dt_3 (A_1/t_1)(A_1/t_2)(A_1/t_3) + \cdots$$

which we abbreviate as:

$$\Phi_1(s_0, s_1) = I + \int_\gamma A_1 + \int_\gamma (A_1)^2 + \int_\gamma (A_1)^3 + \cdots .$$

This can also be written as a limit of products (multiplied right to left)

$$\Phi_1 = \lim_{\Delta t \to 0} \prod (I + (A_1/t_i)\Delta t)$$

In the case when A_1 is constant, parallel transport $C(\gamma(0)) \to C(\gamma(1))$ is given by e^{A_1}. The inverse is given by $\Phi_1(0, 1) = e^{-A_1}$.

Proposition 7.1. *$A_0(\gamma(t))\Phi_1(t,s) = \Phi_1(t,s)A_0(\gamma(s))$, i.e., $\Phi_1(t,s)$ gives a cochain map*

$$C(\gamma(t)) \leftarrow C(\gamma(s)).$$

Proof. By (9) we have:

$$-\frac{d}{dt}A_0(\gamma(t)) = dA_0(\gamma(t))(\gamma') = A_0(\gamma(t))A_1/t - (A_1/t)A_0(\gamma(t))$$

where both negative signs come from the fact that A_0 is odd. So, $X(t) = A_0(\gamma(t))\Phi_1(t,s)$ is the unique solution of the differential equation

$$\frac{\partial}{\partial t}X(t) = (A_1/t)X(t)$$

with initial condition $X(s) = A_0(\gamma(s))$. So, $X(t)$ must also be equal to the other solution of this differential equation which is $X(t) = \Phi_1(t,s)A_0(\gamma(s))$. □

Let

$$\Delta^2 = \{(x,y) \in \mathbb{R}^2 \mid 1 \geqslant x \geqslant y \geqslant 0\}$$

and suppose that $\sigma : \Delta^2 \to B$ is a smooth simplex with vertices $v_0 = \sigma(0,0), v_1 = \sigma(1,0), v_2 = \sigma(1,1) \in B$. Then a chain homotopy

$$\Phi_1(v_0,v_2) \simeq \Phi_1(v_0,v_1)\Phi_1(v_1,v_2)$$

can be obtained by an iterated integral of the form

$$\psi_2(\sigma) = \int_\sigma A_2 + \int_\sigma A_2A_1 + \int_\sigma A_1A_2 + \int_\sigma A_2A_1A_1 + \int_\sigma A_1A_2A_1 + \cdots .$$

The integral over σ is the double integral of the pull-back to Δ^2. The factors of A_1 will just give the parallel transport Φ_1 along paths connecting v_0 and v_2 to the point $v = \sigma(x,y)$

$$\psi_2(\sigma) = \int_{1\geqslant x\geqslant y\geqslant 0} \sigma^*(\Phi_1(v_0,v)A_2(v)\Phi_1(v,v_2)) \in \mathrm{Hom}(C(v_2),C(v_0))$$

where $\Phi_1(v_0,v), \Phi_1(v,v_2)$ are given by parallel transport along paths given by two straight lines each as shown in the Figure.

$$\Phi_1(v,v_2) = \Phi_1(v,\sigma(x,x))\Phi_1(\sigma(x,x),v_2)$$

$$\Phi_1(v_0,v) = \Phi_1(v_0,\sigma(x,0))\Phi_1(\sigma(x,0),v).$$

Proposition 7.2. *$A_0(v_0)\psi_2(\sigma) + \psi_2(\sigma)A_0(v_2) = \Phi_1(v_0,v_2) - \Phi_1(v_0,v_1)\Phi_1(v_1,v_2)$.*

Proof. For $x \in [0,1]$ let $\Phi(x)$ be the parallel transport of $d - A_1$ along the three segment path:

$$\Phi(x) = \Phi_1(v_0,\sigma(x,0))\Phi_1(\sigma(x,0),\sigma(x,x))\Phi_1(\sigma(x,x),v_2).$$

Then

$$\Phi(0) = \Phi_1(v_0,v_2)$$

Figure 1: $\Phi_1(v_0, v), \Phi_1(v, v_2)$ are parallel transport along dark lines.

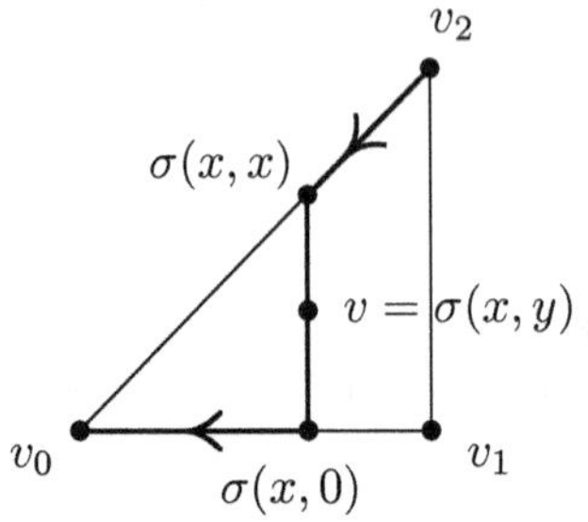

$$\Phi(1) = \Phi_1(v_0, v_1)\Phi_1(v_1, v_2).$$

So the right hand side of the formula we are proving is

$$\Phi(0) - \Phi(1) = -\int_0^1 d\Phi(x). \tag{11}$$

By Proposition 7.1 the left hand side is equal to

$$\int_\sigma \Phi_1(v_0, v)[A_0(v)A_2(v) + A_2(v)A_0(v)]\Phi_1(v, v_2)$$

(The sign in front of $A_2(v)A_0(v)$ is $(-1)^2 = +1$ since the form degree of A_2 is 2.) By (9) this is equal to

$$= \int_\sigma \Phi_1(v_0, v)[-A_1(v)A_1(v) + dA_1(v)]\Phi_1(v, v_2) \tag{12}$$

In (11), we have

$$-\frac{d\Phi(x)}{dx} = \Phi_1(v_0, \sigma(x,0))X(x)\Phi_1(\sigma_1(x,x), v_2)$$

where

$$X(x) = A_1^x\Phi_1(\sigma(x,0),\sigma(x,x)) - \frac{d}{dx}\Phi_1(\sigma(x,0),\sigma(x,x)) - \Phi_1(\sigma(x,0),\sigma(x,x))A_1^x$$

using the notation $\sigma^*(A_1) = A_1^x dx + A_1^y dy$. (The term $\Phi_1(\sigma(x,0),\sigma(x,x))A_1^y$ which occurred with positive sign in the second term and negative sign in the third term was cancelled.) Comparing this to (12) we are reduced to showing that $X(x) = Y(x)$ where

$$Y(x) = \int_{0\leqslant y\leqslant z} dy\, \Phi_1(\sigma(x,0), v)\left(-A_1^x A_1^y + A_1^y A_1^x + \frac{\partial A_1^y}{\partial x} - \frac{\partial A_1^x}{\partial y}\right)\Phi_1(v, \sigma(x,x)).$$

Expressing $\Phi_1(\sigma(x,0),\sigma(x,x))$ as an iterated integral of $A_1^y dy$ we see that the second term of $X(x)$ is equal to the third term of $Y(x)$ (with $\frac{\partial A_1^y}{\partial x}$). The negative sign comes from the fact that we are going backwards along the y direction ($dt = -dy$). The other three terms of $Y(x)$ form the commutator of A_1^x with each factor

$A_1/dt = -A_1^y$ in the iterated integral representation of $\Phi_1(\sigma(x,0),\sigma(x,x))$. This can be more easily seen in the product limit form:

$$\Phi_1(\sigma(x,0),\sigma(x,x)) = \lim_{\substack{n\to\infty \\ \Delta y = x/n}} \prod_1^n (I - A_1^y \Delta y)$$

The commutator of A_1^x with $I - A_1^y\Delta y$ is:

$$A_1^x(I - A_1^y\Delta y) - (I - A_1^y\Delta y)(A_1^x + \Delta_y A_1^x) = -A_1^x A_1^y \Delta y + A_1^y \Delta y A_1^x - \Delta_y A_1^x + o(\Delta y)$$

So, the commutator of A_1^x with $\Phi_1(\sigma(x,0),\sigma(x,x))$ is

$$A_1^x\Phi_1 - \Phi_1 A_1^x = \lim_{\Delta y\to 0} \sum_i \prod_1^{i-1}(I - A_1^y\Delta y)\,[-A_1^x A_1^y\Delta y + A_1^y\Delta y A_1^x - \Delta_y A_1^x] \prod_{i+1}^n (I - A_1^y\Delta y)$$

So, the sum of the remaining two terms of $X(x)$ is equal to the sum of the remaining three terms of $Y(x)$ and we conclude that $X(x) = Y(x)$ proving the proposition. □

The construction that we just explained in detail is a special case of a construction outlined by Chen in [**Che73**], sec. 4.5. Chen constructs mappings

$$\theta_{(n)} : I^{n-1} \to P(\Delta^n, v_n, v_0)$$

from the $n-1$ cube I^{n-1} to the space $P(\Delta^n, v_n, v_0)$ of smooth paths in Δ^n from v_n to v_0 by smoothing a piecewise linear construction very similar to the one we explained. When $n = 2$, this is the 1-parameter family of paths in Δ^2 given in Figure 1 above. The main conceptual difference between Chen's construction and ours is that Chen follows this mapping with a smooth mapping of the n simplex into the space B which sends all vertices of Δ^n to the base point of B. He uses this to obtain a cubical chain complex for ΩB.

A very longwinded description of the higher steps in this process of converting a flat superconnection into a simplicial twisting cochain can be found in [**Igu09**] which uses much of Chen's notation to allow for comparison and to make it easier to understand Chen's work.

In the last section of this paper we will show how the entire process can be done in an easier way.

8. Another method

There is another method for constructing a simplicial twisting cochain from a flat connection. We assume that B is compact and we choose a finite "good cover" for B. (See [**BT82**].) This is a covering of B by contractible open sets U so that all nonempty intersections

$$U_{\alpha_1} \cap U_{\alpha_2} \cap \cdots \cap U_{\alpha_n}$$

are also contractible.

Lemma 8.1. *If U is a contractible open subset of B and D is a flat superconnection on a graded vector bundle V over B then the cohomology of the superconnection complex over U is isomorphic to the cohomology of V using A_0 as differential:*

$$H^n(\Omega(U, V|U), D) \cong H^n(V, A_0)$$

where the isomorphism is given by restriction to any point in U.

Proof. The spectral sequence collapses since its E_2-term is $H^p(U; H^q(V))$. □

This lemma implies that

$$F : U \mapsto (\Omega(U, V|U), D)$$

is a functor from the nerve $\mathcal{N}_\bullet\mathcal{U}$ of the good cover $\mathcal{U}$ of B to the category of cochain complexes over $\mathbb{C}$ and cochain homotopy equivalences. Applying the A_∞ cohomology functor we get a contravariant A_∞ functor H^*F on $\mathcal{N}_\bullet\mathcal{U}$. By Proposition 1.3 we can dualize to get a covariant A_∞ functor $\Phi_1 + \psi$ on $\mathcal{N}_\bullet\mathcal{U}$ with coefficients in $(\Phi, \Phi_1) = (V^*, \nabla^*)$. Subtracting $\Phi_1 = \nabla^*$ we get the twisting cochain ψ satisfying the following.

Theorem 8.2. *The twisted tensor product $C_*(\mathcal{N}_\bullet\mathcal{U}) \otimes_\psi V^*$ is homotopy equivalent to the dual of the superconnection complex. I.e.,*

$$(\Omega(B, V), D) \simeq \mathrm{Hom}(C_*(\mathcal{N}_\bullet\mathcal{U}) \otimes_\psi V^*, \mathbb{C}).$$

Proof. This holds by induction on the number of open sets in the finite good covering $\mathcal{U}$. When the number is 1 we use Lemma 8.1. To increase the number we use Mayer-Vietoris. □

Remark 8.3. By Ed Brown's Theorem 2.3 this implies that, if the superconnection is constructed correctly, the superconnection complex gives the cohomology of the total space of a smooth manifold bundle. This construction also allows us to compare flat superconnections with twisting cochains, giving a K-theory difference class with a well defined higher torsion. But, this is the subject of another paper.

References

[BDKW09] Bernhard Badzioch, Wojciech Dorabiala, John R. Klein, and Bruce Williams, *Equivalence of higher torsion invariants*, arXiv:0904.4684.

[BG01] Jean-Michel Bismut and Sebastian Goette, *Families torsion and Morse functions*, Astérisque (2001), no. 275, x+293.

[BL95] Jean-Michel Bismut and John Lott, *Flat vector bundles, direct images and higher real analytic torsion*, J. Amer. Math. Soc. **8** (1995), no. 2, 291–363.

[BL97] Jean-Michel Bismut and John Lott, *Torus bundles and the group cohomology of* $\mathrm{gl}(N, \mathbf{Z})$, J. Differential Geom. **47** (1997), no. 2, 196–236.

[Bor74] Armand Borel, *Stable real cohomology of arithmetic groups*, Ann. Sci. ENS **7** (1974), 235–272.

[Bro59] Edgar H. Brown, Jr., *Twisted tensor products. I*, Ann. of Math. (2) **69** (1959), 223–246.

[BT82] Raul Bott and Loring Tu, *Differential forms in algebraic topology*, Graduate Texts in Math., vol. 82, Springer-Verlag, New York, 1982.

[Bun00] U. Bunke, *Higher analytic torsion of sphere bundles and continuous cohomology of* $Diff(S^{2n-1})$, math.DG/9802100.

[Che73] Kuo-tsai Chen, *Iterated integrals of differential forms and loop space homology*, Ann. of Math. (2) **97** (1973), 217–246.

[Che75] Kuo-tsai Chen, *Connections, holonomy and path space homology*, Differential geometry (Proc. Sympos. Pure Math., Vol. XXVII, Part 1, Stanford Univ., Stanford, Calif., 1973), Amer. Math. Soc., Providence, R. I., 1975, pp. 39–52.

[Che77] Kuo-tsai Chen, *Iterated path integrals*, Bull. Amer. Math. Soc. **83** (1977), 831–879.

[DWW03] W. Dwyer, M. Weiss, and B. Williams, *A parametrized index theorem for the algebraic K-theory Euler class*, Acta Math. **190** (2003), no. 1, 1–104.

[EM53] Samuel Eilenberg and Saunders MacLane, *On the groups of* $h(\pi, n)$*. I.*, Ann. of Math. (2) **58** (1953), 55–106.

[FH78] F. T. Farrell and W. C. Hsiang, *On the rational homotopy groups of the diffeomorphism groups of discs, spheres and aspherical manifolds*, Algebraic and Geometric Topology (Proc. Sympos. Pure Math., Stanford Univ., Stanford, Calif., 1976), Part 1, Amer. Math. Soc., Providence, R.I., 1978, pp. 325–337.

[Fuk93] Kenji Fukaya, *Morse homotopy,* A^∞*-category, and Floer homologies*, Proceedings of GARC Workshop on Geometry and Topology '93 (Seoul, 1993) (Seoul), Lecture Notes Ser., vol. 18, Seoul Nat. Univ., 1993, pp. 1–102.

[Goe01] Sebastian Goette, *Morse theory and higher torsion invariants I*, math.DG/0111222.

[Goe08] Sebastian Goette, *Torsion invariants for families*, Astérisque 328 (2009), 161-206.

[GI10] Sebastian Goette and Kiyoshi Igusa, *Exotic smooth structures on topological fibre bundles*, arXiv:math/1011.4653.

[Igu87] Kiyoshi Igusa, *The space of framed functions*, Trans. Amer. Math. Soc. **301** (1987), no. 2, 431–477.

[Igu88] Kiyoshi Igusa, *The stability theorem for smooth pseudoisotopies*, K-Theory **2** (1988), no. 1–2, vi+355.

[Igu02] Kiyoshi Igusa, *Higher Franz-Reidemeister Torsion*, AMS/IP Studies in Advance Mathematics, vol. 31, International Press, 2002.

[Igu05] Kiyoshi Igusa, *Higher complex torsion and the framing principle*, Mem. Amer. Math. Soc. **177** (2005), no. 835, xiv+94.

[Igu08] Kiyoshi Igusa, *Pontrjagin classes and higher torsion of sphere bundles*, Groups of diffeomorphisms, Advanced Studies in Pure Math., vol. 52, Tokyo, 2008, pp. 22–29.

[Igu08b] Kiyoshi Igusa, *Axioms for higher torsion invariants of smooth bundles*, J. Topol. **1** (2008), no. 1, 159–186.

[Igu09] Kiyoshi Igusa, *Iterated integrals of superconnections*, arXiv:0912.0249 [math.AT].

[IK93] Kiyoshi Igusa and John Klein, *The Borel regulator map on pictures II. An example from Morse theory*, K-Theory **7** (1993), no. 3, 225–267.

[Ma97] Xiaonan Ma, *Formes de torsion analytique et familles de submersions*, C. R. Acad. Sci. Paris Sér. I Math. **324** (1997), no. 2, 205–210.

[Mil86] Edward Y. Miller, *The homology of the mapping class group*, J. Differential Geom. **24** (1986), no. 1, 1–14.

[Mor84] Shigeyuki Morita, *Characteristic classes of surface bundles*, Bull. Amer. Math. Soc. (N.S.) **11** (1984), no. 2, 386–388.

[Mum83] David Mumford, *Towards an enumerative geometry of the moduli space of curves*, Arithmetic and geometry, Vol. II, Birkhäuser Boston, Boston, MA, 1983, pp. 271–328.

[Qui85] Daniel Quillen, *Superconnections and the Chern character*, Topology **24** (1985), 89–95.

[Sta63] J.D. Stasheff, *Homotopy associativity of H-spaces I,II*, Trans. AMS **108** (1963), 275–292, 293–312.

[Sug60] Masahiro Sugawara, *On the homotopy-commutativity of groups and loop spaces*, Mem. Coll. Sci. Univ. Kyoto Ser. A Math. **33** (1960/1961), 257–269.

[Sus81] A. A. Suslin, *On the equivalence of K-theories*, Comm. Algebra **9** (1981), no. 15, 1559–1566.

[Swe69] M.E. Sweedler, *Hopf Algebras*, W.A. Benjamin, Inc., New York, 1969.

[Vol71] I. A. Volodin, *Algebraic K-theory as an extraordinary homology theory of the category of associative rings with unit*, Math. USSR Izv. 5 (1971), 859–887.

This article may be accessed via WWW at `http://tcms.org.ge/Journals/JHRS/`

Kiyoshi Igusa
`igusa@brandeis.edu`

Department of Mathematics, Brandeis University
Waltham, MA 02454

Journal of Homotopy and Related Structures, vol. 6(2), 2011, pp.239–288

TENSOR PRODUCTS OF REPRESENTATIONS UP TO HOMOTOPY

CAMILO ARIAS ABAD, MARIUS CRAINIC AND BENOIT DHERIN

(*communicated by James Stasheff*)

Abstract

We study the construction of tensor products of representations up to homotopy, which are the A_∞ version of ordinary representations. We provide formulas for the construction of tensor products of representations up to homotopy and of morphisms between them, and show that these formulas give the homotopy category a monoidal structure which is uniquely defined up to equivalence.

Contents

The first author was partially supported by NWO Grant "Symmetries and Deformations in Geometry", SNF Grant 200020-121640/1, the Forschungskredit of the Universität Zürich and the ESI in Vienna. The second author was partially supported by NWO Grant "Symmetries and Deformations in Geometry". The third author was partially supported by NWO Grant "Generating functions and Poisson manifolds" and SNF Grant PA002-113136.
Received February 24, 2011, revised August 21, 2011; published on November 27, 2011.
2000 Mathematics Subject Classification: 18G55
Key words and phrases: Homotopy invariant algebraic structures, monoidal categories

1. Introduction

This work is motivated by the study of the cohomology of classifying spaces of Lie groupoids. For a Lie group G, Bott [**4**] constructed a spectral sequence converging to the cohomology of the classifying space BG with first page

$$E_1^{pq} = H_{\text{diff}}^{p-q}(G, S^q(\mathfrak{g}^*)), \tag{1}$$

the differentiable cohomology with coefficients in the symmetric powers of the coadjoint representation. If the Lie group G is compact then the first page of the spectral sequence vanishes outside of the diagonal, and one obtains that the cohomology of BG is isomorphic to the invariant polynomials on the Lie algebra. The Cartan model for equivariant cohomology can be seen as a generalization of this computation for classifying spaces of groupoids associated to compact group actions on manifolds. In [**6**] Getzler constructed a model for equivariant cohomology of non compact groups, generalizing Bott's spectral sequence to the case of general group actions. Behrend [**5**] extended Getzler's model to the case of stacks that can be represented by "flat groupoids". For general Lie groupoids the situation is more subtle because the "adjoint representation" is no longer a representation in the usual sense. Instead, one has to work with the notion of representation up to homotopy. We have shown in [**1**] that the Bott spectral sequence does exist for arbitrary Lie groupoids, provided one has a well-behaved operation of taking symmetric powers of representations up to homotopy. In the present paper we study the existence and the uniqueness of tensor products of representations up to homotopy.

For a small category $\mathcal{C}$, the notion of representation up to homotopy is the A_∞ version of the usual notion of representation. In terms of A_∞ structures, one associates to $\mathcal{C}$ the differential graded category $\mathbb{R}\mathcal{C}$ whose objects are those of $\mathcal{C}$, and whose morphisms are the linear span over $\mathbb{R}$ of those of $\mathcal{C}$. Then, a representation up to homotopy of $\mathcal{C}$ is an A_∞ functor from $\mathbb{R}\mathcal{C}$ to the dg-category of differential graded vector spaces. We will be interested in the case where $\mathcal{C} = G$ is a Lie groupoid and require the structure operators to be smooth in the appropriate sense. We would like to point out that the assumption that $\mathcal{C}$ is a Lie groupoid does not play any role in the construction of tensor products. We chose this level of generality only because our original motivation comes from studying this case. However, the whole construction applies to arbitrary categories and, more generally, to twisting cochains over a simplicial set (see [**17**]). These more general twisting cochains arise in the integration of infinitesimal representations up to homotopy to global ones, which can also be thought of as a version of the Riemann-Hilbert correspondence, see [**2, 3, 8**].

The works of Loday [**10**], Markl-Schnider [**14**], Saneblidze-Umble [**19**] and Stasheff [**17**] explain that the construction of tensor products of higher homotopy algebraic structures amounts to the construction of certain "diagonal maps" in some appropriate family of polytopes. In the case of A_∞-morphisms, the right family of polytopes is the multiplihedra (see [**7**]). Since we consider the case in which the domain category is strict, the polytopes controlling our problem become much simpler, indeed, they are cubes. See also Sugawara [**20**] and Forcey [**7**] for an account of this. This simplification of the combinatorics allows us to study not only tensor products of representations up to homotopy, but also of morphisms between them.

Here is a short account of the results of this paper. We provide explicit and universal formulas for tensor products that are unital and strictly associative or strictly symmetric, while showing that any two tensor product operations are equivalent. There is a particularly simple choice for this tensor product, given by the Serre diagonal, which is associative but not symmetric. Since we are interested in the symmetric case, and do not want to make any arbitrary choices, we treat all possible tensor product operations on an equal footing. We explain how to take tensor products of morphisms between representations up to homotopy, which correspond to natural transformations between the A_∞ functors. We prove that once a choice has been made for taking tensor products of objects, there is a natural way to take tensor products of morphisms. We show that these constructions produce monoidal structures on the homotopy category of the representations up to homotopy, and that this monoidal structure is unique up to equivalence.

The category of representations up to homotopy of a Lie groupoid is naturally a *dg*-category and it seems natural to ask whether the monoidal structure on the homotopy category can be lifted to this *dg*-category by making choices of tensor products of all lengths in a coherent way. This is an interesting question that we do not address here.

We conclude this introduction with an outline of the paper.

In §2 we review the definitions of representations up to homotopy, the morphisms between them, and the homotopies between the morphisms.

The purpose of §3 is to isolate the algebraic structure that controls the problem of tensoring representations up to homotopy. We show that a representation up to homotopy of G on a complex of vector bundles E is the same thing as a Maurer-Cartan element in a certain DGA (differential graded algebra) $\bar{\mathcal{A}}_E$ associated to G and E. However, for the purpose of handling tensor products, the structure of DGA is not fine enough: $\bar{\mathcal{A}}_{E\otimes F}$ cannot be expressed in terms of the DGA's $\bar{\mathcal{A}}_E$ and $\bar{\mathcal{A}}_F$. For that reason, we introduce the finer notion of DB-algebra and we describe a functor $\bar{K}$ from the category of DB-algebras to the category of complete DGA's.

The DGA $\bar{A}_E$ comes from a canonical DB-algebra $\mathcal{A}_E$ and, this time, $\mathcal{A}_{E\otimes F}$ is related to the tensor product of the DB-algebras $\mathcal{A}_E$ and $\mathcal{A}_F$. Thus, one can state the problem of constructing tensor products of representations up to homotopy in the language of DB-algebras.

In §4 we construct a DB-algebra Ω that is universal with respect to Maurer-Cartan elements in the sense that a morphism of DB-algebras $\Omega \to A$ is the same as a Maurer-Cartan element in the complete differential graded algebra $\bar{K}(A)$.

In §5 we show that the problem of constructing tensor products of representations up to homotopy corresponds to finding certain Maurer-Cartan elements in some universal differential graded algebra. We prove the existence and uniqueness of these tensor products, and provide explicit formulas for strictly associative or strictly symmetric one. We show that a tensor product can not enjoy both of these properties at the same time. We also explain that the tensor product can be chosen so that the product of unital representations remains unital.

In section §6 we explain how to take tensor products of morphisms between representations. We show that any two choices are homotopic. We prove that the homotopy category $\mathcal{D}(G)$ has a monoidal structure that is uniquely defined up to natural equivalence.

Hochschild cohomology appears in §7. There we prove that the universal algebra $\bar{K}(\Omega)$ comes with a canonical Hochschild cocycle of degree zero, and we use this cocycle to specify a construction of the tensor product of morphisms.

The appendix contains general facts about Maurer-Cartan elements in complete differential graded algebras, morphisms between them and their relationship to Hochschild cohomology.

Acknowledgements. We thank Andre Henriques, Jean Louis Loday and Bruno Vallette for various conversations we had at different stages of this work. We specially thank Ieke Moerdijk for his constructive comments during the process of writing this paper. C.A.A. and B.D. also thank Calder Daenzer for suggesting we think about these questions in his cabin in Lake Tahoe.

2. Representations up to homotopy

In this section, we recall the definition of the category of representations up to homotopy. As mentioned in the introduction, we work over general Lie groupoids G (see e.g. [**11**] for the basics), but the reader may assume for simplicity that G is a group.

Hence throughout this paper the letter G will stand for a Lie groupoid (which we also identify with the space of arrows) over the base smooth manifold M (the space of units). The source and target maps will be written as $s, t : G \longrightarrow M$. A representation up to homotopy of G consists of the following data:

1. A graded vector bundle E over M.
2. A differential ∂ on E; that is, a degree-one vector bundle morphism

$$\partial : E^{\bullet} \longrightarrow E^{\bullet+1}$$

with $\partial \circ \partial = 0$.

3. A smooth operator that associates to each $g \in G$ a chain map
$$\lambda_g : E_{s(g)} \longrightarrow E_{t(g)}, \ e \mapsto \lambda_g(e),$$
which we will refer to as the *quasi-action.* Here, *quasi* refers to the fact that it may fail to respect the composition operation.
4. A smooth operation that associates to each pair (g, h) of composable arrows a homotopy between $\lambda_g \lambda_h$ and λ_{gh}; i.e., a linear map that lowers the cochain degree by one,
$$R_2(g, h) : E_{s(h)} \longrightarrow E_{t(h)}$$
with the property that
$$\lambda_g \lambda_h - \lambda_{gh} = \partial(R_2(g, h)), \tag{2}$$
where the last expression makes use of the induced differential in the Hom-bundle:
$$\partial(R_2(g, h)) = [\partial, R_2(g, h)] = \partial \circ R_2(g, h) + R_2(g, h) \circ \partial.$$
5. Similar higher-order operations R_k, in which each R_k measures the failure of higher-coherence equations for $\partial, \lambda, R_2 \dots, R_{k-1}$. In order to have more uniform notation, we will often write $R_0 = \partial$, $R_1 = \lambda$.

For the precise definition, we recall that a string of k composable arrows is a k-tuple $(g_1, \dots, g_k) \in G^k$ of arrows satisfying $s(g_i) = t(g_{i+1})$ for $i = 1 \dots, k$.

Definition 2.1. *A* ***representation up to homotopy*** (E, R_k) *of a Lie groupoid* G *is a graded vector bundle* E *over the base* M*, together with a sequence of operations* R_k*,* $k \geqslant 0$*, where* R_k *associates to a string of* k*-composable arrows* $(g_1, \dots, g_k)$ *a linear map*
$$R_k(g_1, \dots, g_k) : E_{s(g_k)} \longrightarrow E_{t(g_1)},$$
of degree $1 - k$*, depending smoothly on the arguments and satisfying the equations*
$$\sum_{j=1}^{k-1} (-1)^j R_{k-1}(g_1, \dots, g_j g_{j+1}, \dots, g_k) = \sum_{j=0}^{k} (-1)^j R_j(g_1, \dots, g_j) \circ R_{k-j}(g_{j+1}, \dots, g_k). \tag{3}$$
The representation up to homotopy (E, R_k) *is said to be* ***unital*** *if the restriction of* R_1 *to the unit space* M *is the vector bundle identity map* id_E*, and if the higher components* R_k *vanish when one of the arguments is a groupoid unit.*

We will denote the vector bundle morphism R_0 by ∂^E or simply by ∂ when no confusion arises. We will say that (E, ∂) is the complex underlying the representation up to homotopy E, or that the operators $\{R_k\}_{k \geqslant 1}$ define a representation up to homotopy on the complex (E, ∂). With this notation, the equations above read:
$$\sum_{j=1}^{k-1} (-1)^j R_{k-1}(g_1, \dots, g_j g_{j+1}, \dots, g_k) + \sum_{j=1}^{k-1} (-1)^{j+1} R_j(g_1, \dots, g_j) \circ R_{k-j}(g_{j+1}, \dots, g_k)$$
$$= \partial \circ R_k(g_1, \dots, g_k) + (-1)^k R_k(g_1, \dots, g_k) \circ \partial.$$

We turn now to the definition of morphisms between representations up to homotopy:

Definition 2.2. *A* ***morphism*** *from a representation up to homotopy* (E, R_k) *to another one* $(E', R_k^{'})$ *is a sequence* $\Phi = \{\Phi_k\}_{k \geqslant 0}$*, where* Φ_k *is an operator that associates to a string of* k*-composable arrows* $(g_1, \ldots, g_k)$ *a linear map*

$$\Phi_k(g_1, \ldots, g_k) : E_{s(g_k)} \longrightarrow E'_{t(g_1)}$$

of degree $-k$*, depending smoothly on the arguments, such that*

$$\sum_{i+j=k}(-1)^j \Phi_j(g_1, \ldots, g_j) \circ R_i(g_{j+1}, \ldots, g_k) = \sum_{i+j=k} R_j^{'}(g_1, \ldots, g_j) \circ \Phi_i(g_{j+1}, \ldots, g_k) \quad (4)$$
$$+ \sum_{j=1}^{k-1}(-1)^j \Phi_{k-1}(g_1, \ldots, g_j g_{j+1}, \ldots, g_k).$$

The composition of morphisms is given by the formula

$$(\Phi \circ \Psi)_k(g_1, \ldots, g_k) = \sum_{i+j=k} \Phi_i(g_1, \ldots, g_i) \circ \Psi(g_{i+1}, \ldots, g_k).$$

We will denote by $\mathrm{Rep}^\infty(G)$ the resulting category of representations up to homotopy of G. Note that a morphism Φ is an isomorphism if and only if Φ_0 is an isomorphism of (graded) vector bundles. We will also need the following stronger notion of isomorphism.

Definition 2.3. *We say that two representations up to homotopy* (E, ∂, R_k) *and* $(E', \partial', R_k^{'})$ *are* ***strongly isomorphic*** *if* $E = E'$, $\partial = \partial'$ *and there exists a morphism* Φ *with* $\Phi_0 = Id_E$*. In this case,* Φ *will be called a strong isomorphism.*

There is also a natural notion of homotopy between morphisms:

Definition 2.4. *Let* Φ *and* Ψ *be morphisms of representation up to homotopy from* (E, R_k) *to* $(E', R_k^{'})$*. A* ***homotopy*** *between* Φ *and* Ψ *consists of a sequence* $h = \{h_k\}$*, where* h_k *is an operator that associates with a string of* k*-composable arrows* $(g_1, \ldots, g_k)$ *a linear map*

$$h_k(g_1, \ldots, g_k) : E_{s(g_k)} \longrightarrow E_{t(g_1)},$$

of degree $-k-1$ *depending smoothly on the arguments, and such that*

$$\begin{aligned}\Phi_k - \Psi_k = {} & \partial \circ h_k(g_1, \ldots, g_k) + (-1)^k h_k(g_1, \ldots, g_k) \circ \partial \\ & + \sum_{i=0}^{k-1}(-1)^i h_i(g_1, \ldots, g_i) \circ R_{k-i}(g_{i+1}, \ldots, g_k) \\ & + \sum_{i=1}^{k}(-1)^i R_i'(g_1, \ldots, g_i) \circ h_{k-i}(g_{i+1}, \ldots, g_k) \\ & + \sum_{i=1}^{k-1}(-1)^{i+1} h_{k-1}(g_1, \ldots, g_i g_{i+1}, \ldots, g_k).\end{aligned} \quad (5)$$

The composition is well defined on homotopy classes of morphisms. The **homotopy category** $\mathcal{D}(G)$ is defined as the category whose objects are representations up to homotopy and whose morphisms are homotopy classes of morphisms between representations up to homotopy.

Let us now describe the problem of constructing tensor products. Given two representations up to homotopy E and F, the tensor product $E \otimes F$ is defined, first of all, as a cochain complex of vector bundles over M with the standard differential

$$\partial(e \otimes f) = \partial(e) \otimes f + (-1)^p e \otimes \partial(f),$$

where p is the degree of e. The first step toward giving this complex the structure of a representation up to homotopy is to define the R_1-term. Thinking of it as a quasi-action, there is again a standard choice, the diagonal one:

$$\lambda_g(e \otimes f) = \lambda_g(e) \otimes \lambda_g(f).$$

However, for higher R's, the problem is more subtle. For instance, when looking for an R_2, we have to make sure that the equation (2) for $E \otimes F$ is satisfied. Already in this case the equation has more than one natural and interesting solution. For instance, if one is interested in a symmetric tensor product, then there is only one solution for R_2:

$$\begin{aligned} R_2(g,h)(e \otimes f) = \frac{1}{2}(&R_2(g,h)(e) \otimes \lambda(gh)(f) + R_2(g,h)(e) \otimes (\lambda(g) \circ \lambda(h))(f) \\ &+\lambda(gh)(e) \otimes R_2(g,h)(f) + (\lambda(g) \circ \lambda(h))(e) \otimes R_2(g,h)(f)). \end{aligned}$$

On the other hand, this specific second component would not work if we wanted the tensor product to be associative, in which case we could choose, for instance, the second component to be

$$R_2(g,h)(e \otimes f) = R_2(g,h)(e) \otimes \lambda(gh)(f) + (\lambda(g) \circ \lambda(h))(e) \otimes R_2(g,h)(f)).$$

For higher values of k, the equations become much more involved. The aim of this paper is to understand the algebraic structure that governs representations up to homotopy, and to use it to classify all possible tensor products of representations up to homotopy and of morphisms between them.

3. Maurer-Cartan elements and DB-algebras

In this section we discuss the algebraic structures that are relevant to the construction of tensor products. First we interpret representations up to homotopy as Maurer-Cartan elements in a certain DGA (Differential Graded Algebra) and then we describe the building pieces of the DGAs involved. This underlying algebraic structure is important when tensoring two representations up to homotopy, and we axiomatize it under the name of DB-algebra (Differential Bar algebra). Hence the main outcome is the construction of functors

$$\{\text{Complexes of vector bundles } (E, \partial) \text{ over } M\} \longrightarrow \{\text{DB-algebras}\} \longrightarrow \{\text{Complete DGAs}\},$$

$$E \mapsto \mathcal{A}_E \mapsto \bar{\mathcal{A}}_E,$$

so that representations up to homotopy on E correspond to Maurer-Cartan elements in $\bar{\mathcal{A}}_E$ and the resulting composition functor behaves well with respect to tensor products. Moreover, the notion of strong isomorphism on the left hand side corresponds to the notion of gauge equivalence between Maurer-Cartan elements on the right hand side. For the general notion of complete DGA's, Maurer-Cartan elements and gauge equivalences, we refer the reader to the appendix.

3.1. Representations up to homotopy as Maurer-Cartan elements

We start by constructing $\bar{\mathcal{A}}_E$. Let G_k be the submanifold of G^k consisting of strings of k composable arrows of G. By convention, $G_0 = M$. We will also denote by s and t the maps $G_k \to M$ given by $s(g_1, \ldots, g_k) = s(g_k)$ and $t(g_1, \ldots, g_k) = t(g_1)$. For a graded vector bundle E over M, we consider the pull-back bundles s^*E and t^*E to G_k, and we form the graded Hom-bundle $\mathrm{Hom}(s^*E, t^*E)$ over G_k. Recall that $\phi : s^*E \longrightarrow t^*E$ has degree l if it maps $s^*(E^\bullet)$ to $t^*(E^{\bullet+l})$. We will consider the resulting spaces of sections

$$\mathcal{A}_E^k(l) := \Gamma(G_k, \mathrm{Hom}^l(s^*E, t^*E)). \tag{6}$$

For $c \in \mathcal{A}_E^k(l)$, we write

$$k(c) = k, \; l(c) = l, \; |c| = k(c) + l(c),$$

and we call $|c|$ the total degree of c. All these spaces together define a bigraded algebra, with the product $\star$ that associates to $c \in \mathcal{A}_E^k(l)$ and $c' \in \mathcal{A}_E^{k'}(l')$ the element $c \star c' \in \mathcal{A}_E^{k+k'}(l + l')$, given by

$$(c \star c')(g_1, \ldots, g_{k+k'}) = (-1)^{k(k'+l')} c(g_1, \ldots, g_k) \circ c'(g_{k+1}, \ldots, g_{k+k'}). \tag{7}$$

When E is a cochain complex, then so is the Hom-bundle, with the differential

$$\partial(\phi) = [\partial, \phi] = \partial \circ \phi - (-1)^l \phi \circ \partial, \tag{8}$$

for $\phi \in \mathrm{Hom}^l(s^*E, t^*E)$. This defines a differential

$$\partial : \mathcal{A}_E^k(l) \longrightarrow \mathcal{A}_E^k(l + 1), \tag{9}$$

induced by the differential of E. On the other hand, the groupoid structure induces a differential along the other degree:

$$d : \mathcal{A}_E^k(l) \longrightarrow \mathcal{A}_E^{k+1}(l),$$

$$d(c)(g_1, \ldots, g_{k+1}) = \sum_{j=1}^{k} (-1)^j c(g_1, \ldots, g_j g_{j+1}, \ldots, g_{k+1}).$$

We denote by $\bar{\mathcal{A}}_E$ the DGA that, in degree n, is given by

$$\bar{\mathcal{A}}_E^n := \Pi_{k+l=n} \mathcal{A}_E^k(l)$$

and whose elements should be thought of as infinite sums $\gamma_0 + \gamma_1 + \gamma_2 + \cdots$ of homogeneous elements $\gamma_i \in \bar{\mathcal{A}}_E^i$, $i \geqslant 0$. The product $\star$ and the total differential

$$d_{\mathrm{tot}}(c) := \partial(c) + (-1)^n d(c)$$

give $\bar{\mathcal{A}}_E$ the structure of a DGA. The signs are chosen so that the differential is a derivation with respect to $\star$. This DGA is a *complete* DGA, in the sense of the appendix, with the filtration:

$$F_p\bar{\mathcal{A}}_E := \Big\{\gamma = \gamma_0 + \gamma_1 + \ldots \in \bar{\mathcal{A}}_E : \gamma_0 = \ldots = \gamma_{p-1} = 0\Big\}.$$

Note that the structure of $\bar{\mathcal{A}}_E$ depends on the differential ∂ and not only on the vector bundle E. The formulas that appear in the definition of representations up to homotopy and of morphisms between them take now the following more compact form, which follows by a direct computation.

Proposition 3.1. *Let G be a Lie groupoid over a manifold M and let (E, ∂) a cochain complex of vector bundles over M. Also, let $\{R_k\}_{k\geqslant 1}$ be a sequence of operators such that $R_k \in \mathcal{A}^k_E(1-k)$. Then, (E, ∂, R_k) is a representation up to homotopy of G if and only if*

$$R_E := R_1 + R_2 + \ldots \in \bar{\mathcal{A}}_E$$

is a Maurer-Cartan element for $\bar{\mathcal{A}}_E$. Moreover, for two such sets of operations $\{R_k\}_{k\geqslant 1}$ and $\{R'_k\}_{k\geqslant 1}$, there is a one-to-one correspondence between:

1. *strong isomorphisms between (E, ∂, R_k) and (E, ∂, R'_k) (Definition 2.3), and*
2. *strong gauge equivalences between the Maurer-Cartan elements $R_E, R'_E \in \bar{\mathcal{A}}_E$ (Definition A.5).*

3.2. DB-algebras

The description of representations up to homotopy in terms of Maurer-Cartan elements is still not very useful when it comes to constructing tensor products. The reason is very simple: given two cochain complexes E and F, the DGA $\bar{\mathcal{A}}_{E\otimes F}$ is not directly related to the tensor product of the DGAs $\bar{\mathcal{A}}_E$ and $\bar{\mathcal{A}}_F$. Looking at the differential of $\bar{\mathcal{A}}_{E\otimes F}$ it becomes clear that there is more structure present in $\bar{\mathcal{A}}_E$ and $\bar{\mathcal{A}}_F$ then just that of DGA. This brings us to the notion of a DB-algebra.

Definition 3.2. *A differential bar-algebra, or* ***DB-algebra****, is a bigraded vector space*

$$\mathcal{A} = \bigoplus_{k\geqslant 0, l\in\mathbb{Z}} \mathcal{A}^k(l)$$

together with:

- *A structure of bigraded associative algebra with the product*

  $$\circ : \mathcal{A}^k(l) \otimes \mathcal{A}^{k'}(l') \longrightarrow \mathcal{A}^{k+k'}(l+l').$$

  For $a \in \mathcal{A}^k(l)$, we write $k(a) = k$, $l(a) = l$, and we define the total degree $|a| = k(a) + l(a)$.
- *A derivation of bidegree $(1, 0)$; i.e., a linear map*

  $$\partial : \mathcal{A}^k(l) \longrightarrow \mathcal{A}^k(l+1)$$

  that satisfies

  $$\partial(a \circ b) = \partial(a) \circ b + (-1)^{l(a)} a \circ \partial b. \qquad (10)$$

- *For each $k \geqslant 1$, there are linear maps*

$$d_i : \mathcal{A}^k(l) \longrightarrow \mathcal{A}^{k+1}(l), \quad i = 1, \dots, k,$$

commuting with ∂ and satisfying

$$d_j d_i = d_i d_{j-1}, \quad \text{if } i < j,$$

and, for $a \in \mathcal{A}^k(l)$,

$$d_i(a \circ b) = \begin{cases} d_i(a) \circ b, & k \geqslant i \\ a \circ d_{i-k}(b), & k < i. \end{cases} \tag{11}$$

A morphism between two DB-algebras is a linear map that preserves both degrees and commutes with all the structure maps. We denote by $\underline{\mathcal{DB}ar}$ the resulting category.

For a general DB-algebra $\mathcal{A}$ we introduce the operators

$$d = \sum_{i=1}^{k} (-1)^i d_i : \mathcal{A}^k(l) \longrightarrow \mathcal{A}^{k+1}(l). \tag{12}$$

From the axioms, it follows that d is a biderivation with respect to $\circ$.

Lemma 3.3. *Let $\mathcal{A}$ be a DB-algebra. Then $\mathcal{A}$, together with the total grading, the signed product*

$$a \star b = (-1)^{k(a)|b|} a \circ b \tag{13}$$

and the total differential

$$d_{tot} = \partial + (-1)^n d : \mathcal{A}^n \longrightarrow \mathcal{A}^{n+1},$$

is a DGA.

Definition 3.4. *Given a DB-algebra $\mathcal{A}$, we denote by $\bar{\mathcal{A}}$ the completion of $\mathcal{A}$ with respect to the filtration by the k-degree. In other words, $\bar{\mathcal{A}}$ is the DGA with*

$$\bar{\mathcal{A}}^n = \prod_{k+l=n} \mathcal{A}^k(l),$$

endowed with $\star$ and d_{tot}. The elements $a \in \bar{\mathcal{A}}^n$ will be written as infinite sums

$$a = a_0 + a_1 + \dots, \quad \text{with } a_k \in \mathcal{A}^k(n-k), \tag{14}$$

and we call a_k the k-th component of a. This construction defines a functor

$$\overline{K} : \underline{\mathcal{DB}ar} \longrightarrow \overline{DGA}$$

from the category of DB-algebras to the category of complete DGAs.

Example 3.5. It is clear now that the DGA $(\bar{\mathcal{A}}_E, d_{tot}, \star)$ from the previous subsection comes from a DB-algebra $(\mathcal{A}_E, \partial, \circ, d_i)$:

- the underlying bigraded space is $\bigoplus_{k \geqslant 0,\, l \in \mathbb{Z}} \mathcal{A}_E^k(l)$,
- the product $\circ$ is the unsigned version of $\star$:

$$(c \circ c')(g_1, \dots, g_{k+k'}) = c(g_1, \dots, g_k) \circ c'(g_{k+1}, \dots, g_{k+k'}), \tag{15}$$

- the differential ∂ is the Hom-bundle differential defined in (9), and
- the operators d_i are given by the formulas:

$$d_i(c)(g_1, \ldots, g_{k+1}) = c(g_1, \ldots, g_i g_{i+1}, \ldots, g_{k+1}).$$

3.3. The tensor product of DB-algebras

The category $\underline{\mathcal{DB}ar}$ has a natural tensor product operation that will be denoted by $\boxtimes$. Given two DB-algebras $\mathcal{A}$ and $\mathcal{B}$, their tensor product $\mathcal{A} \boxtimes \mathcal{B}$ is defined as follows. As a bigraded vector space,

$$(\mathcal{A} \boxtimes \mathcal{B})^k(l) = \bigoplus_{i+j=l} \mathcal{A}^k(i) \otimes \mathcal{B}^k(j).$$

For $a \in \mathcal{A}^k(i)$ and $b \in \mathcal{B}^k(j)$, we will denote by $a \boxtimes b$ the resulting tensor in $\mathcal{A} \boxtimes \mathcal{B}$. The differential ∂ and the operators d_i are given by

$$\partial(a \boxtimes b) = \partial(a) \boxtimes b + (-1)^{l(a)} a \boxtimes \partial(b), \ d_i(a \boxtimes b) = d_i(a) \boxtimes d_i(b),$$

while the multiplication $\circ$ by

$$(a \boxtimes b) \circ (a' \boxtimes b') = (-1)^{l(b)l(a')} (a \circ a') \boxtimes (b \circ b').$$

The previous definition is designed so that the construction $E \mapsto \mathcal{A}_E$ behaves well with respect to tensor products.

Proposition 3.6. *For any two complexes of vector bundles E and F over M, the canonical map*

$$m_{E,F} : \mathcal{A}_E \boxtimes \mathcal{A}_F \longrightarrow \mathcal{A}_{E \otimes F},$$

$$m_{E,F}(c \boxtimes c')(g_1, \ldots, g_k) = c(g_1, \ldots, g_k) \otimes c'(g_1, \ldots, g_k)$$

is a morphism of DB-algebras.

For later use we mention here that, for any DB-algebra $\mathcal{A}$, there is a natural action of the group S_m on $\mathcal{A}^{\boxtimes m}$. For $\sigma \in S_m$, the associated automorphism of $\mathcal{A}^{\boxtimes m}$ is denoted by $\hat{\sigma}$. To define $\hat{\sigma}$, it suffices to describe it when $\sigma = \tau_{i,i+1}$ is a transposition that interchanges the positions i and $i+1$; in this case:

$$\hat{\sigma}(a_1 \boxtimes \ldots \boxtimes a_m) = (-1)^{ll'} a_1 \boxtimes \ldots \boxtimes a_{i-1} \boxtimes a_{i+1} \boxtimes a_i \boxtimes \ldots \boxtimes a_m,$$

for $a_i \in \mathcal{A}^k(l)$, $a_{i+1} \in \mathcal{A}^{k'}(l')$. It is not difficult to see that this defines an action of S_m on $\mathcal{A}^{\boxtimes m}$ by automorphisms of DB-algebras.

4. The Maurer-Cartan DB-algebra

As explained in Proposition 3.1, representations up to homotopy structures on a complex of vector bundles correspond to Maurer-Cartan elements on the associated DGA. This observation allows one to translate the problem of constructing tensor products of representations up to homotopy to that of constructing Maurer-Cartan elements on the DGAs associated to tensor products of DB-algebras. Clearly, this problem can be treated at a universal level. This brings us to the Maurer-Cartan DB-algebra, which is the universal DB-algebra for Maurer-Cartan elements.

Definition 4.1. *For a DB-algebra $\mathcal{A}$, we denote by $MC_1(\bar{\mathcal{A}})$ the set of Maurer-Cartan elements of $\bar{\mathcal{A}}$ whose zeroth component vanishes. A* ***Maurer-Cartan algebra*** *is a DB-algebra Ω, together with a Maurer-Cartan element $L \in MC_1(\bar{\Omega})$ with the property that for any DB-algebra $\mathcal{A}$, the map*

$$Hom_{\underline{\mathcal{DB}ar}}(\Omega, \mathcal{A}) \longrightarrow MC_1(\bar{\mathcal{A}}), \phi \mapsto \phi(L)$$

is a bijection.

Theorem 4.2. *The Maurer-Cartan DB-algebra exists and is unique up to isomorphisms of DB-algebras. Moreover, for each k, $H^l(\Omega^k(\bullet), \partial) = 0$ for all $l \neq 0$.*

The uniqueness follows by standard arguments. The aim of this section is to provide several explicit descriptions of Ω, proving in particular the theorem above. The main conclusion of this section is the resulting reformulation of the notion of representation up to homotopy in terms of Ω:

Corollary 4.3. *Given a Lie groupoid G over M and a complex of vector bundles (E, ∂), there is a 1-1 correspondence between sequences of operations $R = \{R_k\}_{k \geqslant 1}$ making (E, ∂, R_k) into a representation up to homotopy of G and morphisms of DB-algebras*

$$k_{E,R} : \Omega \longrightarrow \mathcal{A}_E.$$

The map $k_{E,R}$, also denoted k_E, will be called the **characteristic map** of the representation up to homotopy (E, R).

4.1. Description in terms of trees

In our construction of Ω, instead of proceeding abstractly and use generators and relations, we follow a pictorial approach. We start by explaining the main idea. Due to the expected universal property of Ω, a representation up to homotopy (E, R_E) is represented by its characteristic map $k_E : \Omega \longrightarrow \mathcal{A}_E$- uniquely determined by the fact that it sends the component L_n of L to the operation R_E^n. Hence general elements A of Ω should encode certain operations R_E^A on E which arise by combining all the given operations R_E^n. There are various such operations one can think of. For instance, one has the following:

$$(g_1, g_2, g_3, g_4, g_5, g_6) \mapsto R_E^2(g_1, g_2 g_3) \circ R_E^1(g_4 g_5) \circ R_E^1(g_6). \tag{16}$$

The idea is to encode such operations graphically, by forests of height two. For instance, the operation above is encoded by:

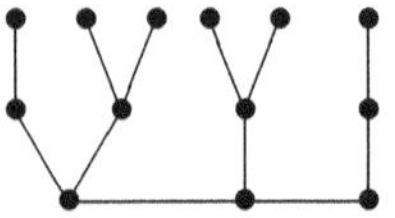

and one should think of the six leaves as labelled by the six elements $g_1, \ldots, g_6$.

Definition 4.4. *We denote by* **T** *the set of isomorphism classes of planar rooted trees whose leaves all have height 2. We denote by* **S** *the set of short forests; that is, the set of finite tuples $(T_1, \ldots, T_n)$ of trees in* **T**.

We represent a short forest by joining the roots of the T_i's by a horizontal line. For instance,

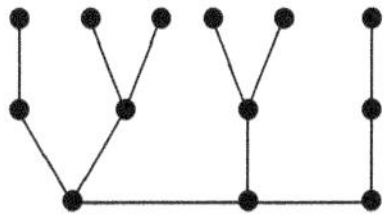

stands for

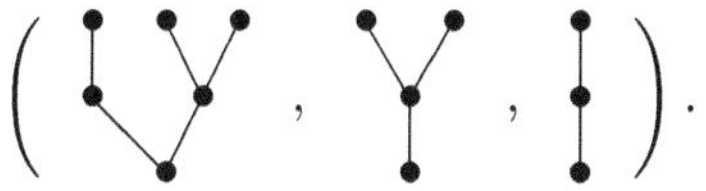

Next, we introduce a bigrading on $\mathbf{S}$.

Definition 4.5. *A branch of a short forest $F \in \mathbf{S}$ is an edge that goes from a root to a vertex that is not a root. For any short forest F, we define*

$$\begin{aligned} k(F) &= \#\ \textit{of leaves of } F, \ \textit{called the order of } F, \\ b(F) &= \#\ \textit{of branches of } F, \\ r(F) &= \#\ \textit{of roots of } F, \\ l(F) &= r(F) - b(F), \ \textit{called the degree of } F. \end{aligned}$$

We denote by $\mathbf{S}^k(l)$ the set of short forests of order k and degree l.

Example 4.6. For the tree mentioned above,

$F =$ 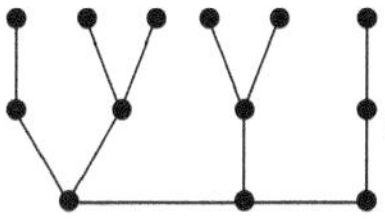 , $k(F) = 6$, $b(F) = 4$, $r(F) = 3$, $l(F) = -1$.

The fact that $F \in S^6(-1)$ corresponds to the fact that the operation (16) belongs to $\mathcal{A}_E^6(-1)$.

Definition 4.7. *We denote by $(\mathbf{\Omega}, \circ)$ the free algebra over $\mathbb{R}$ generated by the trees in $\mathbf{T}$ or, equivalently, the linear span over $\mathbb{R}$ of $\mathbf{S}$ with the product given by the concatenation.*

Pictorially, $F \circ F'$ is the forest obtained by joining the roots by an edge, as in the following example:

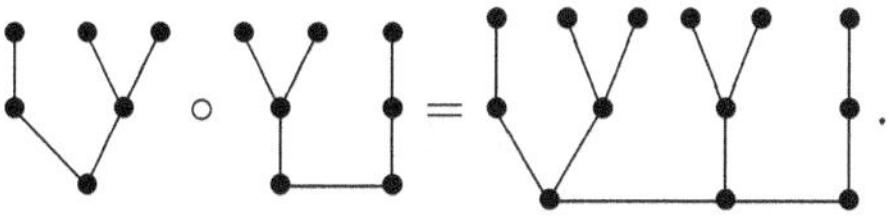

The bigrading on $\mathbf{S}$ induces a similar bigrading on $\mathbf{\Omega}$ and allows us to talk about the spaces $\mathbf{\Omega}^k(l)$.

Definition 4.8. *For each $i = 1, \ldots, k$, we define the operator*

$$d_i : \mathbf{\Omega}^k(\bullet) \to \mathbf{\Omega}^{k+1}(\bullet),$$

which acts by replacing the i-th leaf of a forest, counting from the left, by two leaves.

For each l, we define $\partial : \mathbf{\Omega}^\bullet(l) \to \mathbf{\Omega}^\bullet(l+1)$ by

$$\partial(F) = \sum_{j=1}^{l} (-1)^{j+1} (\partial_j^1 F - \partial_j^0 F), \tag{17}$$

where $\partial_j^1 F$ is obtained by separating the jth pair of adjacent branches (counted from left to right) and $\partial_j^0 F$ by collapsing the jth pair of adjacent branches (counted from left to right).

Finally, we denote by L_n the tree in Ω that has one root, n branches, and n leaves:

$$L_1 = \quad, \quad L_2 = \quad, \quad L_3 = \quad, \quad L_4 = \quad \ldots,$$

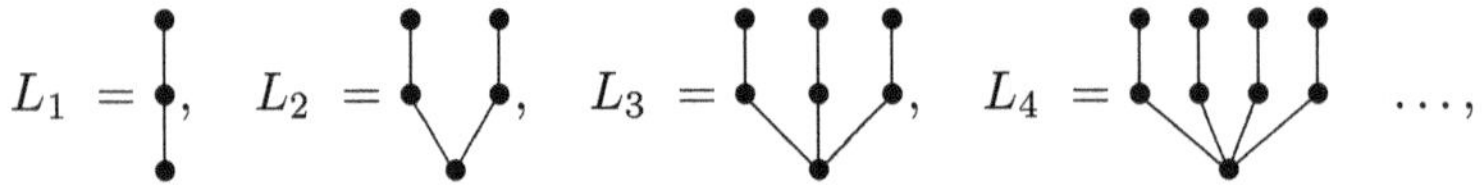

and we set

$$L := L_1 + L_2 + \ldots \in \bar{\Omega}.$$

Example 4.9. Here is an example of the action, on the short forests, of the d_i operators,

$$d_2 \left(\right) = $$

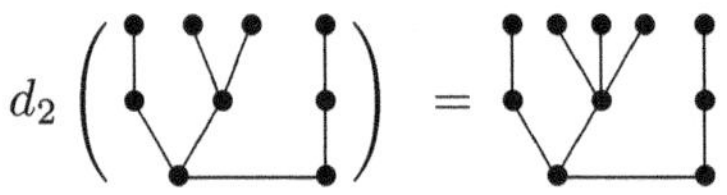

and, here, of the differential,

$$\partial \left(\right) = \left(- \right) - \left(- \right).$$

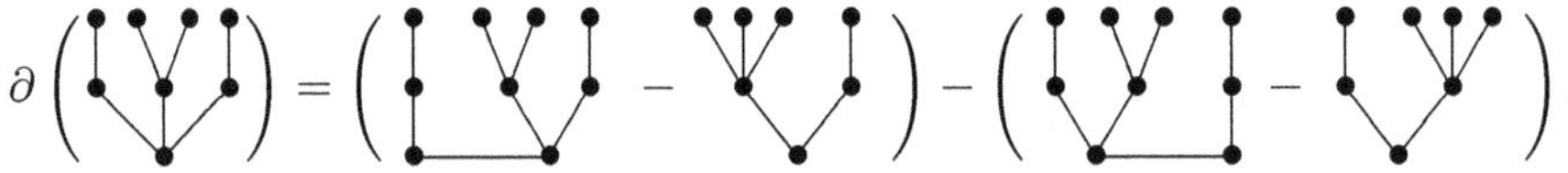

4.2. Description in terms of words in three letters

The set $\mathbf{S}^{k+1}$ of short forests with $k + 1$ leaves can be naturally identified with the set of words in three letters $\{a, b, c\}$ of length k as follows. To a short forest F with $k + 1$ leaves $l_1, \ldots, l_{k+1}$, numbered from left to right, we associate the word $e_1 \ldots e_k$, where:

$$e_i = \begin{cases} a & \text{if } l_i \text{ and } l_{i+1} \text{ belong to the same branch,} \\ b & \text{if } l_i \text{ and } l_{i+1} \text{ belong to different branches of the same tree,} \\ c & \text{if } l_i \text{ and } l_{i+1} \text{ belong to different trees.} \end{cases}$$

The unique short forest in $S(1)$,

$$e := \quad,$$

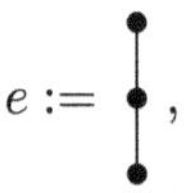

corresponds to the empty word. The following figure illustrates this correspondence:

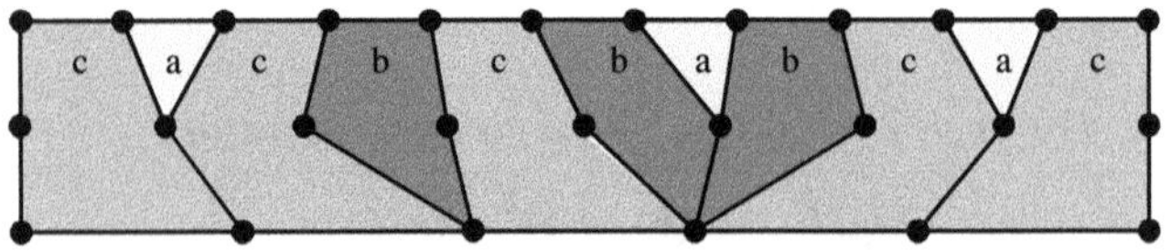

This construction identifies Ω, as a vector space, with the free unital algebra $F_e\langle a, b, c \rangle$ on the generators a, b and c with unit e. In terms of short forests:

$$a = \quad, \quad b = \quad, \quad c = \quad, \quad e = \quad.$$

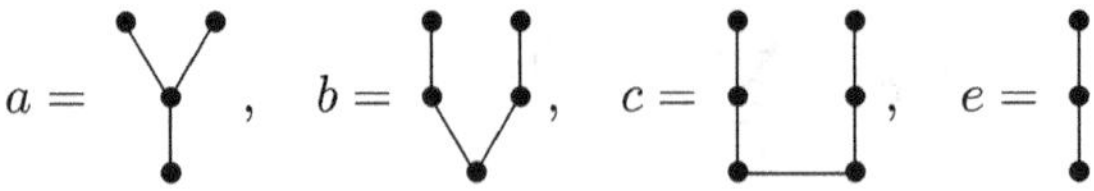

The natural product coming from the concatenation of words will be denoted by $\diamond$. Note that $\diamond$ does not coincide with the product $\circ$ defined on Ω. In terms of forests, $F \diamond F'$ is obtained by identifying the rightmost branch and leaf of F with the leftmost branch and leaf of F', as in the following example:

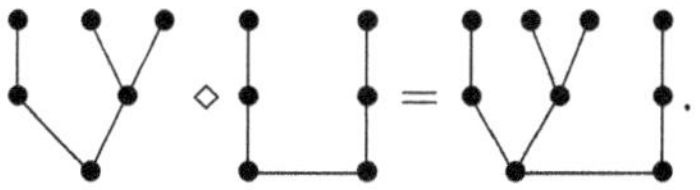

In terms of words on three letters, the algebraic structure of Ω has the following description:

- The product on Ω is given by:

$$T \circ T' = T \diamond c \diamond T'.$$

- The operator d_i acts according to the formula:

$$d_i(e_1 \dots e_k) = e_1 \dots e_{i-1} a e_i \dots e_k.$$

- The operator ∂ is the unique derivation with respect to $\diamond$; i.e., with the property that

$$\partial(F \diamond F') = \partial(F) \diamond F' + (-1)^{l(F)} F \diamond \partial(F'), \tag{18}$$

  given on generators by

$$\partial\,(e) = \partial\,(a) = \partial\,(c) = 0 \text{and} \partial(b) = c - a. \tag{19}$$

- $L_n = b^{\diamond n-1}$ for $n \geqslant 2$, and $L_1 = e$.

The only statement in the list above that requires a proof is the derivation property (18). Denote by $\hat{\partial}$ the operator defined on the generators as in (19) and extended by derivation as in (18). We need to prove that $\hat{\partial} = \partial$. Consider a word $T = e_1 \diamond \dots \diamond e_k$ with $e_i \in \{a, b, c\}$; then:

$$\hat{\partial}(T) = \sum_{e_i = b} (-1)^{l(w_i)} e_1 \diamond \dots \diamond e_{i-1} \diamond (c - a) \diamond e_{i+1} \diamond \dots \diamond e_k, \tag{20}$$

where $l(w_i)$ is the degree of $w_i = e_1 \diamond \dots \diamond e_{i-1}$. On the other hand,

$$e_1 \diamond \dots \diamond e_{i-1} \diamond c \diamond e_{i+1} \diamond \dots \diamond e_k = \partial_j^0(T),$$

$$e_1 \diamond \dots \diamond e_{i-1} \diamond a \diamond e_{i+1} \diamond \dots \diamond e_k = \partial_j^1(T).$$

Thus, from the formula in Definition 4.8, we conclude that $\hat{\partial} = \partial$.

4.3. Description in terms of faces of cubes

The DB-algebra Ω can also be constructed in terms of the faces of cubes. Namely, the set of words in three letters can be naturally identified with the set of faces of the geometric cubes $[0, 1]^k \subset \mathbb{R}^k$ as follows. To a word $F = e_1 \dots e_k$, we associate the face

$$\phi(F) = \psi(e_1) \times \dots \times \psi(e_k) \subset [0, 1]^k,$$

with the convention that

$$\psi(e)=\begin{cases}\{0\} & \text{if } e=a,\\ \{1\} & \text{if } e=c,\\ [0,1] & \text{if } e=b.\end{cases}$$

This simply says that the cells of the cube I^k are products of the cells of the interval, which we label as follows: $\{0\}=a,(0,1)=b$ and $\{1\}=c$. Thus, we identify the cells of I^k with words of length k in the letters $\{a,b,c\}$. Note that the dimension of a cell is the number of times that b appears in the corresponding word. The following figure illustrates this bijection:

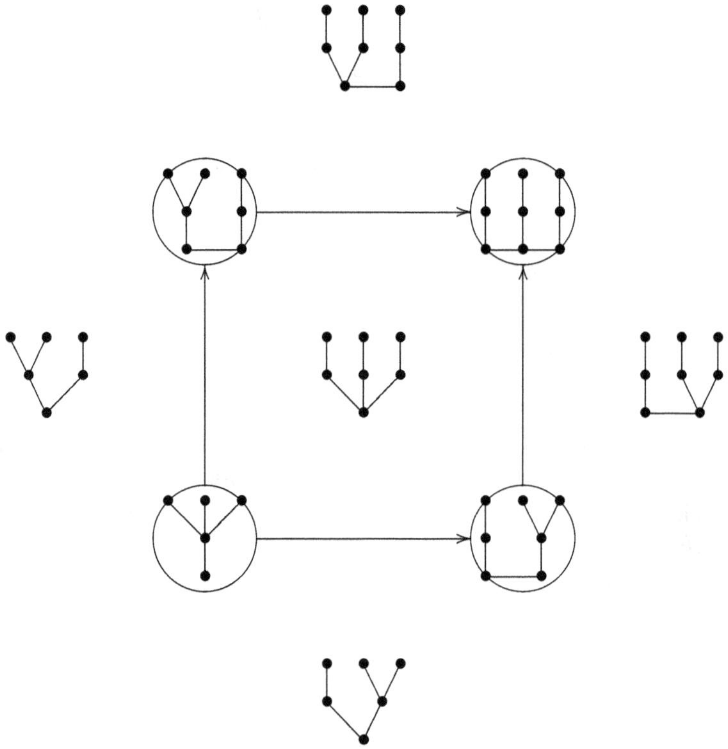

In this correspondence, a forest of degree $-l$ with $k+1$ leaves is sent to a l-dimensional face of the k-dimensional cube. Thus, we obtain the identification

$$(\Omega^{k+1}(\bullet),\partial)\cong(C_\bullet(I^k),\partial),\ k\geqslant 0, \tag{21}$$

where $C_\bullet(I^k)$ is the cellular chain complex computing the homology of the k-dimensional cube with respect to the natural cell decomposition and negative grading of the cells. Also, one easily shows that:

- The product $F_1\diamond F_2$ corresponds to the Cartesian product of cells $\phi(F_1)\times\phi(F_2)$. This also shows that the product $F_1\circ F_2$ corresponds to the operation $\phi(F_1)\times\{1\}\times\phi(F_2)$.

- The operator ∂ corresponds to the boundary operator in $C(I^k)$.
- L_k corresponds to the highest degree cell in I^{k-1}.
- The operators d_i correspond to the various ways of embedding the k-cube into the $(k+1)$-cube as a k-face having the origin as one of its vertices. This is illustrated in the figure below:

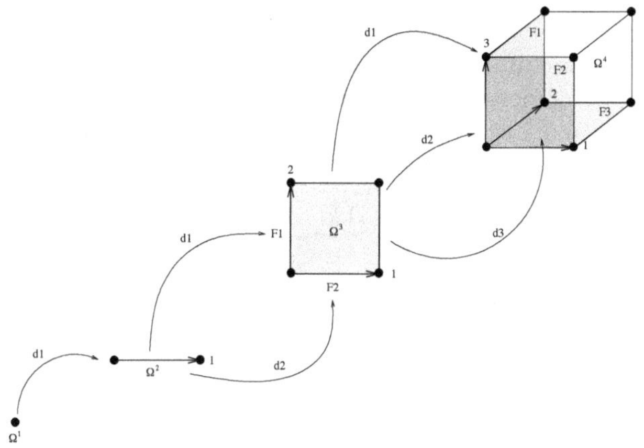

4.4. Proof of Theorem 4.2

We will now prove that $(\mathbf{\Omega}, \circ, d_i, \partial)$, together with

$$L = L_1 + L_2 + \dots,$$

satisfies the universal property of the Maurer-Cartan DB-algebra. The fact that $\mathbf{\Omega}$ is a DB-algebra is quite straightforward now. For instance, to check that $\partial^2 = 0$, one uses the description of ∂ as a derivation with respect to $\diamond$, and one is left with checking this equation on the elements a, b and c. We should also prove that ∂ is a derivation with respect to the product $\circ$. Using the expression of $\circ$ in terms of $\diamond$, we find:

$$\begin{aligned}\partial(T \circ T') &= \partial(T \diamond c \diamond T') = \partial(T) \diamond c \diamond T' + (-1)^{l(T)} \partial(c \diamond T') \\ &= \partial(T) \diamond c \diamond T' + (-1)^{l(T)} c \diamond \partial(T') = \partial(T) \circ T' + (-1)^{l(T)} \circ \partial(T').\end{aligned}$$

Next, to see that L is a Maurer-Cartan element, one has to show that, for each k,

$$\partial(L_k) = \sum_{j=1}^{k-1} (-1)^{j+1} L_j \circ L_{k-j} + \sum_{j=1}^{k-1} (-1)^j d_j(L_{k-1}),$$

which follows immediately from:

$$\begin{aligned}\partial_j^1 L_k &= b^{\diamond j} \diamond c \diamond b^{\diamond(k-j)} = L_j \circ L_{k-j}, \\ \partial_j^0 L_k &= b^{\diamond j} \diamond a \diamond b^{\diamond(k-j)} = d_j L_{k-1}.\end{aligned}$$

For the universality property of $(\mathbf{\Omega}, L)$ it is enough to remark that every forest $F \in \mathbf{\Omega}$ can be written uniquely as

$$F = d_{i_1} \dots d_{i_m}(L_{k_1} \circ \dots \circ L_{k_s}), \tag{22}$$

with $i_1 > \cdots > i_m$.

Finally, the statement about the cohomology follows from the identification with the cellular complexes of the cubes.

5. Tensor products of representations up to homotopy

The main conclusion of the previous section is that, with Ω at hand, representations up to homotopy are characterized by their classifying maps (see Corollary 4.3). With this point of view, the construction of tensor products of representations up to homotopy amounts to the construction of diagonal maps on the DB-algebra Ω. For each $m > 0$, we consider the DB-algebra

$$\Omega_m = \underbrace{\Omega \boxtimes \ldots \boxtimes \Omega}_{m-\text{times}},$$

as well as the associated complete DGA $\bar{\Omega}_m$. In what follows, $e \in \Omega$ stands for the component L_1 of the universal Maurer-Cartan element L.

Definition 5.1. *A **universal Maurer-Cartan element** of length m is any Maurer-Cartan element ω of $\bar{\Omega}_m$ with the property that its degree 1 component is*

$$\omega_1 = e^{\boxtimes m} \tag{23}$$

We denote by $\mathcal{MC}_m$ the set of such Maurer-Cartan elements.

Due to the universal property of Ω, elements $\omega \in \mathcal{MC}_m$ can be identified with morphisms in $\underline{\mathcal{DB}ar}$

$$\Delta_\omega : \Omega \longrightarrow \Omega_m,$$

such that $\Delta_\omega(e) = e^{\boxtimes m}$. This last condition will allow us to recover the usual diagonal tensor product of (strict) representations.

Definition 5.2. *A universal Maurer-Cartan element ω is said to be*

- ***symmetric** if $\hat{\sigma}(\omega) = \omega$ for all $\sigma \in S_m$ (for the action of S_m, see subsection 3.3),*
- ***associative** when $m = 2$ and the induced map $\Delta_\omega : \Omega \longrightarrow \Omega \boxtimes \Omega$ is coassociative.*

Coming back to representations up to homotopy, it is now clear that universal Maurer-Cartan elements induce tensor product operations. For instance, using the DB-morphisms Δ_ω and $m_{E,F} : \mathcal{A}_E \boxtimes \mathcal{A}_F \longrightarrow \mathcal{A}_{E\otimes F}$ of Proposition 3.6, we now define:

Definition 5.3. *Given $\omega \in \mathcal{MC}_2$ and two representations up to homotopy E and F, we define $E \otimes_\omega F$ as the representation up to homotopy with the characteristic map*

$$k_{E\otimes_\omega F} = m_{E,F} \circ (k_E \boxtimes k_F) \circ \Delta_\omega.$$

From the second part of Proposition 3.1 and the naturality of the construction we deduce the following:

Corollary 5.4. *The operations $\otimes_\omega$ have the following properties:*

(1) Any strong gauge equivalence between $\omega, \omega' \in \mathcal{MC}_2$ induces a strong isomorphism between $E \otimes_\omega F$ and $E \otimes_{\omega'} F$ (see Definitions 2.3 and A.5).

(2) If ω is associative or symmetric, then so is the operation $\otimes_\omega$.

Similarly, any universal Maurer-Cartan element ω of length m induces a tensor product operation on m-arguments $\otimes_\omega(E_1, \ldots, E_m)$, defined by

$$k_{\otimes_\omega(E_1,\ldots,E_m)} = m_{E_1,\ldots,E_m} \circ (k_{E_1} \boxtimes \ldots \boxtimes k_{E_m}) \circ \Delta_\omega.$$

As before, a gauge equivalence between Maurer-Cartan elements induces a strong isomorphism between the corresponding tensor products. Also, if ω is symmetric, so is the associated tensor product.

Corollary 5.5. *A symmetric universal Maurer-Cartan element ω of length m induces a symmetric power operation $E \mapsto S^m E$ on representations up to homotopy.*

In this section we study the universal Maurer-Cartan elements. First of all, we clarify their existence and uniqueness.

Theorem 5.6. *For each $m > 0$, we have the following:*

1. *Symmetric universal Maurer-Cartan elements of length m exist.*
2. *Any two universal Maurer-Cartan elements of length m (symmetric or not) are strongly gauge equivalent, and any two gauge equivalences are homotopic.*

Hence the resulting tensor product and symmetric power operations are uniquely defined up to strong isomorphisms. However, as the next theorem shows, the tensor product operation does not posses all the properties one would hope for. Namely:

Corollary 5.7. *For $m = 2$,*

1. *There exist universal Maurer-Cartan elements that are associative.*
2. *There is no universal Maurer-Cartan element that is both associative and symmetric.*

In this section we also discuss a more special class of universal Maurer-Cartan elements, called rigid, which behave well with respect to the additional product $\diamond$ on Ω. One advantage of the rigid Maurer-Cartan elements is that they can be described completely. More importantly, the resulting tensor products preserve the unitality. To describe them, we begin by extending the product $\diamond$ from Ω to Ω_m by the usual formula

$$(a_1 \boxtimes a_2) \diamond (b_1 \boxtimes b_2) = (-1)^{l(a_2)l(b_1)}(a_1 \diamond b_1) \boxtimes (a_2 \diamond b_2),$$

giving $(\Omega_m, \diamond)$ the structure of a unital algebra with unit $e^{\boxtimes m}$.

Definition 5.8. *A universal Maurer-Cartan element $\omega \in \mathcal{MC}_m$ is called* ***rigid*** *if its characteristic map $\Delta_\omega : \Omega \to \Omega_m$ is a map of unital algebras with respect to the product $\diamond$.*

Here are the main properties of the rigid Maurer-Cartan elements.

Theorem 5.9. *For each $m \geqslant 2$,*

1. *The set of all rigid Maurer-Cartan elements $\omega \in \mathcal{MC}_m$ is in 1-1 correspondence with the set $\mathcal{R}_m$ of elements $x \in \Omega^2_m(-1)$ that satisfy the following equation*
$$\partial(x) = c^{\boxtimes m} - a^{\boxtimes m}. \tag{24}$$
The correspondence is characterized by
$$\omega = e^{\boxtimes m} + \sum_{k \geqslant 1} x^{\diamond k} \tag{25}$$
2. *There exists and is unique a symmetric rigid Maurer-Cartan element in $\mathcal{MC}_m$.*
3. *If $\omega \in \mathcal{MC}_m$ is rigid, then the induced tensor product $\otimes_\omega$ of unital representations up to homotopy is unital.*

In the case $m = 2$ we deduce the following.

Corollary 5.10. *The set of rigid Maurer-Cartan elements in $\mathcal{MC}_2$ coincides with the one-parameter family $\{\omega_t\}_{t\in\mathbb{R}}$ given by*
$$\omega_t = e \boxtimes e + \sum_{k \geqslant 1} (B + At)^{\diamond k},$$
where
$$B = b \boxtimes a + c \boxtimes b, \quad A = b \boxtimes c + a \boxtimes b - b \boxtimes a - c \boxtimes b.$$
Among these, $\omega_{\frac{1}{2}}$ is the only symmetric one, and ω_0 and ω_1 are the only associative ones.

Example 5.11. For ω_0, the second component is B and its third component is $B \diamond B$. In terms of trees it is given by

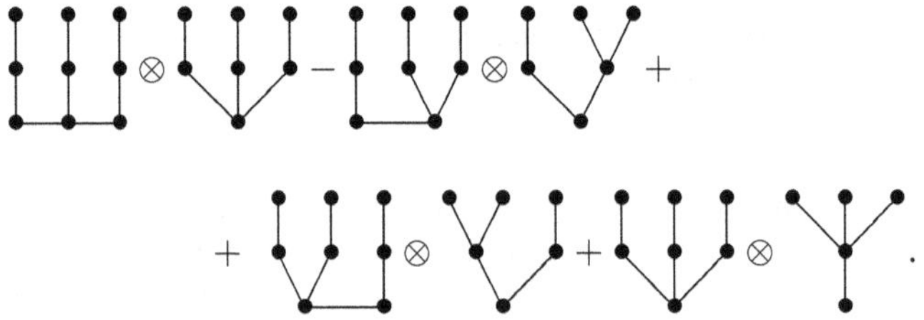

It yields, for the third component of the tensor product of the representations up to homotopy (E, R_k) and (F, R_k), the expression
$$\begin{aligned} R_3(g_1, g_2, g_3) = &\Big(R_1(g_1) \circ R_1(g_2) \circ R_1(g_3)\Big) \otimes R_3(g_1, g_2, g_3) - \\ &-\Big(R_1(g_1) \circ R_2(g_2, g_3)\Big) \otimes \Big(R_2(g_1, g_2 g_3)\Big) \\ &+\Big(R_2(g_1, g_2) \circ R_1(g_3)\Big) \otimes R_2(g_1 g_2, g_3) \\ &+R_3(g_1, g_2, g_3) \otimes R_1(g_1 g_2 g_3). \end{aligned}$$

Remark 5.12. Given $\omega \in \mathcal{MC}_2$, we can interpret the restriction of the characteristic map Δ_ω to Ω^{n+1} as a diagonal on the cubical complex of the n-cube (see paragraph 4.3). If we take the rigid and associative universal Maurer-Cartan element ω_0, the resulting (associative) coincides with the Serre diagonal present in any cubical complex ([**16**]).

The rest of this section is devoted to the proofs of the theorems stated above.

5.1. Proof of Theorem 5.6

That universal Maurer-Cartan elements exist can be derived abstractly using Proposition A.3 from the appendix (for $r = 3$). However, the existence will also follow from the results on rigid Maurer-Cartan elements. Note also that starting with any Maurer-Cartan element $\omega \in \mathcal{MC}_m$, one can produce a symmetric one by averaging:

$$\mathrm{Av}(\omega) = \frac{1}{m!} \sum_{\sigma \in S_m} \hat{\sigma}(\omega).$$

For the second part of the theorem we use Proposition A.6 of the appendix applied to $r = 2$ and to the complete DG-algebra associated to Ω_m. Denoting by $\ldots \subset F_2 \subset F_1 \subset F_0$ the associated filtration of $\bar{\Omega}_m$, we have $F_k/F_{k+1} = \Omega^k(\bullet)$. Note that any universal Maurer-Cartan element is congruent to $e^{\boxtimes m}$ modulo F_2 and the induced differential on F_2/F_3 becomes ∂. On the other hand, the cochain complex $(\Omega^k_m(\bullet), \partial)$ is the the tensor product of the m copies of the complex $(\Omega^k(\bullet), \partial)$. Hence, from the last part of Theorem 4.2, it has trivial cohomology in non-zero degrees. In conclusion, for the cohomology of F_k/F_{k+1} indexed by the total degree (as needed in the theorem of the appendix), we obtain

$$H^i(F_k/F_{k+1}) = 0 \;\; \forall\, i \neq k$$

and we can apply Proposition A.6.

5.2. Proof of Theorem 5.9

For the first part of the theorem, let $\Delta_\omega : \Omega \to \Omega_m$ be the map associated to a rigid element $\omega \in \mathcal{MC}_m$. We verify immediately that $x = \Delta_\omega(b)$ satisfies Equation (24) by applying ∂ on both sides. For the converse, we need to show that given x, there is a unique way to extend it to a rigid Maurer-Cartan element ω. Recall that the kth component ω_k of ω is $\Delta_\omega(L_k)$. Since $L_k = b^{\diamond(k-1)}$ and Δ_ω is required to respect $\diamond$, we conclude that if ω exists, then $\omega_k = x^{\diamond(k-1)}$ for $k \geqslant 2$ and $\omega_1 = e^{\boxtimes m}$. Now we only need to prove that such ω is a Maurer-Cartan element. Note that for $x \in \Omega_m^{k(x)}$ and $y \in \Omega_m^{k(y)}$ with $k(x) > 0$ and $k(y) > 0$, the following identities hold:

$$x \circ y = x \diamond c^{\boxtimes m} \diamond y,$$
$$d_{k(x)}(x \diamond y) = x \diamond a^{\boxtimes m} \diamond y.$$

Using also that ∂ is a derivation with respect to $\diamond$, we have

$$\begin{aligned}\partial(x^{\diamond k-1}) &= \sum_{i=1}^{k-1}(-1)^{i+1}\big(x^{\diamond(i-1)}\diamond(c^{\boxtimes m}-a^{\boxtimes m})\diamond x^{\diamond(k-i-1)}\big),\\ &= \sum_{i=1}^{k-1}(-1)^{i+1}\big(x^{\diamond(i-1)}\circ x^{\diamond(k-i-1)}-d_i x^{\diamond(k-2)}\big),\end{aligned}$$

hence ω is a Maurer-Cartan element.

For the existence of the second part of the theorem, note that the symmetric element

$$x_m=\frac{1}{m!}\sum_{\sigma\in S_m}\sum_{j=0}^{m-1}\hat{\sigma}\left(\underbrace{c\boxtimes\cdots\boxtimes c}_{j\text{ times}}\boxtimes b\boxtimes\underbrace{a\boxtimes\cdots\boxtimes a}_{m-j-1\text{ times}}\right)$$

belongs to $\mathcal{R}_m$. This follows by direct computation (recall that $\partial(b)=c-a$ and ∂ kills a and c). For the uniqueness, note first that the elements of the form

$$X_j=c^{\boxtimes j}\boxtimes b\boxtimes a^{\boxtimes m-j-1},\quad 0\leqslant j\leqslant m-1$$

span a vector space consisting of representatives of the orbits the action of the permutation group on $\Omega^2_m(-1)$. Since ∂ commutes with the action of the permutation group, averaging gives a 1-1 correspondence between symmetric solutions of the equation (24) a and solutions of type $X=\sum_i a_iX_i$ of the same equation, with a_i-some coefficients. It is now easy to see that the resulting equation on the a_i's has the unique solution $a_i=1$.

The last part of the theorem requires a more conceptual understanding of the unitality of representations up to homotopy and is postponed to the final subsection of this section.

5.3. The case $m=2$: the proof of Corollary 5.7 and of Corollary 5.10

We start with the proof of Corollary 5.10. In view of Theorem 5.9, in order to prove that ω_t is a rigid Maurer-Cartan element, it is enough to show that $B+At\in\mathcal{R}_2$. A simple computation shows that:

$$\begin{aligned}\partial(B)&=c^{\boxtimes m}-a^{\boxtimes m}\\ \partial A&=0.\end{aligned}$$

For the uniqueness note that, in general, $\mathcal{R}_m$ is an affine space with underlying vector space consisting of ∂-cocycles in $\Omega^2_m(-1)$. Due to the identification with the cellular complex of the cubes, this vector space coincides with $Z^1(C(I^m))$. When $m=2$, this is easily seen to be 1-dimensional, hence the family $\{\omega_t\}$ exhausts all the rigid elements.

For the last part of the corollary, let $\Delta_t:\Omega\to\Omega\boxtimes\Omega$ be the characteristic map of the rigid element $\omega_t\in\mathcal{MC}_2$. For each $t\in\mathbb{R}$, ω_t produces an associative tensor product of representations up to homotopy if and only if Δ_t is coassociative. Since Δ_t respects $\diamond$, we only need to check coassociativity on the generators a,b c and e.

This always holds for on a, c and e. A direct computation shows that

$$(\Delta_t \boxtimes \mathrm{id})(B + At) = (\mathrm{id} \boxtimes \Delta_t) \circ (B + At)$$

if and only if $t \in \{0, 1\}$. A similar argument shows that Δ_t is symmetric if and only if $t = \frac{1}{2}$.

Turning to Corollary 5.7, we are left with proving the last part. Assume that $\omega \in \mathcal{MC}_2$ is symmetric and we show that it cannot be associative. We claim that $\Delta_\omega(b)$ must belong to $\mathcal{R}_2$. Indeed, since Δ_Ω is a DB-morphism and $\partial(b) = c - a$, $c = e^2$, $a = d_1(e)$, we have

$$\partial(\Delta_\omega(b)) = \Delta_\omega(c) - \Delta_\omega(a) = \Delta_\omega(e)^2 - d_1(\Delta_\omega(e)) = c^{\boxtimes 2} - a^{\boxtimes 2}.$$

Hence, from the uniqueness of symmetric elements of $\mathcal{R}_m$, $\Delta_\omega(b)$ must coincide with A. But, a simple computation similar to the one above show that But a simple computation shows that

$$(\Delta_\omega \boxtimes \mathrm{id})(A) \neq (\mathrm{id} \boxtimes \Delta_\omega)(A),$$

hence ω cannot be associative.

5.4. End of proof of Theorem 5.9: unitality

In this paragraph, we look at tensor products of unital representations up to homotopy, proving in particular the last part of Theorem 5.9. We start by expressing the unitality in terms of the characteristic map $k_E : \Omega \to \mathcal{A}_E$ of the representation up to homotopy. This brings us to the question of the unitality of the DB-algebras Ω and $\mathcal{A}_E$ themselves. So far, Ω has been nonunital (Ω^k is nontrivial only for $k \geqslant 1$). On the other hand, $\mathcal{A}_E$ does have a unit (the vector bundle identity map $\mathrm{id}_E \in \mathcal{A}_E^0$)- but we did not use it so far. Unital representations up to homotopy force us to consider unital DB-algebras: throughout this section, we regard $\mathcal{A}_E$ as a unital DB-algebra with unit id_E, and we formally adjoint a unit to Ω in degree 0, which we denote by 1 and interpret as the empty tree. The characteristic map becomes unital by imposing

$$k_E(1) = \mathrm{id}_E\,.$$

Remark 5.13. The set of Maurer-Cartan elements $\mathcal{MC}_1(\bar{\mathcal{A}})$ are not concerned by the unitality of the DB-algebra $\mathcal{A}$, since we require these elements to have no component in degree zero. For instance, the universal element $L = L_1 + L_2 + L_3 + \cdots$ in the nonunital Ω remains the same in its unital version.

Definition 5.14. *Let $\mathcal{A}_E$ be the DB-algebra associated to a representation up to homotopy (E, R_k) of a Lie groupoid G over M. For each $k \geqslant 1$, we define the following operators*

$$s_i : A_E^k \longrightarrow A_E^{k-1}, \quad i = 1, \ldots, k$$

by the formula

$$s_i(c)(g_1, \ldots, g_{k-1}) = c(g_1, \ldots, g_{i-1}, x_i, g_i, \ldots, g_{k-1}),$$

where $x_i = s(g_{i-1}) = t(g_i)$. For $k = 1$, we define $s_1(c)$ to be the restriction of c to the unit space M of the groupoid. We extend these operators to the powers $\mathcal{A}_E^{\boxtimes n}$ diagonally.

Clearly, (E, R_k) is unital if and only if $s_1(R_1) = \mathrm{id}_E$ and $s_i(R_k) = 0$ for $k > 1$ and $i =, 1 \ldots, k$. Now let us define the corresponding operators at the universal level:

Definition 5.15. *For $k > 1$, we define the operators*

$$s_i : \Omega^k \to \Omega^{k-1},\ i = 1, \ldots, k,$$

as follows. Let $F \in \Omega^k$ be a short forest. We denote by F_i the forest with $k-1$ leaves obtained from F by deleting its i-th leaf. Then,

1. *if F_i is a short forest, we set $s_i(F) = F_i$.*
2. *if F_i is not a short forest (its i-th leaf is now of height 1), writing $F = F_1 \circ L_1 \circ F_2$, we set*

$$s_i(F) = \begin{cases} F_1 \circ F_2, & \textit{if} \quad k(F_1) = i - 1, \\ 0, & \textit{otherwise.} \end{cases}$$

For $k = 1$, we set $s_1(L_1) = 1$. We extend the s_i's diagonally to Ω_m.

Example 5.16. Applying s_2 to the short forests

$F =$, $G =$, $H =$,

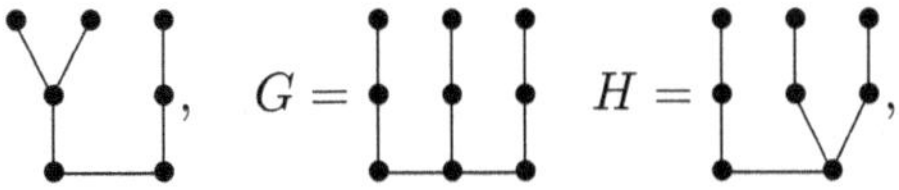

respectively, yields the short forests c for F and G, and 0 for H.

Deleting the i-th leaf of a short forest $F \in \Omega^k$ corresponds, at the universal level, to plugging a groupoid unit in the i-th argument slot of the operator $k_E(F)$ in $\mathcal{A}_E^k$. In particular, we see that the s_i's commute with the characteristic map $k_E : \Omega \to \mathcal{A}_E$ of a representation up to homotopy if and only if the representation is unital. Namely, our definitions yield that $s_1 \circ k_E(L_1)$ is the restriction of R_1 to the unit space and that $k_E \circ s_1(L_1)$ is the identity id_E. For $n > 1$, we have that $s_i \circ k_E(L_n)$ is the operator R_n restricted to the unit space at its i-th slot and, on the other hand, that $k_E \circ s_i(L_n) = 0$.

Now we are ready to characterize the universal Maurer-Cartan elements whose associated tensor products preserve the unitality of the representations.

Definition 5.17. *We will say that a universal Maurer-Cartan element $\omega \in \mathcal{MC}_m$ is unital if its characteristic map Δ_ω commutes with the s_i's.*

Lemma 5.18. *Let $\omega \in \mathcal{MC}_m$ be a unital Maurer-Cartan element. Then the tensor product of m unital representations up to homotopy with respect to ω is also unital.*

Proof. The characteristic map

$$k_{\otimes_\omega(E_1,\ldots,E_m)} = m_{E_1,\ldots,E_m} \circ (k_{E_1} \boxtimes \ldots \boxtimes k_{E_m}) \circ \Delta_\omega,$$

of the tensor product $\otimes_\omega(E_1, \ldots, E_m)$ of the unital representations $E_1, \ldots, E_m$ is a composition of three maps, each of which commutes with the s_i's. Hence the composition itself also commutes s_i's and the resulting representation up to homotopy is unital. □

Proposition 5.19. *For any $x \in \Omega^2_m(-1)$,*

$$s_i(x^{\diamond k}) = 0, \quad k > 0, \quad 1 \leqslant i \leqslant k+1.$$

In particular, if $x \in \mathcal{R}_m$, then the associated rigid Maurer-Cartan element ω (of Theorem 5.9) is unital.

Proof. For the purpose of this proof, we will say that an element $y \in \Omega_m$ is normalized if $s_i(y) = 0$ for all i. The set of elements of the form $e_1 \boxtimes \cdots \boxtimes e_m$, where $e_i \in \{a, b, c\}$ and b occurs exactly once, form a basis of $\Omega^2_m(-1)$. Such a string is normalized if one of its factors is normalized; since b is normalized, all elements $x \in \Omega^2_m(-1)$ are normalized. Next, suppose that all the powers $x^{\diamond k}$ are normalized for $k < n$. Then one immediately sees that $x^{\diamond n}$ is normalized if and only if

$$s_2(x^{\diamond n}) = 0.$$

On the other hand, since

$$s_2(x^{\diamond n}) = s_2(x \diamond x \diamond (x^{\diamond n-2})) = s_2(x \diamond x) \diamond (x^{\diamond n-2}),$$

we conclude that it is enough to prove that $x \diamond x$ is normalized. Let us show that this is the case. First note that $x \diamond x = \frac{1}{2}[x, x]_\diamond$, where $[\,,\,]_\diamond$ is the graded commutator of the associative product $\diamond$. Therefore, it is enough to show that the graded commutator of two elements of the basis of $\Omega^2_m(-1)$ mentioned above is always normalized. Let $v = v_1 \boxtimes \cdots \boxtimes v_m$ be a basis vector with the unique occurrence of b at position j, and let $w = w_1 \boxtimes \cdots \boxtimes w_m$ be another basis vector with b at position l. If $j = l$, we are done, since the factor bb will appear in both terms of the commutator. If $j < l$, we have that

$$[v, w] = v_1 \diamond w_1 \otimes \cdots \otimes v_m \diamond w_m - w_1 \diamond v_1 \boxtimes \cdots \boxtimes w_m \diamond v_m,$$

the minus sign reflecting that the unique b in v "passes over" the unique b in w while carrying out the product. Since both v and w are normalized, we only need to show that s_2 vanishes on the commutator. This follows from the fact that $s_2(x \diamond y) = s_2(y \diamond x)$, for $x, y \in \{a, b, c\}$. This last statement can be checked by direct inspection. □

6. Tensor products of morphisms

In the previous section, we have introduced the sets $\mathcal{MC}_m$ of universal Maurer-Cartan elements, and we have shown that any $\omega \in \mathcal{MC}_m$ induces a tensor product operation $\otimes_\omega$ on the objects of $\mathrm{Rep}^\infty(G)$. In this section, we discuss tensor product operations for the morphisms. We will proceed in a way that is completely similar to the tensor product of representations up to homotopy:

- introduce the notion of a DB-module over a DB-algebra.
- describe morphisms between representations up to homotopy in terms of Maurer-Cartan morphisms.
- introduce the Maurer-Cartan module $\mathcal{T}$ (the analogue of Ω).

- study universal Maurer-Cartan morphism x (between two universal Maurer-Cartan elements ω and η): existence and uniqueness.
- show that any such x induces a tensor product operation $\otimes_x$ on morphisms between representations up to homotopy.

Moreover, we check that the basic properties of the resulting tensor products (e.g. associativity) hold up to homotopy. The main conclusion will be that the homotopy category $\mathcal{D}(G)$ has a monoidal structure uniquely defined up to natural isomorphism.

6.1. Universal Maurer-Cartan morphisms

We start with the notion of a DB-module.

Definition 6.1. *Given a DB-algebra $\mathcal{A}$, a left* ***DB-module*** *$\mathcal{E}$ over $\mathcal{A}$, or simply a left $\mathcal{A}$-module, is a bigraded vector space*

$$\mathcal{E} = \bigoplus_{l \in \mathbb{Z}, k \geqslant 0} \mathcal{E}^k(l),$$

together with a differential ∂ and operations d_i, as in the definition of DB-algebras, and an operation

$$\circ : \mathcal{A}^k(l) \otimes \mathcal{E}^{k'}(l') \longrightarrow \mathcal{E}^{k+k'}(l+l'), \ (a,x) \mapsto a \circ x.$$

These are required to satisfy the same equations (10) and (11) as in Definition 3.2, with $a \in \mathcal{A}$, $b \in \mathcal{E}$. Similarly, one defines the notion of right DB-module.

We will be interested in $\mathcal{A}$-$\mathcal{B}$-bimodules with $\mathcal{A}$ and $\mathcal{B}$ two DB-algebras. There is a version of the functor

$$\overline{K} : \underline{\mathcal{DB}ar} \longrightarrow \overline{DGA}$$

from the category of $\mathcal{A}$-$\mathcal{B}$-bimodules to the category of complete $\bar{\mathcal{A}}$-$\bar{\mathcal{B}}$-DG-bimodules. Thus, given an $\mathcal{A}$-$\mathcal{B}$-bimodule $\mathcal{E}$, there is a complete $\bar{\mathcal{A}}$-$\bar{\mathcal{B}}$-DG-bimodule $(\bar{\mathcal{E}}, d_{\text{tot}})$, where the left and right actions are defined using the signed operation $\star$ (see equation (13)). Given two Maurer-Cartan elements $\theta \in MC(\bar{\mathcal{A}})$ and $\omega \in MC(\bar{\mathcal{B}})$, we consider the set of Maurer-Cartan $\bar{\mathcal{E}}$-morphisms (see the appendix):

$$\bar{\mathcal{E}}(\omega, \theta) = \{x \in \ \bar{\mathcal{E}}^0 : x \star \omega - \theta \star x = d_{\text{tot}}(x)\}.$$

Let G be a Lie groupoid over M and $(E, \partial), (F, \partial)$ two cochain complexes of vector bundles over M. There is a $\mathcal{A}_F$-$\mathcal{A}_E$ bimodule that we will denote by $\mathcal{E}_{E,F}$ and which is defined as follows. In bidegree (k,l):

$$\mathcal{E}^k_{E,F}(l) := \Gamma(G_k, \text{Hom}^l(s^*E, t^*F)),$$

and the structure maps are defined exactly as for $\mathcal{A}_E$. We can now characterize morphisms of representations up to homotopy in this language.

Proposition 6.2. *There is a natural bijective correspondence between morphisms*

$$\Phi : E \longrightarrow F$$

in $Rep^\infty(G)$ *and elements*

$$x \in \bar{\mathcal{E}}_{E,F}(R_E, R_F).$$

Moreover, two morphisms $\Phi, \Psi : E \longrightarrow F$ *are homotopic if and only if the corresponding elements* $x, y \in \bar{\mathcal{E}}_{E,F}(R_E, R_F)$ *are homotopic in the sense of the appendix.*

Proof. The statements follow from comparing the definitions in the appendix with equation (4) in the definition of morphism, and equation (5) in the definition of homotopy. □

Let $\mathcal{E}$ be an $\mathcal{A}$-$\mathcal{B}$-bimodule, where $\mathcal{A}$ and $\mathcal{B}$ are DB-algebras endowed with Maurer-Cartan elements $\theta \in MC_1(\bar{\mathcal{A}})$ and $\omega \in MC_1(\bar{\mathcal{B}})$. Then, using the characteristic maps associated to θ and ω, $\mathcal{E}$ can be given the structure of an Ω-bimodule by the formulas

$$a \circ x := k_\theta(a) \circ x, \;\; x \circ b := x \circ k_\omega(b), \quad a, b \in \Omega, x \in \mathcal{E}.$$

This Ω-bimodule will be denoted by $\mathcal{E}_{\omega,\theta}$ and the associated DG module by $\bar{\mathcal{E}}_{\omega,\theta}$.

For an Ω-bimodule $\mathcal{S}$, we denote by

$$\mathrm{Hom}_{\underline{\mathcal{DB}ar}}(\mathcal{S}, \mathcal{E}_{\omega,\theta})$$

the space of morphisms of Ω-bimodules. Given $S \in \overline{\mathcal{S}}(L, L)$, where L is the universal Maurer-Cartan element of Ω, there is an induced map

$$\mathrm{Hom}_{\underline{\mathcal{DB}ar}}(\mathcal{S}, \mathcal{E}_{\omega,\theta}) \longrightarrow \bar{\mathcal{E}}(\omega, \theta), \;\; f \mapsto f(S). \tag{26}$$

Definition 6.3. *A* ***universal Maurer-Cartan module*** *is an* Ω*-bimodule* $\mathcal{T}$*, together with an element* $T \in \mathcal{T}(L, L)$*, with the property that*

$$Hom_{\underline{\mathcal{DB}ar}}(\mathcal{T}, \mathcal{E}_{\omega,\theta}) \longrightarrow \bar{\mathcal{E}}(\omega, \theta), \;\; f \mapsto f(T) \tag{27}$$

is bijective for all $(\mathcal{A}, \theta, \mathcal{E}, \mathcal{B}, \omega)$ *as above. Given* $x \in \overline{\mathcal{E}}(\omega, \eta)$*, the associated map will be denoted by*

$$k_x : \mathcal{T} \longrightarrow \mathcal{E}$$

and will be called the ***characteristic map*** *of* x*.*

Theorem 6.4. *The Maurer-Cartan DB-module* $\mathcal{T}$ *exists and is unique up to isomorphism. Moreover, for each* k*,* $H^m(\mathcal{T}^k(\bullet), \partial) = 0$ *for all* $m \neq 0$*.*

Proof. The uniqueness follows from the universal property. For the existence part, we will construct $\mathcal{T}$ explicitly. As a vector space, $\mathcal{T}$ is spanned by expressions of type (A, X, B), where each A and B are short forests in $\mathbf{S}\coprod\{\emptyset\}$, and X belongs to the space of short trees $\mathbf{T}\coprod\{\emptyset\}$. Note that A, B and X may be the empty tree, which we denote by 1. The bimodule structure is described by the following natural formulas:

$$C \circ (A, X, B) = (C \circ A, X, B), \qquad (A, X, B) \circ C = (A, X, B \circ C).$$

We introduce the bigrading on $\mathcal{T}$ by

$$k(A, X, B) = k(A) + k(X) + k(B), \;\; l(A, X, B) = l(A) + (l(X) - 1) + l(B),$$

where we put $k(1) = l(1) = 0$. The operators d_i are defined exactly as in the case of Ω. Finally, the differential ∂ is defined as follows. Denote by T_n the element $(1, L_n, 1) \in \mathcal{T}$ and $T_0 = (1, 1, 1)$. We set

$$\partial(T_n) = \sum_{i=0}^{n-1} (-1)^i T_i \circ L_{n-i} - \sum_{i=1}^{n} L_i \circ T_{n-i} + \sum_{i=1}^{n-1} (-1)^{i+1} d_i(T_{n-1}),$$

and extend ∂ by forcing it to be a derivation and to commute with the operators d_i. In order to prove that $\partial^2 = 0$, it is enough to show that $\partial^2(T_n) = 0$ for all n, and this can be checked by a simple computation. The universal Maurer-Cartan morphism in $\mathcal{T}$ is

$$T = (1,1,1) + \left(1, \vcenter{\hbox{$\cdot\!\!\cdot$}}, 1\right) + \left(1, \vcenter{\hbox{$\bigvee$}}, 1\right) + \left(1, \vcenter{\hbox{$\bigvee\!\!\bigvee$}}, 1\right) + \cdots.$$

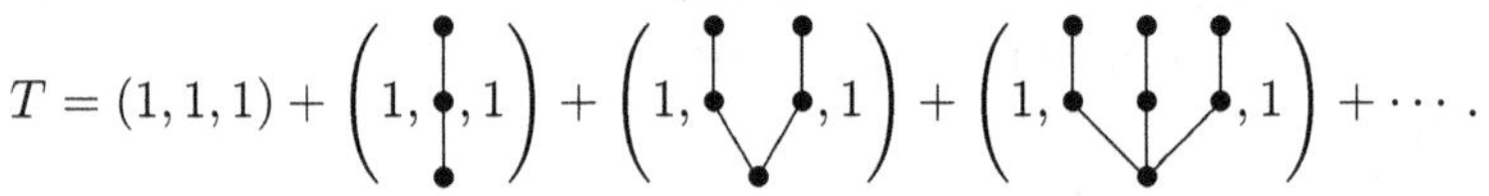

By construction, $\mathcal{T}$ satisfies the universal property. Finally, the statement about the cohomology is analogous to that of Ω. $\qquad\square$

Next we consider the Ω_m-bimodule

$$\mathcal{T}_m := \underbrace{\mathcal{T} \boxtimes \ldots \boxtimes \mathcal{T}}_{m \ \text{times}}.$$

Definition 6.5. *Given $\omega, \theta \in MC_m$, a **universal Maurer-Cartan morphism** from ω to θ is any Maurer-Cartan morphism $x \in \bar{\mathcal{T}}_m(\omega, \theta)$ with the property that its degree 0 component is*

$$x_0 = \underbrace{T_0 \boxtimes \ldots \boxtimes T_0}_{m \ \text{times}}.$$

We denote by $\mathcal{MC}_m(\omega, \theta)$ the set of such elements.

Because of the universal property of $\mathcal{T}$, an element $x \in \mathcal{MC}_m(\omega, \theta)$ may be interpreted as a map of Ω-bimodules

$$\Delta_x : \mathcal{T} \longrightarrow \mathcal{T}_{m,\omega,\theta}.$$

Definition 6.6. *Let $\mathcal{E}, \mathcal{E}'$ be Ω-bimodules and $\phi, \phi' : \mathcal{E} \to \mathcal{E}'$ morphisms. We say that ϕ and ϕ' are homotopic if there exists a degree -1 linear map h that commutes with the Ω action and the d_i operators such that $hd + dh = \phi - \phi'$. In this case, h is called a homotopy between ϕ and ϕ'.*

Lemma 6.7. *Let $\mathcal{A}, \mathcal{B}$ be DB algebras and θ, ω Maurer-Cartan elements of $\mathcal{A}$ and $\mathcal{B}$, respectively. Suppose also that $\mathcal{E}$ is an $\mathcal{A}$-$\mathcal{B}$-bimodule and that x and y belong to $\bar{\mathcal{E}}(\omega, \theta)$. Then homotopies between x and y correspond naturally to homotopies between the characteristic maps $k_x : \mathcal{T} \to \mathcal{E}_{\omega,\theta}$ and $k_y : \mathcal{T} \to \mathcal{E}_{\omega,\theta}$.*

Proof. The correspondence sends a homotopy $h : \mathcal{T} \to \mathcal{E}_{\omega,\theta}$ to $h(T) \in \bar{\mathcal{E}}_{\omega,\theta}$. Clearly, since h commutes with the Ω action and the d_i operators, it is determined by the value of $h(T)$. On the other hand, the equation $hd + dh = \phi - \phi'$ corresponds precisely to the equation

$$x - y = d_{tot}h(T) + h(T) \star \omega + \theta \star h(T).$$

□

As before, Maurer-Cartan morphisms induce tensor product operations between morphisms of representations up to homotopy. Given $x \in \mathcal{T}_m(\omega, \theta)$ and morphisms between representations up to homotopy $\Phi_i : E_i \longrightarrow F_i$, $1 \leqslant i \leqslant m$, we define a morphism

$$\Phi = \otimes_x(\Phi_1, \ldots, \Phi_m) : \otimes_\omega(E_1, \ldots, E_m) \longrightarrow \otimes_\theta(F_1, \ldots, F_m)$$

by specifying its characteristic map:

$$k_\Phi := m_\boxtimes \circ (k_{\Phi_1} \boxtimes \ldots \boxtimes k_{\Phi_m}) \circ \Delta_x : \mathcal{T} \longrightarrow \mathcal{E}_{E,F},$$

where $E = \otimes_\omega(E_1, \ldots, E_m), F = \otimes_\theta(F_1, \ldots, F_m)$, and

$$m_\boxtimes : \mathcal{E}_{E_1,F_1} \boxtimes \ldots \boxtimes \mathcal{E}_{E_m,F_m} \longrightarrow \mathcal{E}_{E,F}$$

is defined, as in Proposition 3.6, by:

$$m_\boxtimes(\phi_1 \boxtimes \cdots \boxtimes \phi_m)(g_1, \ldots, g_k) = \phi_1(g_1, \ldots, g_k) \otimes \cdots \otimes \phi_m(g_1, \ldots, g_k).$$

Theorem 6.8. *Let ω, θ be elements of $\mathcal{MC}_m$ and $\Phi_i : E_i \longrightarrow F_i$, for $1 \leqslant i \leqslant m$, morphisms of representations up to homotopy, $E = \otimes_\omega(E_1, \ldots, E_m), F = \otimes_\theta(F_1, \ldots, F_m)$. Then:*

1. *$\mathcal{MC}_m(\omega, \theta)$ is nonempty. Moreover, every two elements in $\mathcal{MC}_m(\omega, \theta)$ are homotopic in the sense of the appendix.*
2. *Any homotopy between $x, y \in \mathcal{MC}_m(\omega, \theta)$ induces a homotopy between the morphisms $\otimes_x(\Phi_1, \ldots, \Phi_m)$ and $\otimes_y(\Phi_1, \ldots, \Phi_m)$.*
3. *For any $x \in \mathcal{MC}_m(\omega, \theta)$, the zeroth component of $\otimes_x(\Phi_1, \ldots, \Phi_m)$ is the tensor product of the zeroth components of the Φ_i's. Also, if Φ_i are strict morphisms, then so is $\otimes_x(\Phi_1, \ldots, \Phi_m)$.*
4. *Symmetric Maurer-Cartan morphisms exist. Moreover, any symmetric x induces a symmetric power operation on morphisms.*
5. *The tensor product of morphisms is well defined on homotopy classes. Namely, if Φ'_i is homotopic to Φ_i then $\otimes_x(\Phi'_1, \ldots, \Phi'_m)$ is homotopic to $\otimes_x(\Phi_1, \ldots, \Phi_m)$.*
6. *For any $x \in \mathcal{MC}_m(\omega, \omega)$, the tensor product of the identity morphisms $\otimes_x(\mathrm{id}_{E_1}, \ldots, \mathrm{id}_{E_n})$ is homotopic to the identity morphism on $\otimes_\omega(E_1, \ldots, E_n)$.*

Proof. The first claim is a direct application of Proposition A.4 from the appendix. The second claim is a consequence of Proposition 6.2. The third part follows from the condition

$$x_0 = T_0 \boxtimes \ldots \boxtimes T_0$$

for elements of $\mathcal{MC}_m(\omega, \theta)$. The statement about the symmetric elements holds, because given a Maurer-Cartan morphism, one can construct a symmetric one by averaging.

Let us now prove that the tensor product of morphisms is well defined on homotopy classes. Since homotopy is a transitive relation, we can assume that $\phi_i = \phi'_i$ for $i > 1$. Now fix a homotopy h_1 between the characteristic maps k_{ϕ_1} and $k_{\phi'_1}$. Then

$$m \circ (h_1 \boxtimes k_{\phi_2} \boxtimes \ldots \boxtimes k_{\Phi_m}) \circ \Delta_x : \mathcal{T} \longrightarrow \mathcal{E}_{E,F}$$

is a homotopy between the characteristic maps of $\otimes_x(\Phi'_1, \dots, \Phi'_m)$ and $\otimes_x(\Phi_1, \dots, \Phi_m)$. By Lemma 6.7, we conclude that the two morphisms are homotopic.

We turn now to the last claim. Consider the natural map of Ω-bimodules $\pi : \mathcal{T} \to \Omega$, defined on generators by $\pi(T_0) = 1$, and $\pi(T_n) = 0$ for $n \geqslant 1$. In particular, π does not vanish only for those triples (A, X, B) such that $X = 1$, in which case we obtain

$$\pi(A, 1, B) = A \circ B.$$

This induces a map of Ω-bimodules

$$\pi^{\boxtimes m} : (\mathcal{T}^{\boxtimes m})_{\omega,\omega} \longrightarrow \Omega_m.$$

Because we are taking tensor products of identity morphisms, the characteristic map of $\otimes_x(\mathrm{id}_{E_1}, \dots, \mathrm{id}_{E_m})$ factors as follows:

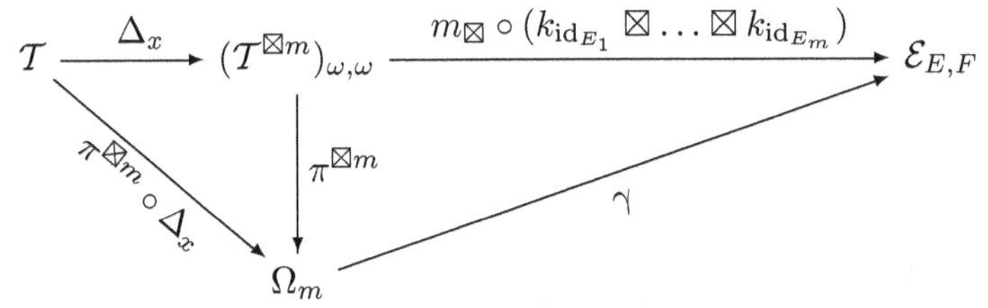

where γ is characterized by the commutativity of the diagram. On the other hand, the characteristic map of the identity morphism of E is given by the composition

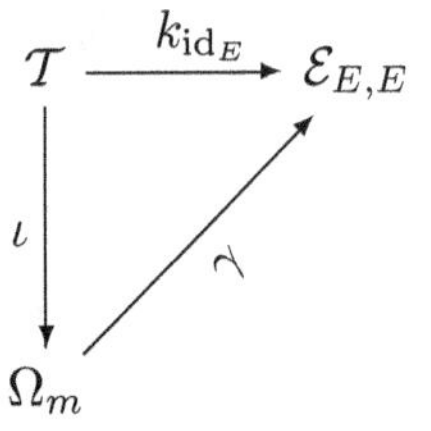

where ι is the map of Ω-bimodules defined on generators as $\iota(T_0) = 1$, and $\iota(T_n) = 0$ for $n > 0$. Thus, it is enough to prove that the maps ι and $\pi^{\boxtimes m} \circ \Delta_x$ are homotopic. In view of the universal property of $\mathcal{T}$ and Lemma 6.7, we only need to prove that the Maurer-Cartan morphisms $x_\iota, x_{\pi^{\boxtimes m} \circ \Delta} \in \bar{\Omega}_2(\omega, \omega)$ associated to ι and $\pi^{\boxtimes m} \circ \Delta_x$ are homotopic in the sense of the appendix. For this, we observe that they coincide modulo $F_1\bar{\Omega}_m$, and since

$$H^{-p}(\Omega^p_m(\bullet), \partial) = 0, \quad \forall p \geqslant 1,$$

we can use Corollary A.6 to conclude the claim. □

6.2. Composition of Maurer-Cartan morphisms

We will now express the composition of morphisms of representations up to homotopy in terms of the Maurer-Cartan DB-module $\mathcal{T}$ in order to show that the homotopy category $\mathcal{D}(G)$ has a monoidal structure. For this, we need to consider the tensor product of DB-modules. Suppose that $\mathcal{A}, \mathcal{B}, \mathcal{C}$ are DB-algebras, $\mathcal{E}$ is an $\mathcal{A}$-$\mathcal{B}$-modules and $\mathcal{E}'$ is an $\mathcal{B}$-$\mathcal{C}$-bimodule. Then one can construct the $\mathcal{A}$-$\mathcal{C}$ DB-bimodule $\mathcal{E} \otimes_B \mathcal{E}'$ as follows. As a bigraded vector space, $\mathcal{E} \otimes_B \mathcal{E}'$ it is the same as the vector space underlying the usual tensor product of graded bimodules over an algebra. The operators d_i are given by the formulas:

$$d_i(v \otimes w) = \begin{cases} d_i(v) \otimes w & \text{if } i \leqslant k_1, \\ v \otimes d_{i-k_1}(w) & \text{if } i > k_1. \end{cases}$$

The operation ∂ is given by:

$$\partial(v \otimes w) = \partial(v) \otimes w + (-1)^{l(v)} v \otimes \partial(w).$$

This construction has the property that:

$$\overline{\mathcal{E}_1 \otimes_{\mathcal{B}} \mathcal{E}_2} \cong \bar{\mathcal{E}}_1 \otimes_{\overline{\mathcal{B}}} \bar{\mathcal{E}}_2.$$

Now let us go back to our initial goal, which is to express the composition of Maurer-Cartan morphisms in terms of $\mathcal{T}$: consider a sequence of Maurer-Cartan morphisms:

$$(\bar{\mathcal{A}}, \omega) \xrightarrow{x_1} (\bar{\mathcal{B}}, \theta) \xrightarrow{x_2} (\bar{\mathcal{C}}, \nu),$$

where $x_1 \in \bar{\mathcal{E}_1}(\omega, \theta)$ and $x_2 \in \bar{\mathcal{E}_2}(\theta, \nu)$.

As explained in the appendix, the composition $x_2 \circ x_1$ is defined as the tensor product $x_1 \otimes_{\bar{\mathcal{B}}} x_2$ in the $\bar{\mathcal{A}}$-$\bar{\mathcal{C}}$-DG-bimodule $\bar{\mathcal{E}_1} \otimes_{\bar{\mathcal{B}}} \bar{\mathcal{E}}_2$. The characteristic map $k_{x_2 \circ x_1}$ of the composition $x_2 \circ x_1$ can be expressed in terms of the characteristic maps k_{x_1} and k_{x_2} as follows:

$$\mathcal{T} \xrightarrow{\Lambda} \mathcal{T} \otimes_\Omega \mathcal{T} \xrightarrow{k_{x_1} \otimes_\Omega k_{x_2}} (\mathcal{E}_1)_{\omega,\theta} \otimes_\Omega (\mathcal{E}_2)_{\theta,\nu} \longrightarrow (\mathcal{E}_1 \otimes_{\mathcal{B}} \mathcal{E}_2)_{\omega,\nu}, \tag{28}$$

where Λ a canonical map of Ω-bimodules that is completely determined by

$$\Lambda(T_n) = \sum_{i=0}^{n} T_i \otimes T_{n-i}.$$

For morphisms of representations up to homotopy,

$$(\bar{\mathcal{A}}_{E_1}, R_1) \xrightarrow{\phi_1} (\bar{\mathcal{A}}_{E_2}, R_2) \xrightarrow{\phi_2} (\bar{\mathcal{A}}_{E_3}, R_3),$$

the morphisms live in the bimodules $\bar{\mathcal{E}}_{E_1,E_2}$ and $\bar{\mathcal{E}}_{E_2,E_3}$. In order to obtain the characteristic map of the composition $k_{\phi_2 \circ \phi_1}$ we need to further compose (28) with the canonical morphism

$$m_\otimes : \mathcal{E}_{E_1,E_2} \otimes_\Omega \mathcal{E}_{E_2,E_3} \longrightarrow \mathcal{E}_{E_1,E_3},$$

which is defined as

$$m_\otimes(\phi \otimes \psi)(g_1, \ldots, g_n) = \psi(g_1, \ldots, g_k) \circ \phi(g_{k+1}, \ldots, g_n).$$

Our next goal is to prove that tensor products and compositions of representation up to homotopy are compatible

Proposition 6.9. *Suppose we have $\omega_1, \omega_2, \omega_3 \in \mathcal{MC}_2$, together with*

$$x_1 \in \mathcal{MC}_2(\omega_1, \omega_2),\ x_2 \in \mathcal{MC}_2(\omega_2, \omega_3),\ x_3 \in \mathcal{MC}_2(\omega_1, \omega_3).$$

Then for any representations up to homotopy E_i and F_i, for $i = 1, 2, 3$, and any morphisms $\phi_j : E_j \to E_{j+1}$ and $\psi_j : F_j \to F_{j+1}$, for $j = 1, 2$, the composition of the tensor products

$$E_1 \otimes_{\omega_1} F_1 \overset{\phi_1 \otimes_{x_1} \psi_1}{\longrightarrow} E_2 \otimes_{\omega_2} F_2 \overset{\phi_2 \otimes_{x_2} \psi_2}{\longrightarrow} E_3 \otimes_{\omega_3} F_3$$

is homotopic to the tensor product of the compositions

$$(\phi_2 \circ \phi_1) \otimes_{x_3} (\psi_2 \circ \psi_1).$$

Proof. There is a morphism of Ω-bimodules:

$$u : (\mathcal{T} \otimes_\Omega \mathcal{T}) \boxtimes (\mathcal{T} \otimes_\Omega \mathcal{T}) \to \overline{\mathcal{E}}_{E_1 \otimes E_3, F_1 \otimes F_3},$$

defined by the formula $u = m \circ (u_E \boxtimes u_F)$ where:

$$u_E : \mathcal{T} \otimes_\Omega \mathcal{T} \to \overline{\mathcal{E}}_{E_1, E_3}$$

is the composition of the map:

$$k_{\phi_2} \otimes_\Omega k_{\phi_1} : \mathcal{T} \otimes_\Omega \mathcal{T} \to \overline{\mathcal{E}}_{E_2, E_3} \otimes_\Omega \overline{\mathcal{E}}_{E_1, E_2}$$

with the natural quotient map:

$$\overline{\mathcal{E}}_{E_2, E_3} \otimes_\Omega \overline{\mathcal{E}}_{E_1, E_2} \to \overline{\mathcal{E}}_{E_1, E_3}.$$

Similarly,

$$u_F : \mathcal{T} \otimes_\Omega \mathcal{T} \to \overline{\mathcal{E}}_{F_1, F_3}$$

is the composition of the map:

$$k_{\psi_2} \otimes_\Omega k_{\psi_1} : \mathcal{T} \otimes_\Omega \mathcal{T} \to \overline{\mathcal{E}}_{F_2, F_3} \otimes_\Omega \overline{\mathcal{E}}_{F_1, F_2}$$

with the natural quotient map:

$$\overline{\mathcal{E}}_{F_2, F_3} \otimes_\Omega \overline{\mathcal{E}}_{F_1, F_2} \to \overline{\mathcal{E}}_{F_1, F_3}.$$

There is also a canonical morphism of Ω-bimodules

$$p : (\mathcal{T} \boxtimes \mathcal{T}) \otimes_\Omega (\mathcal{T} \boxtimes \mathcal{T}) \to (\mathcal{T} \otimes_\Omega \mathcal{T}) \boxtimes (\mathcal{T} \otimes_\Omega \mathcal{T}),$$

given by the formula:

$$(a \boxtimes b) \otimes (c \boxtimes d) \mapsto (-1)^{l(b)l(c)} (a \otimes c) \boxtimes (b \otimes d).$$

A simple computation shows that the characteristic map $k : \mathcal{T} \to \overline{\mathcal{E}}_{E_1 \otimes F_1, E_3 \otimes F_3}$ of $(\Phi_2 \otimes \Psi_2) \circ (\Phi_1 \otimes \Psi_1)$ is given by the composition

$$k = u \circ p \circ (\Delta_{x_1} \otimes \Delta_{x_2}) \circ \Lambda : \mathcal{T} \to \overline{\mathcal{E}}_{E_1 \otimes F_1, E_3 \otimes F_3}.$$

On the other hand, the characteristic map $k' : \mathcal{T} \to \overline{\mathcal{E}}_{E_1 \otimes F_1, E_3 \otimes F_3}$ of $(\Phi_2 \circ \Phi_1) \otimes (\Psi_2 \circ \Psi_1)$ is given by the composition

$$k' = u \circ (\Lambda \boxtimes \Lambda) \circ \Delta_{x_3} : \mathcal{T} \to \overline{\mathcal{E}}_{E_1 \otimes F_1, E_3 \otimes F_3}.$$

We need to prove that the maps k and k' are homotopic. Clearly, it is enough to prove that the maps:

$$p \circ (\Delta_{x_1} \otimes \Delta_{x_2}) \circ \Lambda : \mathcal{T} \to (\mathcal{T} \otimes_\Omega \mathcal{T}) \boxtimes (\mathcal{T} \otimes_\Omega \mathcal{T}),$$

and

$$(\Lambda \boxtimes \Lambda) \circ \Delta_{x_3} : \mathcal{T} \to (\mathcal{T} \otimes_\Omega \mathcal{T}) \boxtimes (\mathcal{T} \otimes_\Omega \mathcal{T}),$$

are homotopic. Let us now prove that this last statement is true for any two maps of Ω-bimodules $a, b : \mathcal{T} \to (\mathcal{T} \otimes_\Omega \mathcal{T}) \boxtimes (\mathcal{T} \otimes_\Omega \mathcal{T})$, provided they take the same value on $\mathcal{T}_0$, which is satisfied in our case. In order to simplify the notation, let us denote the Ω-bimodule $(\mathcal{T} \otimes_\Omega \mathcal{T}) \boxtimes (\mathcal{T} \otimes_\Omega \mathcal{T})$ by $\mathcal{P}$. In view of the universal property of $\mathcal{T}$ we can identify a and b with Maurer-Cartan morphisms $\overline{a}, \overline{b} \in \overline{\mathcal{P}}(L, L)$. Also, Lemma 6.7 implies that it is enough to prove that the Maurer-Cartan morphisms $\overline{a}$ and $\overline{b}$ are homotopic. Since a and b coincide when applied to $\mathcal{T}_0$ we know that

$$\overline{a} \equiv \overline{b} \bmod(F_1(\overline{\mathcal{P}})).$$

By Proposition A.4 applied to the case $r = 1$, we know that it is enough to prove that

$$H^0(F_k(\overline{\mathcal{P}})/F_{k+1}(\overline{\mathcal{P}}), d_{L,L}) = 0, \text{ for all } k \geqslant 1.$$

On the other hand, $F_k(\overline{\mathcal{P}})/F_{k+1}(\overline{\mathcal{P}})$ is naturally isomorphic to $\mathcal{P}^k$ and, because of the shift in degree, we obtain that:

$$H^0(F_k(\overline{\mathcal{P}})/F_{k+1}(\overline{\mathcal{P}}), d_{L,L}) = H^{-k}(\mathcal{P}^k).$$

Thus, all we need to prove is that the cohomology of $\mathcal{P}^k$ vanishes in negative degree. For this we observe that the complex

$$\mathcal{P}^k = ((\mathcal{T} \otimes_\Omega \mathcal{T}) \boxtimes (\mathcal{T} \otimes_\Omega \mathcal{T}))^k$$

is the tensor product of two copies of the complex $(\mathcal{T} \otimes_\Omega \mathcal{T})^k$ and therefore, by Künneth's formula, it is enough to prove that the cohomology of that last complex vanishes in negative degree. This last claim follows from the fact that the cohomology of $\mathcal{T}$ vanishes in negative degree. □

6.3. Monoidal structure on $\mathcal{D}(G)$

In this paragraph, we show that the tensor product operations defined above give the homotopy category $\mathcal{D}(G)$ a monoidal structure. The reader is referred to [**12**] for the basic facts and definitions concerning monoidal categories.

Proposition 6.10. *Let us fix a universal Maurer-Cartan element $\omega \in \mathcal{MC}_2$ and a universal Maurer-Cartan morphism $x \in \mathcal{MC}_2(\omega, \omega)$. The corresponding operations of tensor product defines a functor:*

$$\otimes_{\omega,x} : \mathcal{D}(G) \times \mathcal{D}(G) \longrightarrow \mathcal{D}(G).$$

Moreover, this functor does not depend on x.

Proof. First we observe that Theorem 6.8 guaranties that the tensor product operation is well defined on homotopy classes of morphisms and therefore this map is well defined. Theorem 6.8 and Proposition 6.9 tell us that

$$\mathrm{id}_E \otimes_x \mathrm{id}_F \sim \mathrm{id}_{E \otimes_\omega F},$$

$$(\phi_2 \circ \phi_1) \otimes_x (\psi_2 \circ \phi_1) \sim (\phi_2 \otimes_x \psi_2) \circ (\phi_1 \otimes_x \psi_1).$$

Observe also that $\otimes_{\omega,x}$ does not depend on x since any two $x, x' \in \mathcal{MC}_2(\omega,\omega)$ are homotopic and, thus, the corresponding morphisms coincide in $\mathcal{D}(G)$. □

Proposition 6.11. *Let $\omega, \omega' \in \mathcal{MC}_2$ be two universal Maurer-Cartan elements. Then the bifunctors $\otimes_\omega$ and $\otimes_{\omega'}$ on $\mathcal{D}(G)$ are equivalent. Moreover, for any two representations up to homotopy E_1 and E_2, the isomorphisms*

$$\mathrm{id}_{E_1} \otimes_y \mathrm{id}_{E_2} : E_1 \otimes_\omega E_2 \longrightarrow E_1 \otimes_{\omega'} E_2,$$

represents an isomorphism in $\mathcal{D}(G)$ which is independent of the choice of $y \in \mathcal{MC}_2(\omega, \omega')$.

Proof. We need to check that $(\phi_1 \otimes_{x'} \phi_2) \circ (\mathrm{id}_{E_1} \otimes_y \mathrm{id}_{E_2})$ is homotopic to $(\mathrm{id}_{F_1} \otimes_y \mathrm{id}_{F_2}) \circ (\phi_1 \otimes_x \phi_2)$ for any $x \in \mathcal{MC}_2(\omega,\omega)$ and for any $x' \in \mathcal{MC}_2(\omega',\omega')$ and for any two representation morphisms $\phi_i : E_i \to F_i$, $i = 1, 2$. This is guaranteed by Proposition 6.9, which tells us that

$$(\phi_1 \otimes_{x'} \phi_2) \circ (\mathrm{id}_{E_1} \otimes_y \mathrm{id}_{E_2}) \sim \phi_1 \otimes_y \phi_2 \sim (\mathrm{id}_{F_1} \otimes_y \mathrm{id}_{F_2}) \circ (\phi_1 \otimes_x \phi_2).$$

□

We now prove that the bifunctor $\otimes_\omega$ endows the homotopy category $\mathcal{D}(G)$ with the structure of a monoidal category.

Theorem 6.12. *For every $\omega \in \mathcal{MC}_2$, the functor $\otimes_\omega : \mathcal{D}(G) \times \mathcal{D}(G) \to \mathcal{D}(G)$ gives the category $\mathcal{D}(G)$ a monoidal structure with unit object given by the trivial representation. Moreover, any two choices of ω give naturally equivalent monoidal categories.*

Proof. In order to prove that this functor gives $\mathcal{D}(G)$ a monoidal structure we need to show that there is a unit for the tensor product and that there are unitors and associators that satisfy the pentagon and triangle axiom. The general idea of the proof is simple: one first shows that the associators and unitors are given by universal maps at the level of the DB-algebra Ω. Then the commutativity of the diagrams translates to the statement that certain maps between powers of Ω are homotopic, which follows from the fact that the algebras are contractible.

The unit is given as usual by the trivial representation. Let us first consider the existence of the unitors. We will prove that

$$E \otimes 1 \cong E; 1 \otimes E \cong E,$$

where the maps, which we denote by λ_r and λ_l respectively, are given by the obvious identification of vector bundles.

We denote by k_E the characteristic map of E and consider the map of DB-algebras

$$\pi : \Omega \boxtimes \Omega \to \Omega$$

defined on generators by setting:

$$\pi(a \boxtimes b) = \begin{cases} a \text{ if } l(b) = 0, \\ 0 \text{ otherwise.} \end{cases}$$

Then, the following diagram commutes:

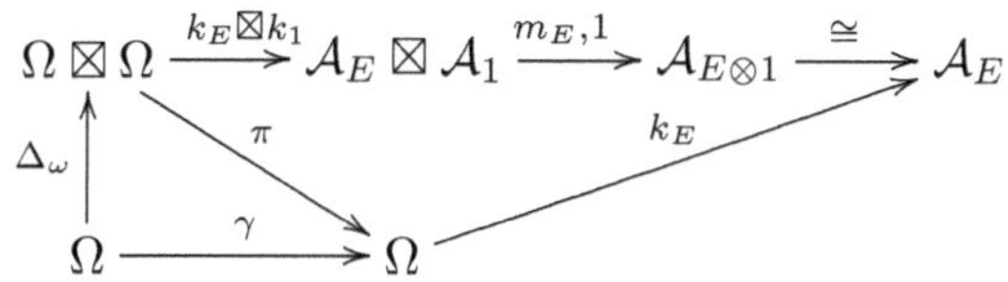

Thus, we see that the characteristic map of $E \otimes_\omega 1$ differs from the characteristic map of E by pre-composing with the map $\gamma : \Omega \to \Omega$. Also, it is clear that $\gamma(L_1) = L_1$. We claim that γ is the identity. By the universal property of Ω it is enough to prove that $\gamma(L_k) = L_k$. We will prove this assertion inductively. Assume that the statement is true for $i < k$. Then:

$$\partial(\gamma(L_k)) = \gamma(\partial(L_k)) = \sum_{j=1}^{k-1}(-1)^{j+1} L_j \circ L_{k-j} + \sum_{j=1}^{k-1}(-1)^j d_j(L_{k-1}) = \partial(L_k).$$

On the other hand, since γ preserves both the degree and the order, we know that $\gamma(L_k)$ is a multiple of L_k. We conclude that they are equal and therefore $E = E \otimes 1$ under the obvious identification. By the same argument we know that $E = 1 \otimes E$.

Let us now construct the associators. The characteristic map $\Delta_\omega : \Omega \to \Omega_2$ can be used to construct two Maurer-Cartan elements $\alpha, \beta \in \mathcal{MC}_3$ with characteristic maps:

$$\Delta_\alpha = (\Delta_\omega \boxtimes \mathrm{id}) \circ \Delta_\omega : \Omega \to \Omega_3,$$
$$\Delta_\beta = (\mathrm{id} \boxtimes \Delta_\omega) \circ \Delta_\omega : \Omega \to \Omega_3.$$

Theorem 6.8 guaranties that there is a universal Maurer-Cartan morphism u from α to β and that any two such are homotopic. For any three representations up to homotopy E_1, E_2, E_3 of G, one can use u to tensor the identity morphisms $E_i \mapsto E_i$ and obtain a map:

$$\hat{u} : (E_1 \otimes_\omega E_2) \otimes_\omega E_3 \to E_1 \otimes_\omega (E_2 \otimes_\omega E_3),$$

which we will call the associator of the monoidal structure. Note that since any two choices of u are homotopic, this map is well defined in $\mathcal{D}(G)$. There is an alternative way to describe the map $\hat{u}$. By Theorem 5.6, we know that α and β are strongly gauge equivalent. By Corollary 5.4 any such gauge equivalence u' induces an isomorphism

$$\hat{u'} : (E_1 \otimes_\omega E_2) \otimes_\omega E_3 \to E_1 \otimes_\omega (E_2 \otimes_\omega E_3).$$

This isomorphism is well defined in the homotopy category because any two such gauge equivalences are homotopic. It is a simple exercise to check that $\hat{u} = \hat{u'}$ in the homotopy category. In what follows we will make use of both constructions of the associator.

We need to prove that the associators define a natural transformation between the functors $((\cdot \otimes_\omega \cdot) \otimes_\omega \cdot)$ and $(\cdot \otimes_\omega (\cdot \otimes_\omega \cdot))$. Namely, we need to prove that given morphisms of representations up to homotopy $\phi_i : E_i \to F_i$ for $i = 1, 2, 3$, the following diagram commutes:

$$\begin{array}{ccc}
(E_1 \otimes_\omega E_2) \otimes_\omega E_3 & \xrightarrow{\hat{u}} & E_1 \otimes_\omega (E_2 \otimes_\omega E_3) \\
\downarrow{\scriptstyle (\phi_1 \otimes \phi_2) \otimes \phi_3} & & \downarrow{\scriptstyle \phi_1 \otimes (\phi_2 \otimes \phi_3)} \\
(F_1 \otimes_\omega F_2) \otimes_\omega F_3 & \xrightarrow{\hat{u}} & F_1 \otimes_\omega (F_2 \otimes_\omega F_3)
\end{array}$$

We will use the second description of the associator map. Consider u' a strong gauge equivalence from α to β and let $x \in \mathcal{MC}_3(\alpha, \alpha)$ and $y \in \mathcal{MC}_3(\beta, \beta)$ be Maurer-Cartan morphisms which are used to define $(\phi_1 \otimes \phi_2) \otimes \phi_3$ and $\phi_1 \otimes (\phi_2 \otimes \phi_3)$, respectively. We can construct two new Maurer-Cartan morphisms as follows. First we use the composition in the Maurer-Cartan category (see Appendix A.1) to define

$$x \circ u' \in \overline{\mathcal{T}_3} \otimes_{\overline{\Omega}_3} \overline{\Omega}_3(\alpha, \beta),$$

and

$$u' \circ y \in \overline{\Omega}_3 \otimes_{\overline{\Omega}_3} \overline{\mathcal{T}_3}(\alpha, \beta).$$

Then, we apply the canonical morphisms of $\overline{\Omega}_3$ modules:

$$\overline{\Omega}_3 \otimes_{\overline{\Omega}_3} \overline{\mathcal{T}_3} \mapsto \overline{\mathcal{T}_3}$$

and

$$\overline{\mathcal{T}_3} \otimes_{\overline{\Omega}_3} \overline{\Omega}_3 \mapsto \overline{\mathcal{T}_3}$$

to $x \circ u'$ and $u' \circ y$ to obtain elements $t, z \in \mathcal{MC}_3(\alpha, \beta)$.

We now observe that the morphisms $\hat{u} \circ (\phi_1 \otimes \phi_2) \otimes \phi_3)$ and $\phi_1 \otimes (\phi_2 \otimes \phi_3) \circ \hat{u}$ are precisely the result of tensoring the morphisms ϕ_1, ϕ_2, ϕ_3 with respect to the Maurer-Cartan morphisms $t, z \in \mathcal{MC}_3(\alpha, \beta)$. In view of Theorem 6.8 parts (1) and (2) we know that t and z are homotopic and therefore the corresponding ways of tensoring the morphisms in the homotopy category coincide.

Let us now prove that the units are compatible with the associators. We need to prove the triangle axiom, which is the commutativity of the following diagram in $\mathcal{D}(G)$:

$$\begin{array}{ccc}
(E_1 \otimes_\omega 1) \otimes_\omega E_2 & \xrightarrow{\hat{u}} & E_1 \otimes_\omega (1 \otimes_\omega E_2) \\
\downarrow{\scriptstyle \lambda_r \otimes \mathrm{id}} & \swarrow{\scriptstyle \mathrm{id} \otimes \lambda_l} & \\
E_1 \otimes_\omega E_2 & &
\end{array}$$

By the last part of Theorem 6.8 we know that $\mathrm{id}_{E_1} \otimes \mathrm{id}_{E_2}$ is homotopic to the identity morphism on $E_1 \otimes E_2$. Therefore, it suffices to prove that the following commutes:

$$\begin{array}{ccc}
(E_1 \otimes_\omega 1) \otimes_\omega E_2 & \xrightarrow{\hat{u}} & E_1 \otimes_\omega (1 \otimes_\omega E_1) \\
\big\downarrow {=} & & \big\downarrow {=} \\
E_1 \otimes_\omega E_2 & \xrightarrow{\mathrm{id}_{E_2} \otimes \mathrm{id}_{E_2}} & E_1 \otimes_\omega E_2
\end{array}$$

Let us consider the characteristic map $\Delta_u : \mathcal{T} \to \mathcal{T}_{3,\alpha,\beta}$ and the map of Ω-bimodules: $\mu : \mathcal{T}_{3,\alpha,\beta} \to \mathcal{T}_{2,\omega,\omega}$ defined on generators by the formula:

$$\mu(a_1 \boxtimes a_2 \boxtimes a_3) = \begin{cases} a_1 \boxtimes a_3 \text{ if } l(a_2) = 0, \\ 0 \text{ otherwise.} \end{cases}$$

We will show that μ is a map of right modules, the other case follows by a symmetric argument. For $a \in \Omega_3$ and $T \in \mathcal{T}_3$ one easily checks that:

$$\mu(Ta) = \mu(T)\left((\pi \boxtimes \mathrm{id})(a)\right),$$

where $\pi : \Omega_2 \to \Omega$ is the map defined above, for which we proved that $\pi \circ \Delta_\omega = \mathrm{id}$. Now we take $F \in \Omega$ and compute:

$$\begin{aligned}
\mu(TF) &= \mu(T(\Delta_\omega \boxtimes \mathrm{id})(\Delta_\omega)(F)) = \mu(T)(\pi \boxtimes \mathrm{id})(\Delta_\omega \boxtimes \mathrm{id})(\Delta_\omega)(F)) \\
&= \mu(T)\Delta_\omega(F) = \mu(T)F.
\end{aligned}$$

Thus, μ is indeed a map of Ω bimodules. We now observe that the characteristic map of the morphism

$$\hat{u} : (E_1 \otimes_\omega 1) \otimes_\omega E_2 \to E_1 \otimes_\omega (1 \otimes_\omega E_2)$$

factors through the map $\mu : \mathcal{T}_{3,\alpha,\beta} \to \mathcal{T}_{2,\omega,\omega}$ which implies that the diagram above commutes. We conclude that the unit is compatible with the associators.

Let us now prove that these associators satisfy the pentagon axiom. We need to prove that for any four representations up to homotopy E_1, E_2, E_3, E_4 the following composition is the identity.

$$\begin{array}{ccc}
((E_1 \otimes_\omega E_2) \otimes_\omega E_3) \otimes_\omega E_4 & \xleftarrow[\hat{u}^{-1}]{} & (E_1 \otimes_\omega E_2) \otimes_\omega (E_3 \otimes_\omega E_4) \\
\big\downarrow \hat{u} \otimes \mathrm{id} & & \big\uparrow \hat{u}^{-1} \\
(E_1 \otimes_\omega (E_2 \otimes_\omega E_3)) \otimes_\omega E_4 & & E_1 \otimes_\omega (E_2 \otimes_\omega ((E_3 \otimes_\omega E_4)) \\
\big\downarrow \hat{u} & \nearrow \mathrm{id} \otimes \hat{u} & \\
E_1 \otimes_\omega ((E_2 \otimes_\omega E_3) \otimes_\omega E_4) & &
\end{array}$$

Observe that the five ways of putting brackets in the tensor product correspond to five elements $\theta_i \in \mathcal{MC}_4$ which one can construct from ω. These elements are given by maps $\Delta_{\theta_i} : \Omega \to \Omega_4$ defined as follows:

$$\begin{aligned}
\Delta_{\theta_1} &= (\Delta_\alpha \boxtimes \mathrm{id}) \circ \Delta_\omega \text{ corresponds to } ((E_1 \otimes_\omega E_2) \otimes_\omega E_3) \otimes_\omega E_4,\\
\Delta_{\theta_2} &= (\Delta_\beta \boxtimes \mathrm{id}) \circ \Delta_\omega \text{ corresponds to } (E_1 \otimes_\omega (E_2 \otimes_\omega E_3)) \otimes_\omega E_4,\\
\Delta_{\theta_3} &= (\mathrm{id} \boxtimes \Delta_\alpha) \circ \Delta_\omega \text{ corresponds to } E_1 \otimes_\omega ((E_2 \otimes_\omega E_3) \otimes_\omega E_4),\\
\Delta_{\theta_4} &= (\mathrm{id} \boxtimes \Delta_\beta) \circ \Delta_\omega \text{ corresponds to } E_1 \otimes_\omega (E_2 \otimes_\omega ((E_3 \otimes_\omega E_4)),\\
\Delta_{\theta_5} &= (\Delta_\omega \boxtimes \Delta_\omega) \circ \Delta_\omega \text{ corresponds to } (E_1 \otimes_\omega E_2) \otimes_\omega (E_3 \otimes_\omega E_4).
\end{aligned}$$

Here, as before $\Delta_\alpha = (\Delta_\omega \boxtimes \mathrm{id}) \circ \Delta_\omega$ and $\Delta_\beta = (\mathrm{id} \boxtimes \Delta_\omega) \circ \Delta_\omega$. Now, the morphisms that appear in the pentagon axiom are induced by Maurer-Cartan morphisms between the elements θ_i. In order to write them down we choose specific characteristic maps $\Delta_x : \mathcal{T} \to \mathcal{T}_{2,\omega,\omega}$, $\Delta_u : \mathcal{T} \to \mathcal{T}_{3,\alpha,\beta}$ and $\Delta_{u^{-1}} : \mathcal{T} \to \mathcal{T}_{3,\beta,\alpha}$ which induce the tensor product of morphisms and the associators in $\mathcal{D}(G)$. The Maurer-Cartan morphisms $\phi_i : \theta_i \to \theta_{i+1}$ for $i = 1, \ldots, 5 \pmod 5$ are determined by maps $\Delta_{\phi_i} : \mathcal{T} \to \mathcal{T}_{4,\theta_i,\theta_{i+1}}$ given by the formulas:

$$\begin{aligned}
\Delta_{\phi_1} &= (\Delta_u \boxtimes \mathrm{id}) \circ \Delta_x,\\
\Delta_{\phi_2} &= (\mathrm{id} \boxtimes \Delta_x \boxtimes \mathrm{id}) \circ \Delta_u,\\
\Delta_{\phi_3} &= (\mathrm{id} \boxtimes \Delta_u) \circ \Delta_x,\\
\Delta_{\phi_4} &= (\mathrm{id} \boxtimes \mathrm{id} \boxtimes \Delta_x) \circ \Delta_{u^{-1}},\\
\Delta_{\phi_5} &= (\Delta_x \boxtimes \mathrm{id} \boxtimes \mathrm{id}) \circ \Delta_{u^{-1}}.
\end{aligned}$$

Let us consider the composition of Maurer-Cartan morphisms:

$$\phi := \phi_5 \circ \phi_4 \circ \phi_3 \circ \phi_2 \circ \phi_1 \in \overline{P}(\theta_1, \theta_1),$$

where P is the Ω-bimodule defined by:

$$P := (\mathcal{T}_{4,\theta_5,\theta_1}) \otimes_\Omega \mathcal{T}_{4,\theta_4,\theta_5} \otimes_\Omega (\mathcal{T}_{4,\theta_3,\theta_4}) \otimes_\Omega (\mathcal{T}_{4,\theta_2,\theta_3}) \otimes_\Omega (\mathcal{T}_{4,\theta_1,\theta_2}).$$

The element ϕ is related to the maps in the pentagon as follows. There is a natural map of Ω-bimodules $\sigma : P \to \mathcal{T}_{4,\theta_1,\theta_1}$ and this defines a universal Maurer-Cartan morphism $\sigma(\phi) \in \mathcal{MC}_4(\theta_1, \theta_1)$. The composition of all the maps in the diagram is precisely given by:

$$\otimes_{\sigma(\phi)}(\mathrm{id}_{E_1}, \ldots, \mathrm{id}_{E_4}) : ((E_1 \otimes_\omega E_2) \otimes_\omega E_3) \otimes_\omega E_4 \to ((E_1 \otimes_\omega E_2) \otimes_\omega E_3) \otimes_\omega E_4.$$

By the last part of Theorem 6.8 we know that this morphism is homotopic to the identity, and therefore equal to the identity in $\mathcal{D}(G)$.

We conclude that any ω defines a monoidal structure on $\mathcal{D}(G)$. In a similar manner one can show that the natural equivalences defined in Proposition 6.11 are compatible with the associators and therefore are equivalences of monoidal categories. □

7. Canonical tensor products on morphisms

In this section we point out another universal property of the universal Maurer-Cartan module, a property which reveals relationships with Hochschild cohomology and non-commutative differential forms. In particular, using the universal derivation, we show that any universal Maurer-Cartan element $\omega \in \mathcal{MC}_m$ comes together with a canonical (and explicit) universal Maurer-Cartan endomorphism

$$x_\omega \in \mathcal{T}_m(\omega, \omega).$$

As an immediate consequence, once a Maurer-Cartan element $\omega \in \mathcal{MC}_m$ is fixed, there is a canonical way of taking tensor products of morphisms. This is important when one is forced to work in the category $\mathrm{Rep}^\infty(G)$ instead of the derived category (e.g. in the search of infinitesimal models for the cohomology of classifying spaces). Note that, at the level of $\mathrm{Rep}^\infty(G)$, the resulting $\otimes_\omega$ is a "functor up to homotopy".

We begin with the description of the universal derivation.

Proposition 7.1. *The Ω-bimodule $\mathcal{T}$ admits a unique biderivation $\delta : \Omega \to \mathcal{T}$ of bidegree $(0, -1)$, which is compatible with the d_i's and which sends L_n to T_n.*

Moreover, δ does not commute with ∂, instead, for $A \in \Omega$:

$$\delta(\partial(A)) + \partial(\delta(A)) = T_0 A - A T_0.$$

In other words, $\delta : \Omega \longrightarrow \mathcal{T}$ is a linear map which sends elements of bidegree (k, l) into those of bidegree $(k, l-1)$, commutes with the d_i's and satisfies the derivation condition

$$\delta(A \circ B) = \delta(A)B + (-1)^{l(A)} A \circ \delta(B).$$

Since Ω is generated as a DB-algebra by the L_n's, the proposition is straightforward.

Remark 7.2. Let us explain how derivations come into the picture (even before $\mathcal{T}$!), starting from the notion of morphisms between Maurer-Cartan elements. Let $\mathcal{A}$ and $\mathcal{B}$ be two DB-algebras, $\mathcal{E}$ be an $\mathcal{A}$-$\mathcal{B}$-DB-module, $\theta \in MC(\bar{\mathcal{A}})$ and $\omega \in MC(\bar{\mathcal{B}})$. Let us try to understand elements $x \in \mathcal{E}(\omega, \theta)$ directly in terms of Ω; one would like to re-interpret the components x_k of x as images of the elements $L_k \in \Omega$ of a certain map

$$\delta_x : \Omega \longrightarrow \mathcal{E}.$$

The equations that the x_k's must satisfy (involving ω and θ) indicate that δ_x should be required to be a derivation on the bimodule $\mathcal{E}_{\omega,\theta}$. Adding the extra-condition that δ_x commutes with the d_i's, δ_x will be determined uniquely. In turn, the fact that $x \in \mathcal{E}(\omega, \theta)$ is equivalent to the equation

$$\delta_x(\partial(A)) + \partial(\delta_x(A)) = x_0 A - A x_0$$

for all $A \in \Omega$. Note also that δ_x does not make use of x_0, hence we are really looking at triples $(\mathcal{E}, \delta, x_0)$ with such properties. Among these, $(\mathcal{T}, \delta, T_0)$ shows up as the universal one. See also Remark 7.5 below.

In order to simplify formulas (and rather intricate signs) we will need some formalism. Let us first introduce some terminology.

Consider the category $\underline{\mathbb{VS}}_B$ of **B-vector spaces**, whose objects are collections $V = \{V^k, d_i\}$ consisting of vector spaces V^k (one for each integer $k \geqslant 0$) and maps $d_i : V^k \longrightarrow V^{k+1}$ for $1 \leqslant i \leqslant k$ satisfying $d_j d_i = d_i d_{j-1}$ for $i < j$. A morphism from V to W consists of families of maps from V^k to W^k, commuting with all the operators d_i; such morphisms form the hom-spaces $\mathrm{Hom}_B(V,W)$. As in the case of simplicial vector spaces, one can realize $\underline{\mathbb{VS}}_B$ as the category of contravariant functors from a small category $\mathbb{B}$ to the category $\underline{\mathcal{V}}$ of vector spaces.

Associated to $\underline{\mathcal{V}}_B$ is the category $\mathrm{Gr}(\underline{\mathbb{VS}}_B)$ of graded objects of $\underline{\mathbb{VS}}_B$ and $\mathrm{Ch}(\underline{\mathbb{VS}}_B)$ of cochain complexes in $\underline{\mathbb{VS}}_B$. Given X and Y graded objects in $\underline{\mathbb{VS}}_B$, one defines the graded hom $\underline{\mathrm{Hom}}^*_B(X,Y)$ whose degree l-part is

$$\underline{\mathrm{Hom}}^l_B(X,Y) = \prod_p \mathrm{Hom}_B(X(p), X(p+l)).$$

When X and Y are complexes in $\underline{\mathbb{VS}}_B$, then $\underline{\mathrm{Hom}}^*_B(X,Y)$ has a natural differential:

$$\partial(f) = \partial \circ f - (-1)^{l(f)} f \circ \partial,$$

where $l(f)$ is the degree of f. Note that the internal hom of $\mathrm{Ch}(\underline{\mathbb{VS}}_B)$ is the space of zero-cocycles of $\underline{\mathrm{Hom}}^*_B$.

The category $\underline{\mathbb{VS}}_B$ comes with a tensor product operation $\otimes$ which makes it into a monoidal category: for V and W in $\underline{\mathbb{VS}}_B$, $V \otimes W$ is defined by

$$(V \otimes W)^k = \bigoplus_{k_1+k_2=k} V^{k_1} \otimes W^{k_2},$$

with the operators d_i given by the formulas:

$$d_i(v \otimes w) = \begin{cases} d_i(v) \otimes w & \text{if } i \leqslant k_1, \\ v \otimes d_{i-k_1}(w) & \text{if } i > k_1, \end{cases}$$

where k_1 is the degree of v. There is an obvious notion of tensor products of morphisms in $\underline{\mathbb{VS}}_B$, and the unit is the base field concentrated in degree zero. This tensor product operation extends to $\mathrm{Gr}(\underline{\mathbb{VS}}_B)$ and $\mathrm{Ch}(\underline{\mathbb{VS}}_B)$ in the standard way:

$$(X \otimes Y)(l) = \bigoplus_{l_1+l_2=l} X(l_1) \otimes X(l_2)$$

It also extends to the graded-hom's using the standard sign conventions:

$$\otimes : \underline{\mathrm{Hom}}^l_B(X,X') \times \underline{\mathrm{Hom}}^{l'}_B(Y,Y') \longrightarrow \underline{\mathrm{Hom}}^{l+l'}_B(X \otimes Y, X' \otimes Y'),$$

$$(f \otimes g)(x \otimes y) = (-1)^{l(g)l(x)} f(x) \otimes g(y).$$

Note that a DB-algebra is the same as a monoid in the monoidal category $\mathrm{Ch}(\underline{\mathbb{VS}}_B)$. With this in mind, there is a B-version of Hochschild cohomology. Given a DB-algebra $\mathcal{A}$ and an $\mathcal{A}$-bimodule $\mathcal{E}$, we consider the vector spaces

$$C^{p,l}(A,\mathcal{E}) := \underline{\mathrm{Hom}}^l_B(\mathcal{A}^{\otimes p}, \mathcal{E}),$$

and the space of Hochschild cochains:

$$C^n(\mathcal{A},\mathcal{E}) = \bigoplus_{p+l=n} C^{p,l}(\mathcal{A},\mathcal{E}).$$

We define the differentials by the same formulas as in the case of DG-algebras, but taking into account only the l-degree. More precisely, the horizontal differential

$$b : C^{p,l}(\mathcal{A},\mathcal{E}) \longrightarrow C^{p+1,l}(\mathcal{A},\mathcal{E}),$$

is given by

$$b(c)(a_1,\dots a_{p+1}) = (-1)^{l(a_1)l} a_1 c(a_2,\dots,a_{p+1}) + \sum_{i=1}^{p} (-1)^i c(a_1,\dots,a_i a_{i+1},\dots,a_{p+1}) + (-1)^{p+1} c(a_1,\dots,a_k) a_{p+1}.$$

The vertical differential

$$d_v : C^{p,l}(\mathcal{A},P) \longrightarrow C^{p,l+1}(\mathcal{A},P),$$

is given by

$$d_v(c)(a_1,\dots,a_p) = d(c(a_1,\dots,a_p)) - \sum_{i=1}^{p} (-1)^{\epsilon_i} c(a_1,\dots,\delta(a_i),\dots,a_p),$$

where $\epsilon_i = l + l(a_1) + \ldots + l(a_{i-1})$. These two differentials commute and we will denote the resulting total complex by $C^*_B(\mathcal{A},\mathcal{E})$. For any $\zeta \in C^{p,l}_B(\mathcal{A},\mathcal{E})$ we define $\bar{\zeta} \in C^{p,l}(\bar{\mathcal{A}},\bar{\mathcal{E}})$ by

$$\bar{\zeta}(a_1,\dots,a_n) = (-1)^{\epsilon}\zeta(a_1,\dots,a_n), \quad \text{where } \epsilon = \sum_{1\leqslant i<j\leqslant n} k(a_i)|a_j|.$$

Note that this expression is well defined even if the a_i are infinite sums, because the map ζ preserves the k degree.

Lemma 7.3. *The map*

$$C^*_B(\mathcal{A},\mathcal{E}) \longrightarrow C^*(\bar{\mathcal{A}},\bar{\mathcal{E}}), \ \zeta \mapsto \bar{\zeta}$$

is a morphism of cochain complexes.

Proof. That the horizontal differentials commute is straightforward. For the other direction, recall that the differential d_{tot} in $\overline{\mathcal{A}}$ is given by:

$$d_{tot} = \partial + (-1)^n d,$$

where d is the alternating sum of the operators d_i. The formula for the vertical differential in $C^*(\bar{\mathcal{A}},\bar{\mathcal{E}})$ decomposes in two pieces, one corresponding to ∂ and one corresponding to $(-1)^n d$. One can easily check that the first part corresponds to the vertical differential in $C^*_B(\mathcal{A},\mathcal{E})$, while the second part vanishes on $\bar{\zeta}$, because ζ commutes with the operators d_i. □

The derivation $\delta : \Omega \to \mathcal{T}$ together with the component T_0 can now be interpreted as a canonical Hochschild cochain of degree zero:

$$\zeta^{\mathrm{u}} := \delta + T_0 \in C^0(\Omega,\mathcal{T}).$$

Using cup-product operations, one obtains new cochains in $C^0(\Omega_m,\mathcal{T}_m)$ as follows. To simplify notations, we consider the case $m = 2$. Consider the two cocycles:

$$\zeta^1 = \delta \boxtimes \mathrm{Id} + T_0 \boxtimes 1 \in C(\Omega_2, \mathcal{T} \boxtimes \Omega),$$

$$\zeta^2 = \mathrm{Id} \boxtimes \delta + 1 \boxtimes T_0 \in C(\Omega_2, \Omega \boxtimes \mathcal{T}),$$

where $\mathrm{Id} = \mathrm{Id}_\Omega$, 1 is the unit of Ω and the operations

$$(-) \boxtimes \mathrm{Id} : \underline{Hom}^*_B(\Omega, \mathcal{T}) \longrightarrow \underline{Hom}^*_B(\Omega \boxtimes \Omega, \mathcal{T} \boxtimes \Omega),$$

$$\mathrm{Id} \boxtimes (-) : \underline{Hom}^*_B(\Omega, \mathcal{T}) \longrightarrow \underline{Hom}^*_B(\Omega \boxtimes \Omega, \Omega \boxtimes \mathcal{T}),$$

are defined with the usual sign conventions. Using the composition

$$\circ : (\mathcal{T} \boxtimes \Omega) \otimes (\Omega \boxtimes \mathcal{T}) \longrightarrow \mathcal{T} \boxtimes \mathcal{T} = \mathcal{T}_2,$$

we now form the cup-product

$$\zeta := \zeta^1 \cup \zeta^2 \in C(\Omega_2, \mathcal{T}_2).$$

Using the construction from the last part of the appendix, we define

$$x_\omega := \bar{\zeta}(\omega).$$

Lemma A.8 gives us the following.

Proposition 7.4. *For any $\omega \in \mathcal{MC}_2$, x_ω is an universal Maurer-Cartan morphism from ω to itself.*

Remark 7.5. Here is one final remark on the structures involved. In this paper we have thought of a representation up to homotopy as a cochain complex of vector bundles (E, ∂) together with the extra-data $\{R_k : k \geqslant 1\}$; the relevant algebraic structure was that of DB-algebra and Maurer-Cartan elements with vanishing 0-component. One can follow a slightly different route, which has some advantages when it comes to the universal Maurer-Cartan module: think of a representation up to homotopy as a graded vector bundle together with the extra-data $\{R_k : k \geqslant 0\}$. The relevant algebraic structure is that of B-algebra, which is defined exactly as that of DB-algebra but giving up on the differential ∂ and requiring unitality. In terms of the formalism discussed above, the resulting category $\underline{\mathcal{B}ar}$ of B-algebras coincides with the category $\mathrm{GrAlg}(\underline{\mathbb{VS}}_B)$ of (unital) graded algebras associated to the monoidal category $\underline{\mathbb{VS}}_B$. Then, for a graded vector bundle E, $\mathcal{A}_E$ is a B-algebra and representations up to homotopy on E correspond to Maurer-Cartan elements of $\mathcal{A}_E$ (with no restriction on the zero-component). As analogues of Ω and $\mathcal{T}$, one looks at

- $\Omega_\alpha \in \underline{\mathcal{B}ar}$ together with a Maurer Cartan element $L_\alpha \in \mathrm{MC}(\bar{\Omega}_\alpha)$ which is universal among pairs $(\mathcal{A}, \omega)$ consisting of a Maurer Cartan element in a B-algebra.
- $\mathcal{T}_\partial$ which has the same universal property as $\mathcal{T}$, but for bimodules over B-algebras.

It is not surprising that one can explicitly construct Ω_α out of Ω by adjoining to it a unit and a formal element α of bidegree $(0,1)$:

$$\Omega_\alpha^k(l) = \Omega^k(l) + \alpha \circ \Omega^k(l-1),$$

except in bidegrees $(0,0)$ and $(0,1)$ where

$$\Omega_\alpha^0(0) = \mathbb{Q}, \Omega_\alpha^0(1) = \mathbb{Q}\alpha.$$

One defines the algebra structure on Ω_α by requiring

$$\alpha^2 = 0, \quad \alpha \circ a - (-1)^{l(a)} a \circ \alpha = \partial(a),$$

while the d_i's are defined by

$$d_i(a + \alpha \circ b) = d_i(a) + \alpha \circ d_i(b).$$

Finally, one sets $L_\alpha = \alpha + L$.

For $\mathcal{T}_\alpha$ the situation is similar but a bit simpler:

$$\mathcal{T}_\alpha = \mathcal{T} + \alpha \circ \mathcal{T}$$

and the differential ∂ of $\mathcal{T}$ is encoded in the bimodule structure of $\mathcal{T}_\alpha$:

$$x \circ \alpha = (-1)^{l(x)}(\alpha \circ x - \partial(x)),$$

for $x \in \mathcal{T}$.

The analogue $\delta_\alpha : \Omega_\alpha \longrightarrow \mathcal{T}_\alpha$ of the derivation δ has a nicer universal property: it is universal among all derivations on Ω_α-bimodules. Using the straightforward B-version of Hochschild cohomology and non-commutative forms, we see that $\mathcal{T}_\alpha$ must be the space of non-commutative 1-forms associated to the B-algebra Ω_α. This also gives another description of $\mathcal{T}_\alpha$ (and then of $\mathcal{T}$) out of Ω_α:

$$\mathcal{T}_\alpha = \Omega_\alpha \otimes \overline{\Omega}_\alpha, \quad \text{where } \overline{\Omega}_\alpha = \Omega_\alpha / 1 \cdot \mathbb{R},$$

and where a tensor $a \otimes b$ should be interpreted as a non-commutative 1-form $a\delta_\alpha(b)$. Since $\mathcal{T}$ can be recovered as the subspaces of elements which are not of type $\alpha \circ x$, the derivation property of δ_α shows that $\mathcal{T}$ is spanned by the following types of elements:

$$A\delta_\alpha(T)B,$$

where A is either 1 or an element of Ω, similarly for B, and T is either a tree or α. This corresponds to our original description of $\mathcal{T}$ in terms of trees and forests. It is interesting to point out that the appearance of $\emptyset$ in that description encodes two types of elements: 1 (on the forest side) and $\delta(\alpha)$ (on the tree side).

A. Appendix

A.1. Maurer-Cartan elements

In this appendix we put together some definitions and results that are used in the paper.

We begin with some standard notions regarding Maurer-Cartan elements in Differential Graded Algebras (DGAs).

1. A *Maurer-Cartan element* in a DGA (A, d) is an element $\gamma \in A$ of degree one satisfying $d(\omega) + \omega^2 = 0$. We denote by $MC(A)$ the set of all Maurer-Cartan elements.
2. A *gauge equivalence* between ω and $\theta \in MC(A)$ is an invertible element $u \in A$ of degree zero satisfying $u\omega u^{-1} - \theta = (du)u^{-1}$.
3. Given two Maurer-Cartan elements θ and ω of two DGAs (A, d) and (B, d), respectively, and a DG A-B bimodule (P, d), a *Maurer-Cartan P-morphism* from ω to θ is an element $x \in P$ of degree zero satisfying $x\omega - \theta x = d(x)$. We denote by $P(\omega, \theta)$ the set of such P-morphisms.
4. With the same notations, we say that $x, y \in P(\omega, \theta)$ are *homotopic* if there exists $h \in P$ of degree -1 such that $x - y = dh + h\omega + \theta h$. We denote by $P[\omega, \theta]$ the set of all homotopy classes of P-morphisms from ω to θ.

Altogether, one obtains a category whose objects are DGAs endowed with a Maurer-Cartan element where the morphisms from (B, ω) to (A, θ) are pairs (P, x) consisting of a DG A-B bimodule P and an element $x \in P(\omega, \theta)$. If (Q, y) is another morphism from (C, η) to (B, ω), then their composition is defined as

$$(P, x) \circ (Q, y) = (P \otimes_B Q, x \otimes y).$$

It is easy to check that $x \otimes y$ satisfies the required equation and also that this operation is compatible with the notion of homotopy. In particular, one obtains a quotient of this category in which homotopic morphisms become equal.

We will now concentrate our attention on complete DGAs and complete DG modules. By a filtered algebra we mean an algebra A together with a filtration

$$\cdots \subset F_2A \subset F_1A \subset F_0A = A,$$

satisfying

$$F_pA \cdot F_qA \subset F_{p+q}A.$$

Note that, in particular, F_pA is an ideal in A for all p, hence we can consider the quotient algebras, which fit into a tower

$$A/F_1A \leftarrow A/F_2A \leftarrow \cdots .$$

We denote by $\bar{A}$ the inverse limit of this tower. Note that $\bar{A}$ has a natural filtration, with $F_p\bar{A}$ being the inverse limit of

$$F_pA/F_{p+1}A \leftarrow F_pA/F_{p+2}A \leftarrow \cdots .$$

Moreover, there is a canonical map $c : A \longrightarrow \bar{A}$ that is a map of filtered algebras.

Definition A.1. *A complete algebra is an algebra A together with a filtration $F_\bullet A$, such that the canonical map $c : A \longrightarrow \bar{A}$ is an isomorphism.*

A complete DGA is a DGA (A, d) that also has the structure of a complete algebra, such that each space F_pA of the filtration is a subcomplex of (A, d_A).

A similar discussion applies to modules over filtered algebras. Given A as above, a filtered left A-module P is required to carry a filtration

$$\cdots \subset F_2P \subset F_1P \subset F_0P = P,$$

satisfying

$$F_pA \cdot F_qP \subset F_{p+q}P.$$

The completion $\bar{P}$ of P is the inverse limit of P/F_pP (a left $\bar{A}$-module). If A is a complete algebra, we say that P is a complete (left) A-module if the canonical map from P to $\bar{P}$ is an isomorphism. If (A, d) is a complete DGA, a complete (left) DG module over (A, d) (or simply A-module) is a DG module (P, d) that also has the structure of complete A-module such that each F_pP is a subcomplex of (P, d). Right modules and bimodules are treated similarly.

In general, for a filtered algebra A, $\bar{A}$ is complete and is called the completion of A.

Example A.2. If $A = \oplus_{k,l} A^k(l)$ is a (differential) bigraded algebra, then it can also be viewed as a (differential) graded algebra with $A = \oplus_n A^n$, where

$$A^n = \oplus_{k+l=n} A^k(l).$$

In this case, A carries a natural filtration with

$$F_pA = \oplus_{k \geqslant p} A^k(l).$$

The resulting completion $\bar{A}$ is given by

$$\bar{A}^n = \Pi_{k+l=n} A^k(l).$$

A.2. The case of complete DGAs

We now consider the existence problem for Maurer-Cartan elements whose class modulo F_rA (for some $r \geqslant 1$) is given. Let $\gamma \in A$ be of degree one, and suppose that we look for a Maurer-Cartan element ω that is equivalent to γ modulo F_rA. This condition forces:

$$d\gamma + \gamma^2 \equiv 0 \ \mathrm{mod} F_rA. \tag{29}$$

Any $\gamma \in MC(A)$ induces a new differential on A:

$$d_\gamma(a) = d(a) + [\gamma, a] = d(a) + \gamma a - (-1)^{|a|} a\gamma.$$

This differential descends to a differential d_γ on all the quotients $F_pA/F_{p+1}A$. Actually, for d_γ to be a differential on the quotients, one does not need the full Maurer-Cartan condition on γ but only (29) for $r = 1$. Hence given any γ of degree 1 satisfying (29) for some $r \geqslant 1$, it makes sense to talk about the cohomology of $(F_pA/F_{p+1}A, d_\gamma)$.

Proposition A.3. *Let A be a complete DGA. Then for any degree one element γ satisfying (29) and*

$$H^2(F_pA/F_{p+1}A, d_\gamma) = 0, \quad \forall \ p \geqslant r,$$

there exists $\omega \in MC(A)$ such that $\omega \equiv \gamma$ mod F_rA.

Proof. We will inductively construct $\omega_r, \omega_{r+1}, \ldots$ with the property that

$$d\omega_k + \omega_k^2 = 0, \quad \mathrm{mod}\ F_kA,$$

in such a way that

$$\omega_k = \omega_{k-1} \bmod F_{k-1}A, \ \forall\ k \geqslant r+1,\ \omega_r = \gamma.$$

Assuming that ω_k has been constructed, we are now looking for $a \in F_kA$ such that

$$\omega_{k+1} := \omega_k + a$$

satisfies the Maurer-Cartan equation modulo $F_{k+1}A$. Writing out the equation and using that $a^2 \in F_{k+1}A$, the equation to solve is

$$-d_{\omega_k}(a) = (d\omega_k + \omega_k^2) \bmod F_{k+1}A.$$

This can be seen as an equation in $F_kA/F_{k+1}A$. Moreover, on the quotient, $d_{\omega_k} = d_\gamma$ because $\omega_k - \gamma \in F_1(A)$. Hence, due to the cohomological condition in the statement, we only have have to check that the right hand side of the last equation is closed for d_{ω_k}. But its differential (modulo $F_{k+1}A$) is

$$d(\omega_k^2 + d\omega_k) + \omega_k(\omega_k^2 + d\omega_k) - (\omega_k^2 + d\omega_k)\omega_k = 0.$$

In conclusion, we obtain the desired sequence $(\omega_k)_{k\geqslant r}$. Due to completeness of A, we obtain an element $\gamma \in A$ such that $\omega = \omega_k \bmod F_kA$ for all $k \geqslant r$. Since the Maurer-Cartan expression in ω is congruent, modulo F_kA, to the one of ω_k, hence to zero, we deduce (again from the completeness of A) that $\omega \in MC(A)$. By construction, $\omega = \gamma \bmod F_rA$. □

There is an analogous result for Maurer-Cartan morphisms. Given two Maurer-Cartan elements ω and θ of two complete DGAs (A,d) and (B,d), respectively, and let (P,d) be a a complete DG-A-B bimodule. Then the differential d of P can be twisted by ω and θ to define a new differential:

$$d_{\omega,\theta}(x) = d(x) + \theta x - (-1)^{|x|}x\omega.$$

The following is proven exactly as the previous result.

Proposition A.4. *Let $r \geqslant 1$, and assume that*

$$H^1(F_pP/F_{p+1}P, d_{\omega,\theta}) = 0, \quad \forall\ p \geqslant r.$$

Then for any $x \in P$ satisfying

$$x\omega - \theta x = dx \bmod F_rP,$$

one can find $y \in P(\omega,\theta)$ such that $y = x \bmod F_rP$. Moreover, if the same cohomological condition holds in degree zero, then any two such y's are homotopic.

Gauge equivalence in complete DGAs: In the context of complete DGAs, there is a refined notion of gauge transformation that we now explain. We associate a group $G_1(A)$ to a complete DGA A as follows

$$G_1(A) = \{x \in A^0 :\ x \equiv 1 \bmod F_1A\} = 1 + (F_1A)^0.$$

One can see that $G_1(A)$ is a group with respect to the multiplication in A from the power series expression

$$(1-\alpha)^{-1} = \sum_{k\geqslant 0} \alpha^k,$$

where, for $\alpha \in F_1 A$, completeness implies that the right hand side makes sense as an element of A .

Note that, strictly speaking, the definition of $G_1(A)$ requires A to be unital. However, the role of the elements "1" is purely formal. In other words, $G_1(A)$ makes sense even without the unitality condition. Equivalently, taking this as a definition for unital A's, for a general A, one can replace A by the new (complete) DGA A^+ obtained by adding a unit to A. Define $G_1(A)$ as $G_1(A^+)$. In the case that A already has a unit 1_A, the map $1_A + x \mapsto 1 + x$ identifies the two definitions.

Definition A.5. *Given a complete DGA A, a gauge equivalence u between two Maurer-Cartan elements ω and θ of A is called strong if $u \in G_1(A)$. If such a u exists, we say that ω and θ are strongly gauge equivalent.*

Proposition A.6. *Let A be a complete DGA and $\omega, \theta \in MC(A)$, such that*

$$\omega = \theta \ mod \ F_r A,$$

with $r \geqslant 1$. If

$$H^1(F_p A/F_{p+1}A, d_\omega) = 0, \quad \forall \ p \geqslant r,$$

then ω and θ are strongly gauge equivalent.

Proof. This proof is very similar to the one of Proposition A.3. We construct inductively a sequence $u_r, u_{r+1}, \ldots$ of degree zero elements of A with the property that

$$u_k\omega - \theta u_k = du_k \bmod F_k A$$

for all $k \geqslant r$. Moreover, the sequence will be constructed so that

$$u_k = u_{k-1} \bmod F_{k-1}A \ \forall \ k \geqslant r+1, \quad u_r = 1.$$

Assuming that u_k has been constructed, we are looking for $x \in F_k A$ such that

$$(u_k + x)\omega - \theta(u_k + x) = d(u_k + x) \bmod F_{k+1}A.$$

Note that, since $\omega - \theta \in F_r A \subset F_1 A$,

$$\theta x - x\omega = \omega x - x\omega \bmod F_{k+1}$$

whenever $x \in F_k A$. We see that the previous equation can be written as an equation in $F_k A/F_{k+1}A$:

$$d_\omega x = -du_k + u_k\omega - \theta u_k \bmod F_{k+1}A.$$

Because of the hypothesis, it suffices to show that the right hand side is closed in the quotient. Denoting the right hand side by y, and using that

$$d_\omega(y) = d(y) + \theta y + y\omega,$$

the desired equation follows immediately.

With the sequence $u_r, u_{r+1}, \ldots$ constructed, one uses again the completeness of A to obtain $u \in A$ of degree zero such that $u = 1 \bmod F_r A$, $u\omega - \theta u = du$. □

Remark A.7. The gauge equivalence comes from an action of $G_1(A)$ on $MC(A)$, given by the usual gauge formula:

$$u \cdot \omega = u\omega u^{-1} - du \cdot u^{-1} \quad u \in G_1(A), \omega \in MC(A).$$

Our discussion has an infinitesimal counterpart. First of all, "the Lie algebra of $G_1(A)$" is defined as

$$\mathfrak{g}_1(A) := \{\alpha \in A^0 : \alpha \equiv 0 \bmod F_1 A\},$$

with the commutator bracket

$$[\alpha, \beta] = \alpha\beta - \beta\alpha.$$

The exponential map

$$exp : \mathfrak{g}_1(A) \longrightarrow G_1(A)$$

is defined by the usual power series

$$exp(\alpha) = \sum_{k \geqslant 0} \frac{1}{k!}\alpha^k.$$

The completeness of A make sense of $exp(\alpha)$ for $\alpha \in F_1 A$. The action of $G_1(A)$ on $MC(A)$ has an infinitesimal counterpart: an action of the Lie algebra $\mathfrak{g}_1(A)$ on $MC(A)$, which is familiar in the discussion of Maurer-Cartan elements in differential graded Lie algebras:

$$\alpha \cdot \gamma = [\alpha, \gamma] + d\alpha, \quad \alpha \in \mathfrak{g}_1(A), \gamma \in MC(A).$$

A.3. Relation with Hochschild cohomology

Let A be a DGA and M be an A-bimodule. Here we explain that, given a Maurer-Cartan element ω of A, one can associate a Maurer-Cartan morphism to a degree zero P-valued Hochschild cocycle on A. In low degrees, the idea is very simple: by applying a biderivation $D : A \longrightarrow P$ to the Maurer-Cartan equation for a Maurer-Cartan element ω, one obtains an element $D(\omega) \in P(\omega, \omega)$. We now discuss what happens for general degree zero Hochschild cocycles. For each k and l, we denote by $C^{k,l}(A, P)$ the space of all linear maps

$$c : \underbrace{A \otimes \ldots \otimes A}_{k \text{ times}} \longrightarrow P$$

which raises the total degree by l. The horizontal differential

$$b : C^{k,l}(A, P) \longrightarrow C^{k+1,l}(A, P)$$

is given by

$$b(c)(a_1, \ldots a_{k+1}) = (-1)^{|a_1|l} a_1 c(a_2, \ldots, a_{k+1}) + \sum_{i=1}^{k} (-1)^i c(a_1, \ldots, a_i a_{i+1}, \ldots, a_{k+1})$$
$$+ (-1)^{k+1} c(a_1, \ldots, a_k) a_{k+1}.$$

The vertical differential

$$d_v : C^{k,l}(A, P) \longrightarrow C^{k,l+1}(A, P)$$

is given by

$$d_v(c)(a_1, \ldots, a_k) = d(c(a_1, \ldots, a_k)) - \sum_{i=1}^{k} (-1)^{\epsilon_i} c(a_1, \ldots, d(a_i), \ldots, a_k),$$

where $\epsilon_i = l + |a_1| + \ldots + |a_{i-1}|$. These two differentials commute, and we obtain a complex $C^n(A, P) = \bigoplus_{k+l=n} C^{k,l}(A, P)$ with $D = d_v + (-1)^l b$ as total differential. We are interested in 0-cocycles. Such a cocycle is a finite sum

$$\zeta = \zeta_0 + \zeta_1 + \ldots, \quad \text{with } \zeta_k \in C^{k,-k}(A, P), \tag{30}$$

satisfying

$$b(\zeta_i) + (-1)^i d(\zeta_{i+1}) = 0.$$

For any such ζ, we consider the induced polynomial function

$$\hat{\zeta} : A^1 \longrightarrow P^0, \quad \zeta(a) = \zeta_0 + \zeta_1(a) + \zeta_2(a, a) + \ldots \ .$$

The following is straightforward:

Lemma A.8. *For any Hochschild cocycle $\zeta \in C^0(A, P)$ and any $\omega \in MC(A)$, $\hat{\zeta}(\omega) \in P(\omega, \omega)$. Moreover:*

- *If ζ and ζ' are cohomologous, then $\zeta(\omega)$ and $\zeta'(\omega)$ are homotopic.*
- *This construction is compatible with cup-products.*

References

[1] C. Arias Abad and M. Crainic, *Representations up to homotopy and Bott's spectral sequence for Lie groupoids*, arXiv:0911.2859, submitted for publication.

[2] C. Arias Abad and F. Schätz, *The A_∞ de-Rham theorem and the integration of representations up to homotopy*, arXiv:1011:4693, submitted for publication.

[3] J. Block and A. Smith, *A Riemann Hilbert correspondence for infinity local systems*, arXiv:0908.2843.

[4] R. Bott, *On the Chern-Weil homomorphism and continuous cohomology of Lie groups*, Adv. Math. 11, 289–303 (1973).

[5] K. A. Behrend, *On the de Rham cohomology of differential and algebraic stacks*, Adv. Math., 198(2), 583–622, (2005).

[6] E. Getzler, *The equivariant Chern character for non-compact Lie groups*, Adv. Math., 109(1), 88–107, (1994).

[7] S. Forcey, *Quotients of the multiplahedron as categorified associahedra*, Homology, Homotopy Appl. 10 (2008), no. 2, 227–256.

[8] K. Igusa, *Iterated integrals of superconnections*, arXiv:0912.0249.

[9] T. Kaczynski, K. Mischaikow, M. Mrozek, *Computational Homology*, Applied Mathematical Sciences, Volume 157 (2004) Springer.

[10] J.L. Loday, *The diagonal of the Stasheff polytope*, to appear in International Conference in honor of Murray Gerstenhaber and Jim Stasheff (Paris 2007).

[11] K.C.H. Mackenzie, *General theory of Lie groupoids and Lie algebroids*, London Mathematical Society Lecture Note Series, 213. Cambridge University Press, Cambridge, 2005.

[12] S. Mac Lane, *Categories for the Working Mathematician*, Second edition. Graduate Texts in Mathematics, 5. Springer-Verlag, New York, 1998.

[13] M. Markl, *Homotopy algebras are homotopy algebras*, Forum Math. 16 (2004), no. 1, 129–160.

[14] M. Markl and S. Shnider, *Associahedra, cellular W-construction and products of* A_∞*-algebras*, Trans. Amer. Math. Soc. 358, no. 6, 2353–2372, (2006).

[15] M. Markl, S. Shnider and J. Stasheff, *Operads in algebra, topology and physics*, Mathematical Surveys and Monographs, 96. American Mathematical Society, Providence, RI, 2002.

[16] J.P. Serre, *Homologie singuliere des espaces fibres* (French), Ann. of Math. (2) 54, 425–505 (1951).

[17] J.D. Stasheff, *Homotopy associativity of H-spaces*, I, II. Trans. Amer. Math. Soc. 108 (1963), 275-292, 293–312.

[18] J.D. Stasheff, *A twisted tale of cochains and connections*, Georgian Mathematical Journal. Volume 17, Issue 1, Pages 203215.

[19] S. Saneblidze, R. Umble, *Diagonals on the permutahedra, multiplihedra and associahedra*, Homology Homotopy Appl. 6 (2004), no. 1, 363–411.

[20] M. Sugawara, *On the homotopy commutativity of groups and loop spaces*, Mem. Coll. Sci. Univ. Kyotp Ser. A Math. 33, 257–269 (1960).

This article may be accessed via WWW at `http://tcms.org.ge/Journals/JHRS/`

Camilo Arias Abad
`Camilo.Arias.Abad@math.uzh.ch`

Institut für Mathematik, Universität Zürich

Marius Crainic
`m.crainic@uu.nl`

Mathematics Institute, Utrecht University

Benoit Dherin
`B.R.U.Dherin@math.uu.nl`

Mathematics Institute, Utrecht University

www.ingramcontent.com/pod-product-compliance
Lightning Source LLC
Chambersburg PA
CBHW032328210326
41518CB00041B/1693
9781848901186